W0253753

Proceedings in Life Sciences

Respiratory Function in Birds, Adult and Embryonic

Satellite Symposium of the 27th International Congress of Physiological Sciences, Paris 1977

held at the Max-Planck-Institute for Experimental Medicine, Göttingen (FRG), July 28–30, 1977

Organized and Edited by
Johannes Piiper

With 147 Figures

Springer-Verlag
Berlin Heidelberg New York 1978

PROFESSOR DR. J. PIIPER
Max-Planck-Institut
für Experimentelle Medizin
Abteilung Physiologie
Hermann-Rein-Str. 3
3400 Göttingen

This Symposium was supported by the
Deutsche Forschungsgemeinschaft

ISBN-13:978-3-642-66895-1 e-ISBN-13:978-3-642-66894-4
DOI: 10.1007/978-3-642-66894-4

Library of Congress Cataloging in Publication Data Main entry under title: Respiratory function in birds, adult and embryonic. (Proceedings in life sciences) Bibliography: p. Includes Index. 1. Birds-Physiology-Congresses. 2. Respiration-Congresses. 3. Respiratory organs-Birds-Congresses. I. Piiper, Johannes, 1924-. II. International Congress of Physiological Sciences, 27th, Paris, France, 1977. III. Max-Planck-Institut für Experimentelle Medizin. QL698.R47 598.2'12 78-3534

Softcover reprint of the hardcover 1st edition 1978

2131/3130-543210

Preface

This volume contains the papers presented at the symposium on *Respiratory Function in Birds, Adult and Embryonic* held in Göttingen, July 28-30, 1977. Some of the manuscripts represent the papers as delivered, while others are considerably abridged or extended. An essential part of the symposium was discussion at all levels, from formal discussion remarks with slides to dialogs at the lunch table with scribble on paper napkins. However, no attempt was made to reproduce any part of the discussion.

The symposium was attended by 75 scientists, of varied specializations (physiology, anatomy, zoophysiology, ecology, veterinary science), but all actively engaged in research related to some aspects of avian respiration. The aim of the meeting was to survey critically the present knowledge in avian respiration and related areas in order to provide a basis for future research. Due to the limited time available and to the restricted number of invited attendants, not all relevant topics could be given the attention they deserved, but a selection had to be made. A particular emphasis was given to the following aspects of avian respiration: (1) phylogenetic relationships, (2) structure and function of lungs, (3) mechanisms of adjustments to changes in environment and physiological status, (4) respiration during the embryonic period.

The symposium was one of the numerous Satellite Symposia expanding and complementing the International Congress of Physiological Sciences held in Paris, July 18-23, 1977. Although the institution of such Satellite Symposia is a rather recent development, the Satellite Symposia dedicated to comparative respiration research apparently are in the process of becoming a tradition carried on by a group whose members are conscious of forming an international scientific community.

The first meeting of the group, under the very broad title *Comparative Physiology of Respiration in Vertebrates*, complemented the Munich Congress in 1971. It was jointly organized by Pierre Dejours (Strasbourg), George M. Hughes (Bristol) and myself, and took place in Göttingen, August 2-4, 1971*. The next meeting of the group was held in Bhagalpur (Bihar, India), October 29-November 1, 1974 in conjunction with the 26th International Congress in Delhi. The topic of the symposium, organized

* Published in Respir. Physiol. 14, 1-236 (1972).

by George M. Hughes and J.S. Datta Munshi, was *Respiration of Amphibious Vertebrates**. Two consecutive Satellite Symposia followed the Paris Congress in July 1977. The first, organized by Pierre Dejours, Hermann Rahn, and Blake Reeves, was held in Strasbourg, July 24-27, 1977 on *Interaction of Intra- and Extracellular Acid-Base Balance*. Particular emphasis was on recent recearch on the effects of temperature on acid-base balance in diverse organisms (mollusks, crustaceans, all classes of vertebrates)**.

Another historical root of the current symposium is the workshop on *Receptors and Control of Respiration in Birds* held in Göttingen, May 24-25, 1974, organized by Peter Scheid***. From this workshop the topic of CO_2-receptors, subject of several reports in the present symposium, was inherited, and also the symbol of the model duck with lungs and airsacs. However, the symbol had to be completed by adding an egg, with the air cell, to characterize and distinguish this current symposium, in which considerable time was allotted to the respiration of the avian embryo enclosed in the eggshell. I was happy to enjoy the active cooperation of Hermann Rahn, the initiator and promotor of the rapidly expanding research on the development of the avian respiratory function, in the preparation of the "egg session" (Session V).

It is my pleasant duty to acknowledge the generous financial support provided by the Deutsche Forschungsgemeinschaft (German Research Council), Bad Godesberg. Without this support the symposium could not have taken place. Furthermore I should like to thank Mrs. Elfriede Hansen, secretary of the Department of Physiology, and Peter Scheid, my co-ornithophysiologist, for advice and efforts at all stages from the planning to the publication. Moreover, thanks are also due to all members of the staff of the Department of Physiology for their cooperation.

Another particular pleasure is to acknowledge the excellent cooperation with Dr. Dieter Czeschlik and his co-workers, Springer-Verlag Heidelberg, in shaping this volume of proceedings.

Göttingen, April 1978 Johannes Piiper

---

* Published in Hughes, G.M. (ed.): Respiration of Amphious Vertebrates. London-New York-San Francisco: Academic Press, 1976.
** To be published in Respir. Physiol. 30 (1978).
*** Published in Respir. Physiol. 22, 1-216 (1974).

Contents

Participants

1. Tallman
2. Stone
3. Dawes
4. Geiser
5. Scheid
6. Ar
7. Osborne
8. Bauer
9. Black
10. Ultsch
11. Fedde
12. Kunz
13. Hansen
14. King, A.S.
15. King, D.Z.
16. Hughes
17. Munshi
18. Nilsson
19. Dejours
20. Piiper
21. Rahn
22. Hoyt
23. Carey
24. Schmidt-Nielsen
25. Claeson
26. Maloy
27. Bamford
28. Burns
29. Marin
30. Miller
31. Rüfer
32. Butler
33. Farner
34. Visschedijk
35. Rautenberg
36. Burger
37. Estavillo
38. Mochizuki
39. White
40. Abdalla
41. Bech
42. Gans
43. Wright
44. Milsom
45. Gratz
46. Barteczko
47. Meyer
48. Shelton
49. Gillespie
50. Banzett
51. Molony
52. Duncker
53. Peterson
54. Besch
55. Besch, jr.
56. Laughlin
57. Richards
58. Graf
59. Gans
60. Heisler
61. Bartels
62. Perry
63. Pattle
64. Bouverot
65. Johansen
66. Ballintijn
67. Holland
68. Powell
69. Crawford
70. Torre-Bueno
71. Calder
72. Mongin
73. Tullet
74. Kawashiro
75. Brackenbury
76. Berger
77. Tazawa
78. West
79. Jones
80. Sturkie
81. Metcalfe
82. Perry

1
2
3
4
5
6
7
8
9
10
11
12
13
14
15
16
17
18
19
20
21
22
23
24
25
26
27
28
29
30
31
32
33
34
35
36
37
38
39
40
41
42
43
44
45
46
47
48
49
50
51
52
53
54
55
56
57
58
59
60
61
62
63
64
65
66
67
68
69
70
71
72
73
74
75
76
77
78
79
80
81
82

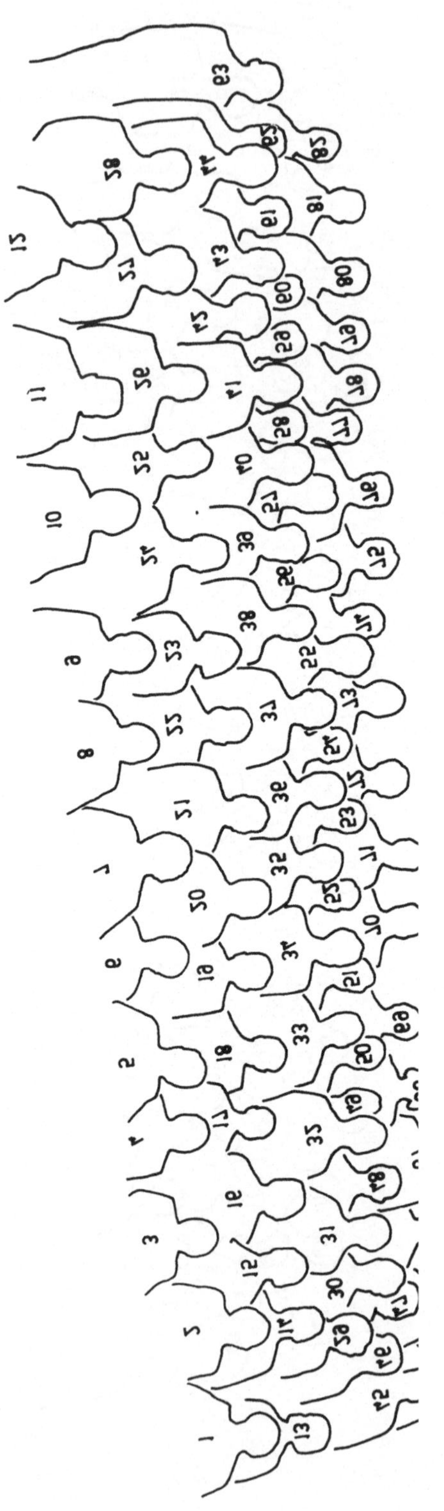

Session I

Evolutionary Relationships from Reptiles to Birds and Mammals

Chairman: K. Johansen

General Morphological Principles of Amniotic Lungs

H.-R. Duncker

Summary

The unicameral lung with a faviform (honeycomb-like) structured wall is the simplest form in living reptiles and is found in most lizards and in snakes. The increasing subdivision of the lung in different reptilian orders has led to an heterogeneously partitioned multicameral lung, which was independently developed in turtles, in monitor lizards, and in crocodiles.
The rigid avian lung, which is the highest evolved form of a heterogeneously partitioned lung, was able to develop because of the special arrangement of the septa which subdivide the body cavity and the development of air sacs. The mammalian lung follows another principle in being homogeneously partitioned, thereby allowing inflation of completely collapsed alveoli.

The true land living vertebrates, the amniotic reptiles, developed very early a great multiplicity of different forms and ways of living, from the running and climbing to the burrowing and flying forms, as well as those forms which returned to water as different swimming orders. Because of this diversity of forms and body sizes, the lungs of the surviving reptiles demonstrate great differences in their structure. This diversity has led to the development of different structural principles in lung construction in the varying forms of reptiles, out of which the basically different respiratory system of birds and of mammals evolved. The following is an attempt to give an outline of these developments.

Today the basic form of the reptilian lung is found in most small lizards and in snakes. Their lungs resemble that of amphibians in being paired, oval-shaped hollow organs lying in the anterior part of the pleuro-peritoneal cavity (Moser, 1902; Marcus, 1937). The well-developed rib cage enables these reptilians, in contrast to amphibians, to ventilate the lungs by respiratory movements and thereby the development of a trachea and a neck region was possible (Fig. 1). This simplest form of the reptilian lung is connected to the mediastinal wall between the left and right anterior part of the pleuro-peritoneal cavity in the region of the lung hilus. There, the lung is fused with the mediastinal wall to varying extents. The large freely movable part of the lung is fixed to this mediastinal septation regularly by a dorsal mesopneumonium over the whole length of the lung, and often also by a shorter ventral mesopneumonium, both of which originate from the hilus region. Apart from this connection the lung is freely movable in the pleuro-peritoneal cavity, cranially touching the pericardium,

Zentrum für Anatomie und Cytobiologie der Justus-Liebig-Universität Giessen, Aulweg 123, D-6300 Lahn-Giessen/West Germany.

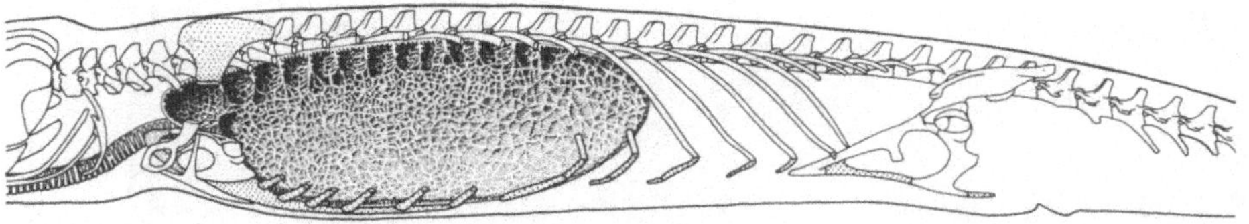

Fig. 1. Green lizard [*Lacerta viridis* (Laurenti)], Unicameral lung, inflated, in its topographical position to the skeleton. Note luminal smooth muscle network

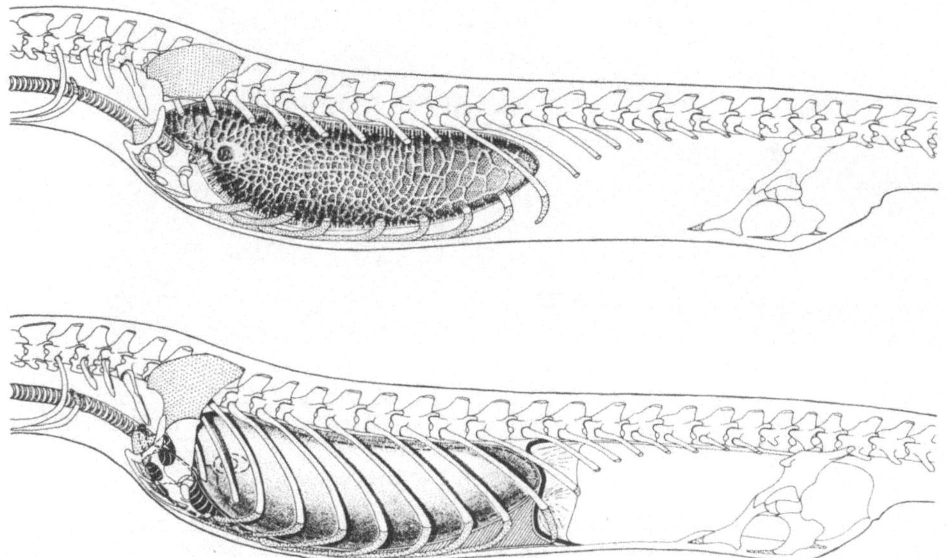

Fig. 2. Teju (*Tupinambis nigropunctatus* Spix). *Top*, Unicameral lung, inflated, with its smooth muscle network converging into one strand to the bronchial opening. *Bottom*, Lung in its topographical position in the body cavity, cranially touching the pericardium, ventrally, the liver and caudally, the posthepatic septum

ventrally and caudally the liver and the intestines, and in the females also the mesosalpinx and the cranial oviduct (Fig. 2).

This oval-shaped lung is unicameral, i.e., it consists of only one chamber. Its wall is developed into a very highly spezialized gas exchange structure. The wall is very regularly partitioned into long, narrow, thin-walled tubes orientated perpendicularly like small honeycomb cells or "faveoli" (Fig. 3). The partitions of this layer of faviform parenchyma contain a capillary net on both sides, making up the exchange surface (Fig. 4). These partitions are held in an expanded or stretched state during the varying inflation stages of the lung by a network of smooth muscle bundles connected to it at the luminal side. This smooth muscle system consists of a first layer of thicker bundles forming a net of heptagonal or hexagonal shape. This net joins regularly on the medial side and in a number of species also on the lateral side of the lung into one large muscle strand going directly to the bronchial opening at the anterior lung tip. A smaller network of smooth muscle cell bundles is expanded between these large meshes. In some species

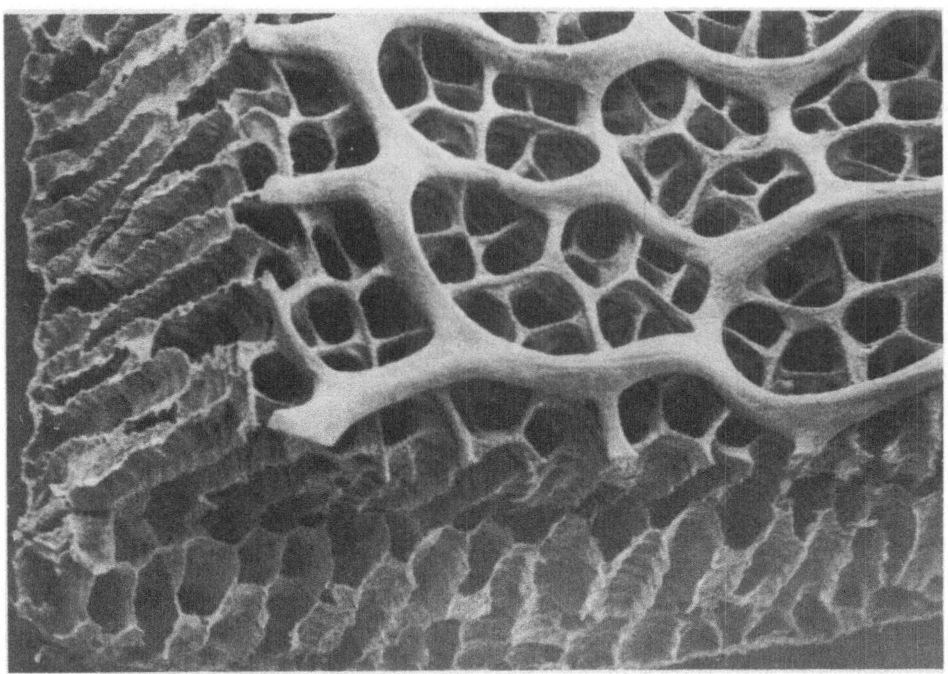

Fig. 3. Teju (*Tupinambis nigropunctatus* Spix). Lung, scanning electron micrograph. Parenchymal wall of the unicameral lung with faveoli and smooth muscle network. Fixed by glutaraldehyde instillation, critical point method, ×50

a system of very small muscle bundles may lie between these secondary net meshes. At the luminal side this muscle network is covered by ciliated epithelium. The extent of the ciliated epithelium varies with species as well as with the size and position of the bundles within the faviform parenchyma. The cilia carry along the longitudinal strands to the bronchial opening.

This faviform parenchyma is present in lizard families and in snakes. It never exeeds 3 to 4 mm in depth. This may be related to the fact that in normal respiratory cycles the gas moves mainly by diffusion in the parenchyma. This is underlined by the apparently greater capillary density at the luminal portion of the partitions. The arterioles often lie in these partitions, usually one-third of the distance from the lung wall. At this level, half of the faveoli terminate, and the other half divide to supply the outer third of the faviform parenchyma. The venules often lie in the partitions at the luminal side. They are collected at the wall of the lung, where all larger branches of the pulmonary vein and artery are located. Here they lie on the medial and dorsal lung surface supplying the other surface by branches.

In small lizards, the right and left lungs are more or less the same size and in the inflated state they occupy 7% to 10% of the total body volume (Perry and Duncker, in preparation). In these species, the wall of the lungs is made up, over its entirety, of the faviform structure, which only decreases slightly in height toward the caudal end of the lung. In a number of species, however, such as burrowing skinks or many snakes, the lungs occupy in the inflated state a higher percentage (up to 30%) of the total body volume. These lungs are more asymmetrically

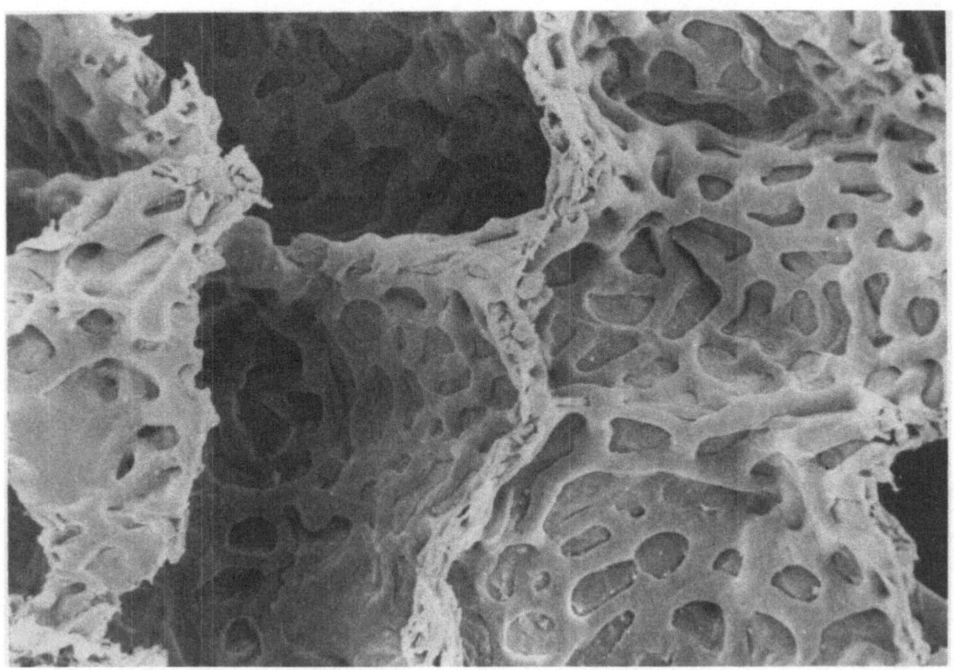

Fig. 4. Algerian skink (*Eumeces algeriensis* Peters). Lung, scanning electron micrograph. Faviform parenchyma of the lung, partitions between the faveoli with the blood capillary network. Fixed by glutaraldehyde instillation, critical point method, ×250

developed, particularly in snakes, where the caudal parts are greatly dilated and more or less lack respiratory tissue. Their wall is reduced to a membranous structure, often without partitions. Especially in large snakes, a pulmonary vascular supply of the caudal dilatation is lacking, and the respiratory tissue is restricted to the cranial part of the lungs. The large caudal dilatation is responsible for nonrespiratory functions such as enlarging the body form in defense reactions, for burrowing as in skinks, or for the maintenance of the body form in large snakes when their intestines are empty. It is remarkable that the compliance of the lungs with dilated caudal extension is so high compared with smaller lungs without dilatations and having solely respiratory functions (Perry and Duncker, in preparation). In these snakes, the lungs are directly attached to the body wall secondarily, preventing a cranial shift of the caudal dilatations.

The structure of these lungs restricts the body size of the animals by virtue of its limited parenchymal thickness. The body mass increases proportionally to the third power of the body length, whilst the exchange surface of the lung increases only to the second power. Therefore, larger reptiles are forced to develop a higher exchange surface area by septation of the unicameral lung. The transition is demonstrated by a number of lizards like iguanids, agamids, and chameleons (Fig. 5). They have one or two septa, subdividing the lung into a small number of chambers: a paucicameral lung. This lung form retains all other structural principles of the faviform unicameral lung, e.g., the vascular supply, the smooth muscle system, and the faviform parenchyma. The bronhcial entrance lies far cranially, and the lung is freely mov-

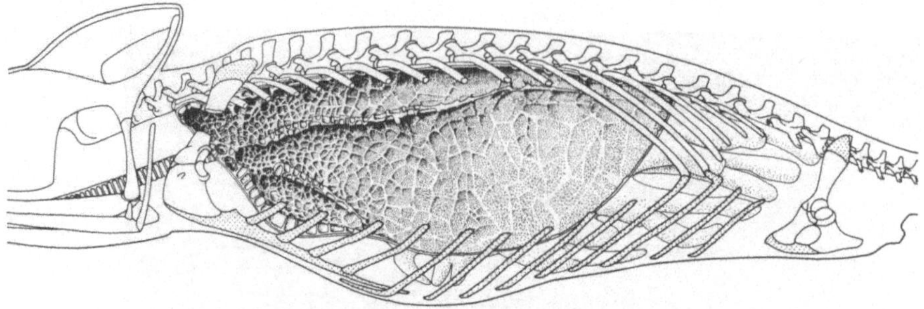

Fig. 5. European chameleon [*Chamaeleo chamaeleon* (L.)]. Paucicameral lung, inflated, in its topographical position in the skeleton. Lung with two septae and smooth muscle network. Ventral and caudal finger-like dilatations

able in the pleuro-peritoneal cavity, fixed only by the two mesopneumonia. Also this lung may be caudally dilated into large nonrespiratory parts as in the chameleon.

From this transitional state of the septated paucicameral lung the lung structure of the monitor lizards, turtles, and crocodiles is understandable. Their lungs are subdivided by an increasing septation into a larger number of chambers: they are multicameral. Together with the development of numerous chambers, one intrapulmonary ductus or bronchus arises to maintain communication of chambers with the extrapulmonary bronchus in all respiratory phases. This ductus or main intrapulmonary bronchus is supported by cartilagineous plates or rings to prevent it from collapsing. One of the best examples of this lung structure is demonstrated by the monitor lizards (Fig. 6). The numerous chambers connected to the intrapulmonary bronchus are very different in size. Medial, dorsal, and dorso-lateral to the bronchus are a large number of small chambers with faviform wall structure exhibiting a large exchange surface (Fig. 7). The ventrally originating chambers that occupy the anterior, the ventral, and the caudal part of the lung, are, in contrast, very large. Only their narrow portion, near the origin of the main bronchus, has a faviform wall structure, whereas the distal parts are more or less dilated and exhibit only a very reduced exchange surface. Accordingly the lungs, in the totally inflated state, occupy up to 30% of the total body volume (Perry and Duncker, in preparation).

Parallel with the development of the multicameral lung, the arrangement of the vascular system changed tremendously. The pulmonary artery and the pulmonary vein and their main branches are totally incorporated into the lung. They run just ventral to the intrapulmonary bronchus, from where the branches which supply the different parts of the lung run in the walls between the chambers. The arrangement of the arterioles and venules supplying the faviform parenchyma seems to be similar to that in the unicameral lung.

Basically the lungs of monitor lizards have the heterogeneous partition (Hughes, 1973) and are totally grown together with the walls of the thoracic cavity. They have lost their pleural space and thus are preexpanded in all respiratory phases (Duncker, 1977). The small dorsal chambers and the very large ventral and caudal chambers are very different in size. Because the surface tension is not only a function of the wet tissue surface with its surfactant film, but also depends on the curvature radius, the large chambers would be overexpanded, whilst the small chambers would collapse if this lung were freely movable in

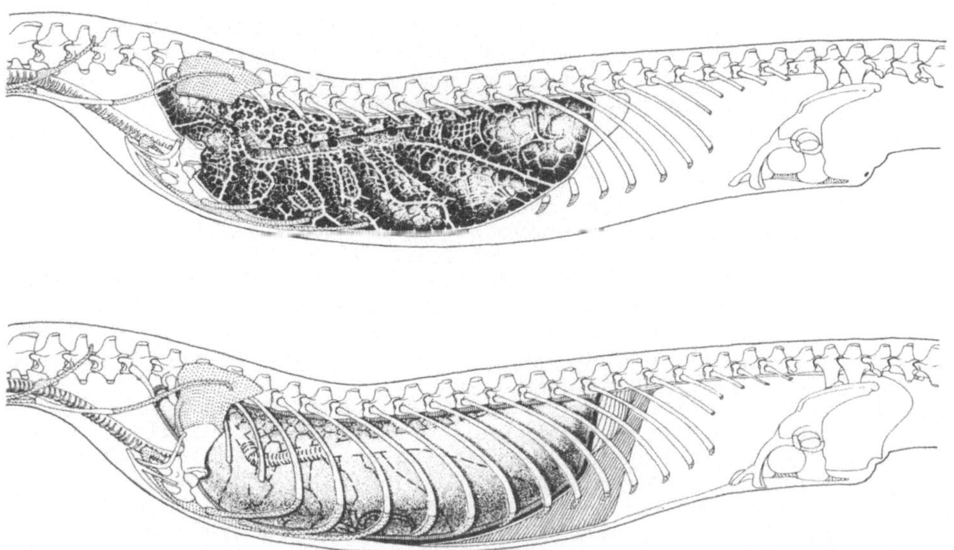

Fig. 6. Monitor lizard [*Varanus exanthematicus* (Bosc)]. *Top*, Multicameral heterogeneous lung, inflated. Small chambers are arranged dorsally around the intrapulmonary bronchus, whereas large thin-walled chambers lie ventrally and caudally. *Bottom*, Lung in its topographical position in the body cavity, adhering to the body wall and to the postpulmonary septum ventrally and caudally. Pericardium in the middle of the lung at the cranial border of the postpulmonary septum, at its caudal side the liver

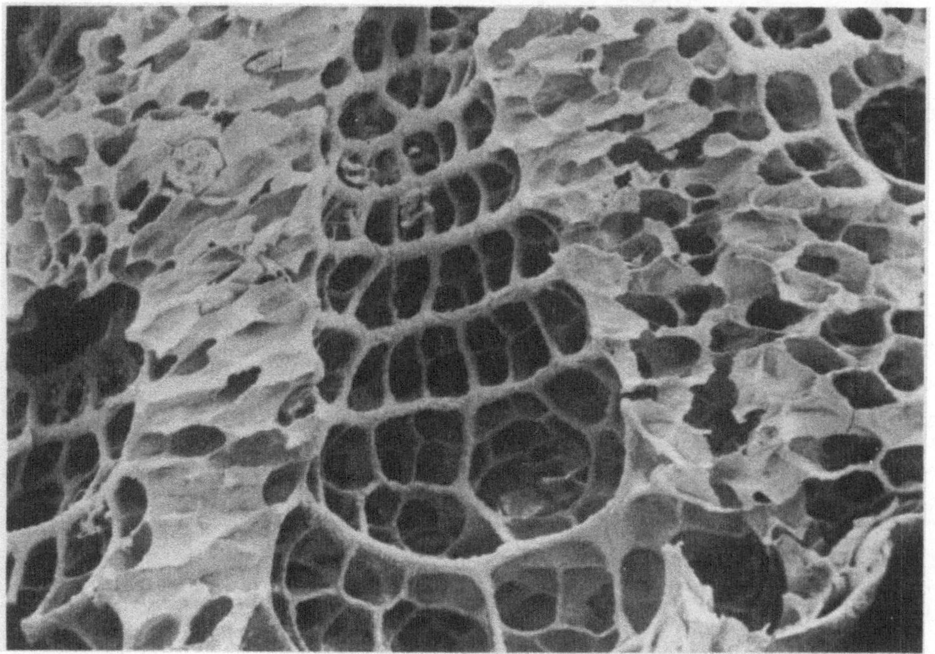

Fig. 7. Monitor lizard [*Varanus exanthematicus* (Bosc)]. Lung, scanning electron micrograph. Part of the small-chambered dorsal region of the heterogeneous lung near the intrapulmonary bronchus with smooth muscle network and densely partitioned chambered wall. Fixed by glutaraldehyde instillation, critical point method, ×30

the pleuro-peritoneal cavity as in lizards described above. Movement is inhibited by the preexpansion of the small-chambered dorsal parts. The lung is ventilated by respiratory movements, which act especially on the dorsal lung parts by the action of the anterior rib cage, as can be readily observed.

The loss of the pleural cleft by the adhesion of the lungs with the body wall is always combined with the development of a postpulmonary septum. It originates from the transverse septum and is closed by the cranial nephric folds (Broman, 1937; Goodrich, 1958; Nelson, 1953), which close the postpulmonary septum dorsally in the form of the cranial mesosalpinx also in adults. This septum separates the lung caudally and ventrally from the liver and the intestines. The lung is also grown together with this septum. Due to total filling of the anterior thoracic cage with the lung, the liver has been pushed caudally together with the postpulmonary septum. Because the pericardium is ontogenetically directly connected to the transverse septum the pericardium also undergoes a descent. Thus, the topographical situation of the heart is a consequence of the lung structure and extension. The descent of the heart determines the position of the lung hilus in the middle of the lung above the cranial pericardium as well as the bronchial and vascular arrangement in the lung (Duncker, 1977).

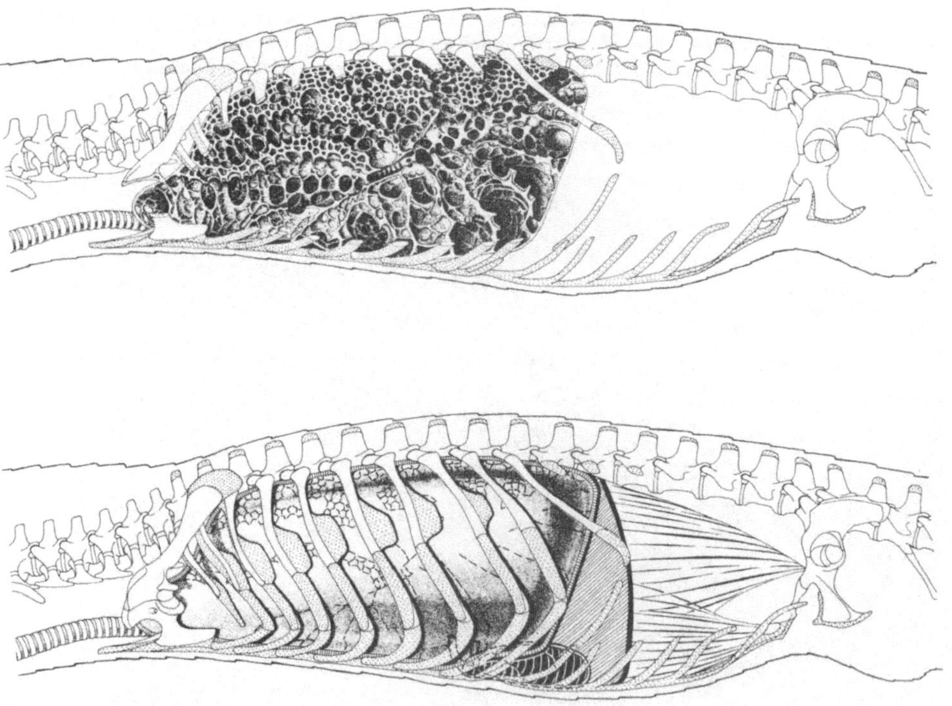

Fig. 8. Caiman [*Caiman crocodilus* (L.)]. *Top*, Multicameral heterogeneous lung, inflated. The multiple lung chambers originate from a short intrapulmonary bronchus, the dorsal chambers being more intensively partitioned than the ventral ones. *Bottom*, Lung in its topographical position in the body cavity, adhering to the body walls and to the postpulmonary septum. Ventral to it, the pericardium. Between postpulmonary and posthepatic septa, the liver. The M. diaphragmaticus connected to the posthepatic septum

The chambers of the heterogeneously subdivided lung are ventilated by the respiratory movements, whereas the exchange of gas in the faveoli of the wall occurs more by diffusion. These structural and functional principles of heterogeneously chambered, multicameral lungs are very similar in crocodiles. From an intrapulmonary bronchus supported by cartilage rings, a great number of chambers originates in three longitudinally arranged series (Fig. 8). They are of different size, the anterior and dorsal being smaller than the ventral and caudal ones, but the differences are not so extreme as in monitor lizards. The walls of the chambers in the crocodile lung are densely partitioned for enlargement of the exchange surface, but in a different manner than in monitor lizards. Instead of possessing a faviform arrangement of the partitions, the parenchyma is subdivided into more bubble- or vesicle-like compartments, into *bullae* (Fig. 9). These bulliform spaces differ in diameter, and are more intensely separated into smaller bullae dorsally than ventrally.

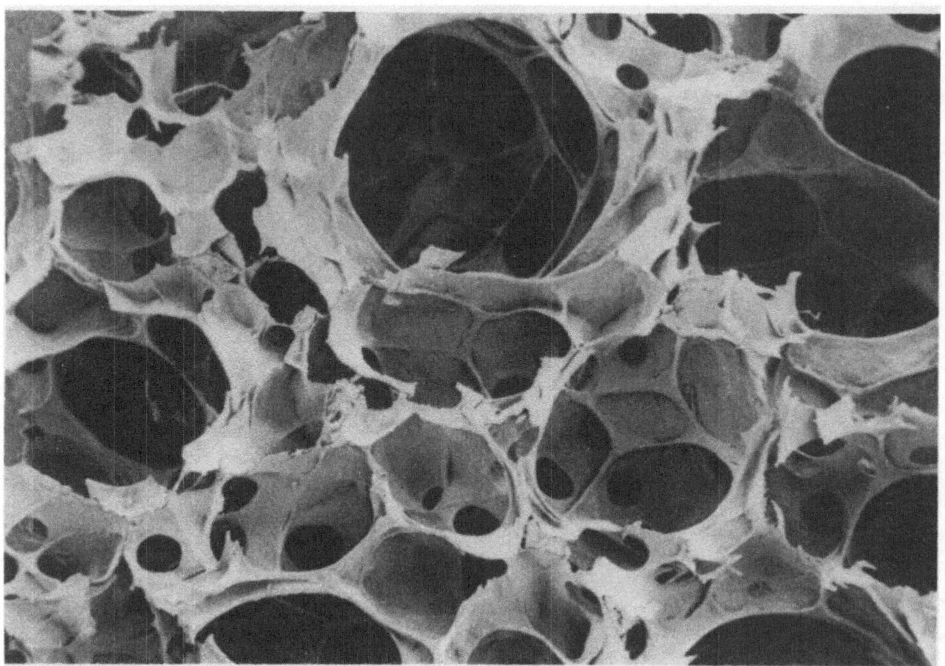

Fig. 9. Caiman [*Caiman crocodilus* (L.)]. Lung, scanning electron micrograph. Heterogeneously partitioned dorsal part of the lung with various sized bulliform spaces. Fixed by glutaraldehyde instillation, critical point method, ×25

Crocodilian lungs also entirely fill the anterior rib cage. The pleural cleft may be totally or partially absent, but in all species the lungs fuse with the postpulmonary septum. Because the lungs fill the entire anterior thoracic cavity, the same topographical relationships exist between lung, heart, and intestines as in monitor lizards. In this way the position of the lung hilus is in the middle of the lung, and the vessels entering it there are also distributed near the bronchus. The respiratory apparatus of crocodilians is characterized by the development of a special muscular apparatus for the inspiratory movements. In addition to the rib cage musculature, the so-called M. diaphragmaticus is present, and serves as a strong inspiratory muscle (Gans and Clark, 1976). It is derived from the internal abdominal musculature and makes up a complete muscle layer in the lateral and ventral abdominal wall (Fig. 8). This diaphragmatic muscle is connected to the posthepatic septum, which lies caudal to the liver. The muscle pulls the septum and the liver backward like a piston and so inflates the lung. Although functionally comparable to the mammalian diaphragm, it is of quite different structural origin.

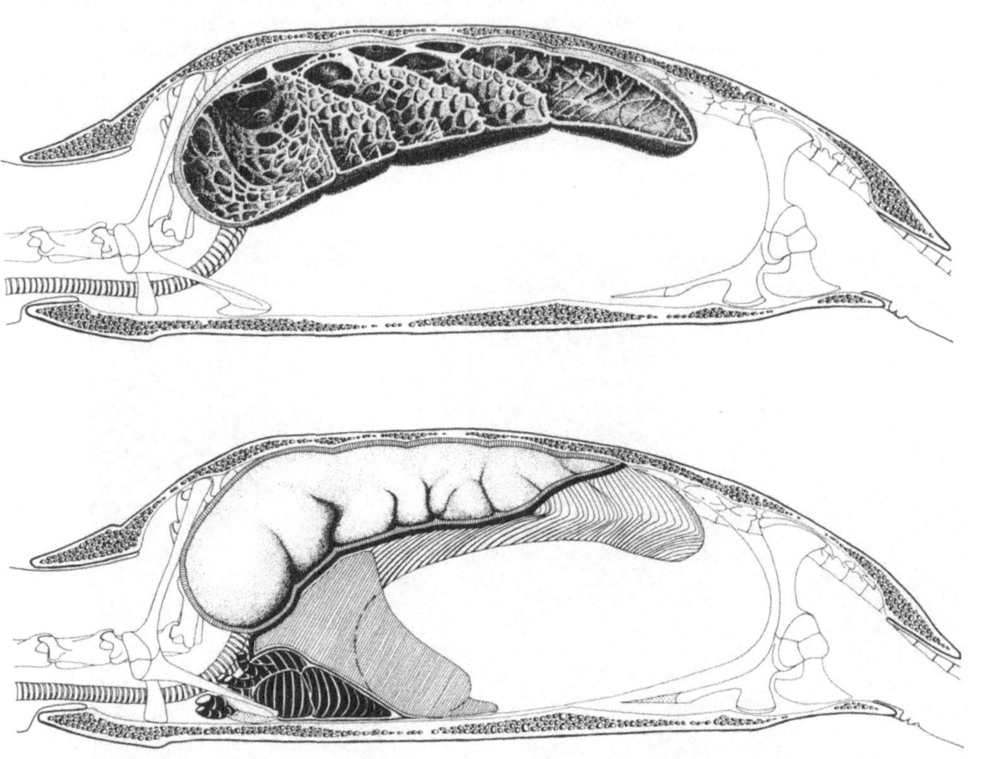

Fig. 10. Red-eared turtle [*Pseudemys scripta elegans* (Wied)]. *Top*, Multicameral heterogeneous lung, inflated. Around the dorsally located duct-like intrapulmonary bronchus smaller and more densely partitioned chambers are developed, laterally and caudally larger and less densely partitioned chambers. *Bottom*, Lung affixed to the carapace and to the underlying postpulmonary septum. Below the cranially located lung hilus, the pericardium; dorsally and caudally to it, the liver

The lungs of turtles are also septated into heterogeneous chambers with smaller and greatly subdivided walls around the dorsal intrapulmonary duct-like bronchus (Fig. 10). The highly developed lungs of marine turtles are densely partitioned. The multiple chambers which are connected to one main bronchus contain a thick parenchyma around a bronchus-like duct. Their lungs are attached to the carapace and to the postpulmonary septum, which separates them from the intestines. The construction of the turtle carapace and the lack of a rib cage requires a totally different mode of respiratory movements from that occurring in most species, namely by the action of the forelegs and shoulder girdle with the assistance of the abdominal musculature and hind limbs (Gans and Hughes, 1967; Gaunt and Gans, 1969). Because of this construction, the lungs do not occupy the total anterior body cavity and thus the liver and the postpulmonary septum are not pushed caudally. The pericardium therefore lies cranial in the thorax near the shoulder girdle, as in the primative lizards, and dorsal to it, the lung hilus is situated.

In turtles the small chambers in the very heterogeneously partitioned, multicameral lungs are prevented from collapse by adhesion of the lung to the body wall. Water turtles demonstrate a dilatation of their caudal and lateral lung chambers for regulating buoyancy. The general organization of this turtle lung resembles that of monitor lizards and crocodiles, especially according to the arrangement of the large vessels which run in the interior of the lung near the interpulmonary bronchus. The septation of the walls of the chambers resembles more that of crocodiles, smooth muscle bundles being arranged around the openings into the single bulliform spaces of varying sizes.

The avian lung demonstrates the highest development of the structural principles of heterogeneous lungs, fixed to the body wall for the maintenance of their functional capacity. The relatively small avian lung is fixed in the dorsal thoracic cavity and separated ventrally from the air sacs and the intestines by the horizontal septum which originates from the postpulmonary septum (Poole, 1909). This avian pulmonary cavity not only allows a preexpansion for the lung, but also insures that the lung maintains a practically constant volume during all respiratory phases. This depends on the dorsal position of the lung and on the action of the costoseptal muscles at the margin of the horizontal septum, which compensate for small volume changes occurring by the respiratory movement of the rib cage. This expansion of the lung prevents the highly developed secondary bronchi of the avian lung from collapsing, thereby functionally guaranteeing the air flow through the lung (Duncker, 1971).

In this volume-constant lung, an air capillary network is able to develop around blood capillaries. The air capillaries, only 3 μm to 10 μm in diameter, cannot be inflated out of a collapsed state because of their high surface tension. Their existence is only guarenteed in the rigid avian lung. Their development increased the exchange surface by a factor of 10 per volume unit compared with mammalian lungs of equal sized species. The meshwork of air and blood capillaries make up the 300-μm to maximally 500-μm wide wall of the parabronchi, the functional unit of the avian lung (Fig. 11). The parabronchi are long, stretched, parallel tubes, connecting the series of secondary bronchi to each another. In the paleopulmo the air flows unidirectionally through this parabronchial system. The specific arrangement of the smooth muscle bundles in the avian parabronchi surrounding the lumen and forming interconnected rings or hexagonal meshes resembles the typical arrangement of smooth muscle in reptilian lungs (Fig. 11). In this highly developed avian parabronchus, the smooth muscle meshwork does not have the function of extending the exchange surface, which is done by the rigid overall expansion of the lung, but they regulate the flow of air through the parabronchi by varying the diameter, thereby regulating the air resistance.

The arrangement of the vascular system has reached its highest development in the avian lung. The long and only poorly interconnected blood capillaries arise from arterioles at the outside of the parabronchi and are collected mainly by venules at the luminal side. This special arrangement of the capillaries and that of the long, unidirectional parabronchi are responsible for the high gas exchange capacity of the avian lung (Duncker, 1974). The surfactant film of the air capillaries is very well developed. Its main function is not to make the air capillaries

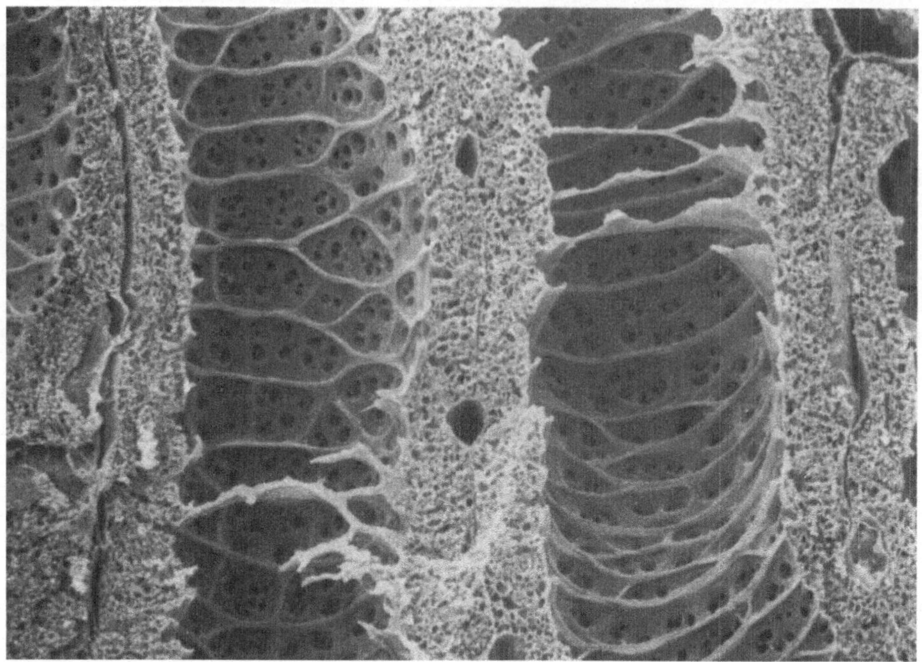

Fig. 11. Khaki Campbell duck (*Anas platyrhynchos* L. f. *domestica*) 6 weeks old. Lung, scanning electron micrograph. Parabronchi in longitudinal section with smooth muscle bundles around the parabronchial lumen. Atria with originating infundibula. The parenchyma of the parabronchi out of the blood and capillary network, incorporating single larger blood vessels. Fixed by glutaraldehyde instillation, critical point method, ×80

expandable, but to prevent transsudation of fluid into them from the blood capillaries, thus preventing an oedema.

This highly developed rigid avian lung can only be ventilated by means of the air sacs. They develop ontogenetically out of bronchi, which penetrate the ventral surface of the lung and invade the post-pulmonary septum, splitting it into the horizontal septum directly beneath the lung, and into the oblique septum, which separates the intestine from the air sacs (Poole, 1909). Only the abdominal air sac in most birds invades the abdominal cavity. The respiratory movements act on the air sacs as bellows which cause the air flow through the lung. This highly complicated unidirectional air flow through the paleopulmo (Scheid et al., 1972) is determined by the way in which the secondary bronchi originate from the main bronchus, and in which the air sacs are connected to the secondary bronchi.

Comparing the avian lung with the higher developed reptilian lungs, we find that the avian lung has taken over some of these structural principles: (1) a single intrapulmonary main bronchus; (2) a similar arrangement of the gross vascular system; (3) the preexpansion by three series of secondary bronchi; (4) the adhesion to the body wall; and (5) the separation from intestines by the postpulmonary septum. However, there are special developments of the avian respiratory system which distinguish it from all reptilian lungs: (1) the interconnecting of the parabronchi from the medioventrobronchi with the parabronchi from the mediodorsobronchi, and from the lateroventrobronchi, making up the functional circuit in the avian lung; (2) the development of air capillaries; and (3) the development of the air sacs by invading and splitting the postpulmonary septum. There is no comparable structure in any other vertebrate, not even in the caudal nonrespiratory dilatations of reptilian lungs.

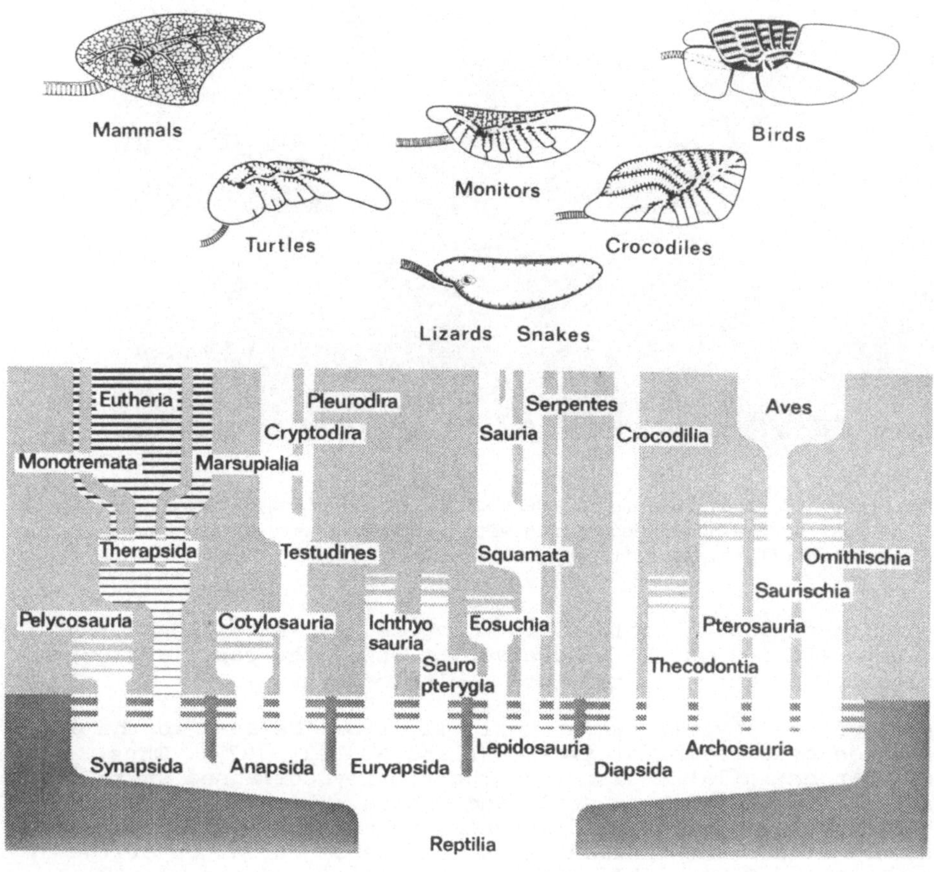

Fig. 12. Phylogenetic tree indicating the relationship between the different reptilian orders with heterogeneous lungs and birds with highest developed heterogeneous lungs, in contrast to mammals with homogeneous lungs

This line of differentiation of the reptilian lung leading to the avian lung, the highest developed vertebrate respiratory system, depends on the heterogeneous lung partitioning. The development to mammals has undergone quite another pattern. Investigation of this development is hampered by the lack of surviving reptilian ancestors of mammals (Fig. 12). Analysis of the mammalian lung shows that in therapsid reptiles the partitioning of the lung for the increase of gas exchange surface must have been homogeneous. Therefore, a pleural space persisted and the convergent development of a postpulmonary septum as the diaphragm made possible a forceful and effective inflation of these homogeneously partitioned lungs. In the further subdivision of the lung chambers, the bulliform spaces are septated into groups of alveoli (Fig. 13). Out of the intrapulmonary bronchi a bronchial tree had to be developed

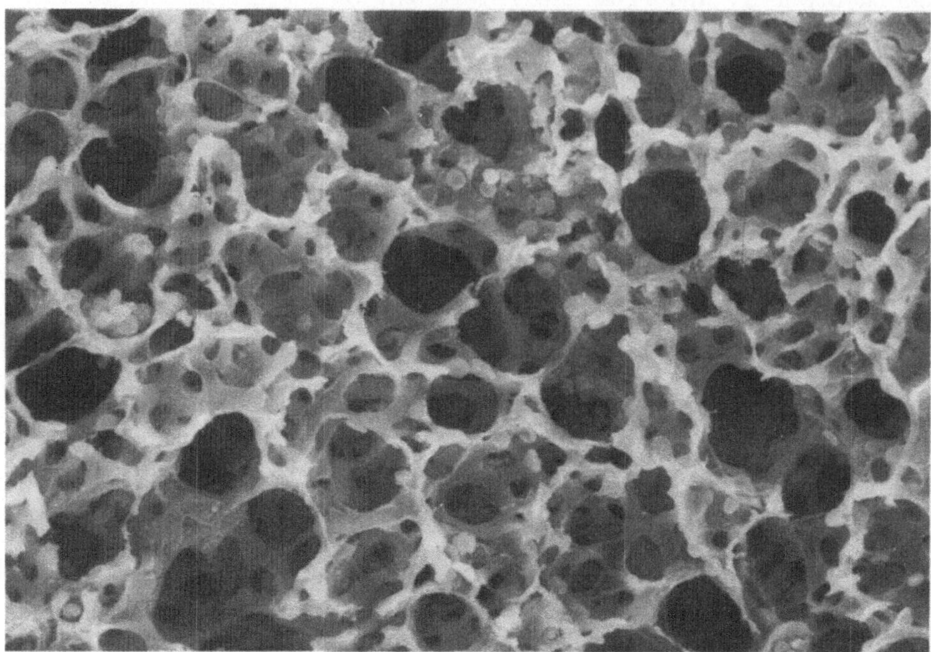

Fig. 13. White-toothed shrew [*Crocidura russula* (Hermann)]. Lung, scanning electron micrograph. Alveoli of the homogeneous lung with multiple interalveolar pores. Fixed by glutaraldehyde instillation, critical point method, ×650

to supply all groups of alveoli. This homogeneously subdivided bronchio-alveolar lung retained in the interalveolar septa the capillary network of reptilian lungs, but combined into a single layer.

The diameter of the alveoli varies maximally from the apex to the diaphragmatic side by a factor of two (Gehr and Weibel, 1977). Thus, the alveoli can be inflated also from the totally collapsed state, as is demonstrated in deep-diving seals and whales. This inflation is made possible by the surfactant film which reduces the surface tension. The dependence of surface tension on the radius of curvature of the alveoli seems to be responsible for the lower limit of the possible alveolar diameter, thus limiting the amount of the exchange surface. These structural conditions of the mammalian lung were the basis for the evolution of the obligate vivipary in metatherian and eutherian mammals. In contrast, the highly evolved avian lung, with its necessarily rigid structure, hinders an expansion even at the moment of birth and therefore birds were forced to maintain the reptilian ovipary.

Acknowledgment. Supported by grants of the Deutsche Forschungsgemeinschaft (DU 50/3, 50/4).

References

Broman, I.: Cölom. In: Bolk, Göppert, Kallius, Lubosch (eds.): Handbuch der vergleichenden Anatomie der Wirbeltiere. Bd. III. Berlin-Wien: Urban u. Schwarzenberg, 1937, 989-1018

Duncker, H.-R.: The lung air sac system of birds. Ergebn. Anat. Entwickl. Gesch. 45, Heft 6, 1-171 (1971)
Duncker, H.-R.: Die Anordnung des Gefäßsystems in der Vogellunge. Verh. Anat. Ges. 68, 517-523 (1974)
Duncker, H.-R.: Coelom-Gliederung der Wirbeltiere - Funktionelle Aspekte. Verh. Anat. Ges. 72 (1977) in press
Gans, C., Clark, B.: Studies on ventilation of *Caiman crocodilus* (Crocodilia: Reptilia). Respir. Physiol. 26, 285-301 (1976)
Gans, C., Hughes, G.M.: The mechanism of lung ventilation in the tortoise *Testudo graeca* Linné. J. Exp. Biol. 47, 1-20 (1967)
Gaunt, A.S., Gans, C.: Mechanics of respiration in the snapping turtle, *Chelydra serpentina* (Linné). J. Morphol. 128, 195-228 (1969)
Gehr, P., Weibel, E.R.: Comparative morphometric analysis of the horse and cow lung. Experientia (1977) in press
Goodrich, E.S.: Subdivisions of the coelom and diaphragm. In: Goodrich (ed.): Studies on the Structure and Development of Vertebrates. London 1930. Reprint New York: Dover Publ., 1958, 613-656
Hughes, G.M.: Comparative vertebrate ventilation and heterogeneity. In: Bolis, Schmidt-Nielsen, Maddrell: Comparative Physiology. Amsterdam: North Holland, 1973, 187-220
Marcus, H.: Lungen. In: Bolk,Göppert, Kallius, Lubosch: Handbuch der vergleichenden Anatomie der Wirbeltiere. Bd. III. Berlin-Wien: Urban u. Schwarzenberg, 1937, 909-988
Moser, F.: Beiträge zur vergleichenden Entwicklungsgeschichte der Wirbeltierlunge (Amphibien, Reptilien, Säuger). Arch. Mikr. Anat. Entw. 60, 587-668 (1902)
Nelsen, D.E.: The development of the coelomic cavities. In: Nelsen: Comparative Embryology of the Vertebrates. New York-Toronto: Blakiston Co., 1953, 857-873
Perry, S.F., Duncker, H.-R.: Reptilian lungs: correlation of volume and static compliance with lung structure (in preparation)
Poole, M.: The development of the subdivisions of the pleuroperitoneal cavity in birds. Proc. Zool. Soc. Lond. 49, 210-235 (1909)
Scheid, P., Slama, H., Piiper, J.: Mechanisms of unidirectional flow in parabronchi of avian lungs: measurements in duck lung preparations. Respir. Physiol. 14, 83-95 (1972)

Ventilation Mechanisms: Problems in Evaluating the Transition to Birds

C. Gans

Summary

Review suggests that the ventilatory system of recent reptiles shows a diversity of structural and mechanical patterns of ventilation. The systems appear to have been selected for low energy expenditure; both behavioral and ecological factors influence the patterns seen. Endothermy apparently involved tighter constraints and hence greater stereotypy.

1. Introduction

Mammals and birds evolved from reptiles more than a hundred million years ago. The six orders of reptiles that survive today, namely turtles (Testudines), crocodilians (Crocodilia), beak-heads (Rhynchocephalia), lizards (Sauria), snakes (Serpentes), and worm-lizards (Amphisbaenia), appear to have separated from one another at or before the time of the origin of birds. Consequently there has been a substantial time for extensive modification of each, and it cannot be assumed that any of the members of an order now retain the structural or the phylogenetic patterns they showed at the time of their divergence.

Each order, furthermore, includes forms highly diverse in their ecology and in almost all aspects of their biology. While examination of this diversity may offer some clues regarding the origin of the ventilatory patterns seen in birds and mammals, only a few of the multiple mechanical patterns have yet been studied. Thus we lack data on most lizards, *Sphenodon*, and the Amphisbaenia, while the functional speculations of Wolf (1933) remain to be tested. In view of the parallel report of Duncker on morphological aspects, I restrict this introduction to the effectors of ventilation.

2. Historical

Recent reptiles are normally biphasic aspiration breathers (Gans and Clark, 1977). Buccal pumping is only seen in traumatized individuals, and during certain defensive displays. Ventilation by aspiration probably evolved before the tetrapod condition; evidence suggests that crossopterygian fishes gave up buccal pumping even before the carboniferous radiation of amphibians (Gans, 1970). Buccal pumping of air was intrinsically inefficient, particularly in animals submerged in water.

Division of Biological Sciences, The University of Michigan, Ann Arbor, MI 48109/ U.S.A.

The larger fossil amphibians, from which the reptiles presumably arose, were clearly aspirators: recent amphibians (frogs, salamanders, and caecilians) are irrelevant in this connection.

3. Mechanical Patterns

Turtles ventilate by changing the anterior and posterior limits of the visceral mass within their shell (which may not be rigid, and indeed has long been known to bend during breathing); they thus indirectly induce volumetric changes of the lungs suspended in the dome of the shell. Turtles may also change the visceral space by other devices such as inward and outward movement of the shoulder or the pelvic girdle (Gans and Hughes, 1967; Gaunt and Gans, 1969).

In crocodilians the pleural cavity lies within the rib cage. This has limited mobility, though the ribs are complexly articulated along their length so that their position and configuration change when the animals shift from land to water. Posteriorly, the pleural cavity is limited by the liver and its radiating ligaments. Contractions of the transverse and diaphragmatic muscles induce anteroposterior shifts of the liver, modify the pulmonary volume, and ventilate the lungs (Gans and Clark, 1976).

Lizards apparently modify the volume of the pleural space by rotation of the ribs between the vertebrae and the sternum as in mammals. The rib cage extends some distance beyond the pleural cavity, although the more posterior visceral mass is supported by abdominal rather than dorsal ribs.

Snakes have an array of ventrally free ribs that extend the full length of the trunk. Their rotation on the vertebral column decreases or increases the visceral space. The ventral skin may also be pulled inward to decrease the volume of the cavity (Rosenberg, 1973).

4. Design Peculiarities

Special features in different forms require amplification and sometimes modification of these oversimplified statements about the bases of reptilian ventilatory mechanics.

In turtles, we have to pay attention to the enormously long trachea and bronchi, which compensate for the extensible head and neck and may thus increase the ventilatory dead space. The ventilatory muscles vary widely in their magnitude and location (George and Shah, 1959). Some turtles have the entire surface of the lung invested with striated muscles. We lack direct observation on the function of such systems.

In crocodilians, the external nares are at the tip of the variously elongate snout, and the secondary palate is enormous, with the internal choanae lying back of the eyes. During ventilation the larynx is adpressed to the internal choanae, and the respiratory dead space is reduced by by-passing the posterior buccal chamber. Nasal valves close the external nares during dives.

Some aquatic lizards show diverse valvular nostrils, while sand living species use them to avoid the penetration of particles. Lateral grooves or folding zones along the trunk may act as volume-adjustment areas in ventilation. A whole spectrum of sternal arrangements and abdominal ribs is observed (cf. Etheridge, 1965), for which we thus far lack adequate functional explanations. The sand-living lizards have behavioral modifications for developing "pressure shadows" (or external air pockets) to accomodate ventilatory movements beneath the sand that presses downward on them (Pough, 1969). The diversity of lung types and pulmonary vascular patterns is as yet poorly characterized, and even less understood (Willnow and Willnow, 1976). Some patterns ascribed to groups such as the chameleons apply only to a few species (Klaver, 1977).

Aquatic snakes may have valvular nostrils, but the internal choanae lie anteriorly. The glottis of snakes is often at the end of a cartilaginously reinforced tube, which is protrusible in forms that feed on prey of large diameter, so that the snakes can continue to breathe during ingestion. Some snakes have a pair of functional lungs and complex vascular patterns. Other snakes have one lung shorter than the other or only a single lung may remain (Underwood, 1967). The lungs generally start at the level of the heart (perhaps 25% down the body). In some species the elongate trachea shows lateral vascularized diverticulae (so-called tracheal lungs). In most snakes, the posterior aspect of the lung is extended as an elongate airsac that may reach to the gonads. Ventilatory movements may thus involve the ribs of only a local section of the trunk, for example, and thus be removed from the site where the prey is constricted or swallowed (McDonald, 1959) or from the cuff of a plethysmograph.

5. Complex Adaptations

It may be easy to design a perfect gas exchanger, but it is impossible to design a perfect organism to fit an unpredictably variable environment. Like all living organisms, reptiles are adapted to current conditions, neither their ventilatory apparatus nor any other system will retain any intermediate historical condition. Consequently, we see adaptive compromises and synergisms among demands for different functions. Two kinds of situation may illustrate the importance of such considerations and the limitations they impose on simple analyses.

First, consider the design constraints for lungs in swimming and diving reptiles. The importance to the animal of gas exchange in the lungs varies with its capacity to exchange gas at other sites, such as the integumentary, buccal, and cloacal surfaces. Maintenance of an integumentary barrier to the passage of water appears to involve an energetic cost. Furthermore, the integumentary permeabilities to water and to respiratory gas are correlated (Wood and Lenfant, 1976). When terrestrial reptiles, such as snakes and turtles, secondarily became aquatic they seem to have relaxed this permeability barrier; however, they obviously found it important to maintain a barrier to salt flux. A barrier to water loss was a problem only when the animals temporarily emerged. Integumentary permeability for gas exchange may independently have been advantageous to a marine organism as it reduced the number of trips that the animal had to make to the surface to replenish its air, thereby limiting its exposure to predators, energy expenditures for the trips, and volumetric change associated with shifts of hydrostatic pressure. The lung could become simplified, and perhaps, smaller, thus reducing the need to counteract its buoyancy. Behavioral changes that limited O_2 use might have become advantageous, though circulatory adjustments may have been required before the shift in absorption sites could be fully utilized. It is difficult to tell whether alteration of the permeability of gas or that of water provided the selective advantage for the change.

A different set of complicating factors is noted when analyzing the effects of absolute size, both in the ontogeny of a single form and among related species. During the growth of an organism, the rate of ventilatory movements is size dependent; indeed, H. Pough's recent work suggests that alveolar gas tensions in snakes are also size-dependent, both in intra- and interspecific comparisons. The location of ventilatory action may differ with age. Thus in hatchling snakes, ventilation proceeds anteriorly only, while in adults the entire set of ribs moves in and out (Pough, pers. comm.). This mechanical change is not obviously size dependent. It may be that packing factors are involved and that the muscles of these animals perhaps become more complex with age, as do those of crocodilians (Gans and Clark, 1976).

6. Four Generalizations

a) Recent Reptiles Represent Multiple Experiments in Evolution

Two of the preceeding sections have shown that there is great diversity in the way in which aspiration breathing is achieved in the major groups of reptiles. Turtles differ among each other, crocodilians are unique, and lizards seem to show significant diversity, as do snakes; we are as yet uncertain about breathing mechanics of amphisbaenians and *Sphenodon*. As far as we can now tell, the mechanics reflect subtly different approaches and the absence of a common structural plan. The pattern of motor control may well be more similar than the sets of muscles involved, indicating that it incorporates greater intrinsic evolutionary flexibility. We see large airsacs and highly trabecular sections, but lack any demonstration that gas flow is more than simply tidal.

b) Reptiles have been Selected for low Metabolic Expenditures

Reptiles differ from birds and mammals, not only in that both their resting and active metabolic rates are about one-tenth of those shown by the latter but also in that they spend a far smaller protion of the day and season as active animals. Estimates (Bennett and Nagy, 1977), based on the use of doubly labeled water in field and laboratory, indicate that some lizards operate at one-thirtieth of the energy cost of a mammal of equivalent mass. This means that energetically marginal environments are more likely to support a reptilian than a mammalian population. Some curious strategies are hence advantageous. For instance, the males of some small snakes have recently been shown to restrict their active hunting activity in early summer, immediately after completion of mating (Feaver, 1977). Further feeding would increase their growth, which might improve chances of successful mating in the following year; however, it seems also to expose them to predation. Their very low basal metabolic rate makes the cost of resting for a protracted period low enough to justify inactivity as a method of avoiding predation.

Reflections of this low metabolism may be seen in the typically very slow breathing movements (particularly of the larger species; even small crocodilians may take four or more seconds to inhale), and in the prolonged apneusis throughout which most gas exchange occurs (Lenfant et al., 1970). Demands on the external gas exchangers of reptiles may be for minimal absolute cost, rather than for relative efficiency.

c) Reptilian Ventilation Incorporates Major Behavioral Effects

Certain reptiles are seemingly immobile and "lethargic"; actually their unresponsiveness is spurious. Immobility may often be considered as a predator-avoidance device; it masks changes in cardiac and ventilatory activity. As the animals notice predators, their ventilatory rate increases, and as their excitement increases, their ventilatory rhythm may shift from biphasic to triphasic (Gans and Clark, 1977).

Withdrawal bradycardia in crocodilians (Gaunt and Gans, 1969; H. Messel, pers. comm.) accompanies diving in response to disturbance (rather than to short-term voluntary dives) and may be accompanied by shifts in the flow pattern of the blood (F. White, this volume). Of course, the pulmonary system also has a role in defense and display. Inflation allows reptiles to wedge themselves into crevices, and it also occurs in response to predators or in sexual displays. Similarly the ventilatory system has a critical role in sound production, both for vocalization (hissing) and when the lung is used as an amplifying device for sounds generated on the integument (Gans and Maderson, 1973). Such functions may have major importance in explaining the lung structures that have been described.

d) Reptilian Ventilation Reflects the Solution of Ecological Problems

A variety of environments have induced adaptive responses that affect ventilatory mechanisms. Examples are the responses of reptiles to riparian or to totally aquatic situations. In some cases, one sees changes in the patterns of muscle utilization, but muscular arrangements also change: aquatic turtles seem to have stronger inhalatory muscles than do terrestrial ones in which the inhalation may be partially achieved by the passive drooping action of the viscera. The undersand environment represents another extreme. Not only are there nostril specializations for filtering out sand particles, but the nostrils may shift to the ventral side of the head, while behavioral postures may permit lizards and snakes to inhale beneath soil and other loose overburden without working to displace overburden coincident with each breath (Norris and Kavanaugh, 1966).

More important, though less obvious, are the indirect effects of environments. Reptiles have adapted to multiple situations by changing their body proportions. Among tetrapods, they show the greatest diversity in their length to diameter ratio. This clearly has implications for the mechanics of lung filling, although ventilatory mechanics were rarely the primary selective force.

7. Question

I cannot avoid noting one perplexing point, namely the function of the smooth muscle in the lungs of many lower tetrapods. This muscle contracts on a frequency much lower than that seen in ventilation. In frogs, the strength of the spontaneous muscular contractions has been shown to increase with CO_2 concentration, and it may show some association with the hysteresis in the pulmonary pressure volume curve (Kato, 1951). Incidental observations suggest that it may be associated with the distribution of gas to different parts of the reptilian lung; however, experimental proof is still lacking.

8. Conclusions

The diversity of reptilian ventilatory arrangements and patterns represent multiple solutions to the problem of ventilation: how air can be aspirated into and forcibly expelled from the lungs. The diversity may result from the effect of numerous synergisms and from compromises with the demands imposed by environments currently or historically occupied by reptiles. The seemingly greater stereotypy of pulmonary mechanics and clearly greater exchange per unit volume seen in birds and mammals may reflect greater emphasis on the demands that must be met by their gas exchange systems. They have higher metabolic rates and maintain such rates over more protracted periods, as required by their adoption of endothermy. This presumably placed a premium on the development of gas exchange systems that are more effective, and hence more sterotyped, than those of ectotherms.

Acknowledgments. I am grateful to those colleagues who participated in some of the studies here referred to and to B.D. Clark and P.J. Regal for assistance and comments. Supported by NSF BMS DEB 76 18289 and a travel grant from the Deutsche Forschungsgemeinschaft.

References

Bennett, A.F., Nagy, K.: Energy expenditure in free-ranging lizards. Ecology 58(3), 697-703 (1977)

Etheridge, R.A.: The abdominal skeleton of lizards in the family Iguanidae. Herpetologica 21(3), 161-166 (1965)

Feaver, P.A.: The demography of the Michigan population of *Natrix sipedon* with discussions of ophidian growth and reproduction. PhD thesis, University Michigan, Ann Arbor, 1977

Gans, C.: Strategy and sequence in the evolution of the external gas exchangers of extothermal vertebrates. Forma et Functio 3, 81-104 (1970)

Gans, C.: Ventilatory mechanisms and problems in some amphibious aspiration breathers (*Chelydra*, *Caiman* - Reptilia). In: Respiration in Lower Vertebrates. Hughes, G.M. (ed.). London: Academic Press, 1976

Gans, C., Clark, B.D.: Studies on ventilation of *Caiman crocodilus* (Crocodilia: Reptilia). Respir. Physiol. 26, 285-301 (1976)

Gans, C., Clark, B.D.: Reptiles ventilate their lungs biphasically. Proc. Int. Union Physiol. Sci. 13, 252 (1977)

Gans, C., Hughes, G.M.: The mechanism of lung ventilation in the tortoise, *Testudo graeca* Linné. J. Exp. Biol. 47, 1-20 (1967)

Gans, C., Maderson, P.F.A.: Sound producing mechanisms in recent reptiles: Review and comments. Am. Zool. 13(4), 1195-1203 (1973)

Gaunt, A.S., Gans, C.: Mechanics of respiration in the snapping turtle, *Chelydra serpentina* (Linné). J. Morph. 128, 195-228 (1969)

Gaunt, A.S., Gans, C.: Diving brachycardia and withdrawal brachycardia in *Caiman crocodilus*. Nature (London) 223, 207-208 (1969)

George, J.C., Shah, R.V.: The structural basis of the evolution of the respiratory mechanism in Chelonie. J. Anim. Morph. Physiol. 6, 1-9 (1959)

Kato, H.: Studies on normal respiration of the frog, Part 1, Normal atmung and CO_2 effect on it. (In Japanese) J. Physiol. Soc. Japan 13, 319-327 (1951)

Klaver, C.J.J.: Comparative lung morphology in the genus *Chamaeleo* Laurenti, 1768 (Sauria: Chamaeleonidae), with a discussion of taxonomic and zoogeographic implications. Beaufortia 25(327), 167-199 (1977)

Lenfant, C., Johansen, K., Petersen, J.A., Schmidt-Nielsen, K.: Respiration in the freshwater turtle, *Chelys fimbriata*. Respir. Physiol. 8, 261-275 (1970)

McDonald, H.S.: Respiratory functions of the ophidian air sac. Herpetologica 15, 193-198 (1959)

Norris, K.S., Kavanaugh, J.L.: The burrowing of the western shovelnosed snake, *Chionactis occipitalis* Hallowell, and the undersand environment. Copeia 4, 650-664 (1966)

Pough, F.H.: Physiological aspects of the burrowing of some lizards (*Uma*, Iguanidae, and other lizards). Comp. Biochem. Physiol. 31, 869-884 (1969)

Rosenberg, H.I.: Functional anatomy and pulmonary ventilation in the garter snake, *Thamnophis elegans*. J. Morphol. 140, 171-184 (1973)

Underwood, G.: A Contribution to the Classification of Snakes. London, British Museum (Natural History), 1967

Willnow, I., Willnow, R.: Bauplan der Lunge von *Sphenodon punctatus*. Eine vergleichend anatomische Betrachtung. Acta Anat. 94, 504-519 (1976)

Wolf, S.: Zur Kenntniss von Bau und Funktion der Reptilienlunge. Zool. Jahrb. Abt. Anat. Ontog. 57(1), 139-190 (1933)

Wood, S.C., Lenfant, J.M.: Respiration. Mechanics, control and gas exchange. In: Biology of the Reptilia, Vol. V. Gans, C., Dawson, W.R. (eds.). London: Academic Press, 1976

Lung Surfactant and Lung Lining in Birds

R. E. Pattle

Summary

Bird lung is equipped with a surfactant which is physically, chemically, and morphologically similar to that of mammalian lung. It also produces relatively large quantities of sheets of "trilaminar substance" (TLS), consisting of an osmiophilic layer resembling a unit membrane, apparently flanked by two nonosmiophilic layers. We can only speculate as to the function of this substance. It is released by partial breakdown of cells in the atrial lining, and is secreted to a lesser extent by cells lining the air capillaries. Further work is needed to elucidate the function of TLS. The writer suggests that it may help to prevent accumulation of liquid in the air capillaries.

1. Introduction; Definitions

Lung surfactant is a complex substance containing lipid and protein, which provides the internal surface of the lung with a film which is capable of reducing the surface tension to less than 1 mN/m, thereby lessening the dangers of collapse and of transudation from the blood. Its presence in bird lung was demonstrated by Pattle (1958). Such substances and films are here called *surpellic* (from Lat. pellere, to push); the term *surface active* is here used in its ordinary physicochemical sense, implying a tendency to concentrate at a surface and lower the surface tension, not in the special sense of surpellic. Qualitative data about surfactant have been easy to obtain, but quantitative measurements are difficult and often untrustworthy. A highly critical approach to the literature is necessary. The early work was reviewed by Pattle (1965). Other reviews, books, and symposia are those of Scarpelli (1968), King and Moloney (1971), Weibel (1973), Villee et al. (1973), Tierney (1974), Clements (1974), Goerke (1974), and Pattle (1976). Much of the interest in surfactant centres on its absence in the respiratory distress syndrome of premature human babies (Avery and Mead, 1959); the "barker" syndrome in the thoroughbred foal appears to have a similar origin (Pattle et al., 1975). A point of major interest in bird lung is the peculiar *trilaminar substance* (TLS) which is not found in other lungs (Figs. 1, 2). The surface features of bird lung can best be appreciated by comparison with other lungs.

Chemical Defence Establishment, Porton Down, Salisbury, Wilts., SP4 OJQ, Great Britain.

2. Lung Surfactant in Mammals

The surfactant was originally noticed (Pattle, 1955) because of the extreme stability which it gives to bubbles derived from the lung. Bubbles of about 50 μm diameter squeezed from a fragment of lung into a hanging drop of air-saturated water remain stable for hours, and from their rates of contraction net surface tensions (at room temperature) of 0.1 mN/m or less can be calculated. When dissolving in deaerated water, lung bubbles 100 μm or so in diameter slowly flatten and suddenly resume spherical shape; a characteristic rhythmical movement known as "clicking" results. Surfactant despersed in liquid can be demonstrated by the stability of very small bubbles (15 μm or less in diameter) (Pattle et al., 1976). Substances more surface active, but less surpellic than lung surfactant (e.g., certain detergents), can prevent lung surfactant from reaching the surface, but cannot dislodge it once it is there (Pattle, 1977).

Chemically, the surfactant consists of a mixture of lipids containing much di-palmitoyl lecithin, with about 10% of protein (King and Clements, 1972). The surface tension of lung bubbles, and presumably of the lung lining, increases if the area is increased. The surface area of a bubble must be increased by 25% to increase its surface tension from a low metastable value to the equilibrium value of about 30 mN/m (Pattle, 1977). Somewhat similar properties are displayed by lung extracts on the Wilhelmy trough (Clements et al., 1958) and by the lung lining in situ (Schürch et al., 1976).

The surfactant lining of the alveoli is difficult to demonstrate in transmission electron micrographs, because the embedding material dissolves or rolls up must of the lining film. Osmiophilic wisps, probably rolled-up lining film, and occasionally patches of osmiophilic multilayer may be seen; the film is better shown in freeze-etched specimens (Roth et al., 1973). The Type II cells of the alveoli contain *lamellated osmiophilic bodies* (LOPBs), which are intracellular depots of surfactant (Gil and Reiss, 1973). They are easily eroded by embedding fluids; this may in part be combated by the improved fixing methods of Schock et al. (1973). Freeze-fracture studies, however, show that the coarse lamellations seen in transmission E.M.s are at least in simian lungs, artefacts, and that the bodies actually consist of solid packs of leaflets with a spacing of about 4.4 nm (Smith et al., 1972). In man and monkeys, the bodies have a concentric structure (Creasey et al., 1974) and the lamellations seen in transmission E.M.s may be real. Geometrical "myelin figures" sometimes found in the alveoli are probably a hydration product of surfactant; after inhalation of silica, they may block the alveolus (Heppleston and Young, 1972).

The E.M.s of Gil and Weibel (1969/70) show signs of a fluid in the alveoli, particularly in the niches. The presence of microvilli, which probably project into liquid rather than into air (see Gil and Weibel, 1969/70, Fig. 4), and the ease with which surfactant-coated bubbles are squeezed from lung fragments, provide circumstantial evidence for the permanent presence of liquid in the alveoli.

3. Surfactant in Reptiles and Amphibia

Pattle and Hopkinson (1963) found that stable bubbles could be obtained from the lungs of the legless lizard *Anguis fragilis* (slow-worm), but that "clicking" was less vigorous than that in mammalian lung bubbles. Stable bubbles are also produced by tortoise lung. Osmiophilic bodies, but no trace of TLS, have been found in the alveolar walls of reptiles by Bargman and Knoop (1961, tortoise lung), Okada et al. (1962b, snake and lizard lungs), and Brooks (1970b, snake lung).

The surfactant in the three subclasses of amphibia has been studied systematically by Pattle et al. (1977). From the lungs of an Anuran (*Rana temporaria*) and a Caecilian (*Ichthyophis orthoplicatus*), they obtained surfactant-lined bubbles, which at room temperature were not quite as stable as those from mammals, but at 3°C were equally stable and exhibited "clicking". In the lining cells of the lungs there were LOPBs like those of mammals, but less easily eroded by embedding fluids. Extrusion of an LOPB by merocrine secretion was noted in *Ichthyophis*. No sign of a TLS like that of birds was found, either inside the cells or in the air space. Similar morphological findings are reported by Barmann and Knoop (1956, frog and clawed toad), Schulz (1959, toad), Okada et al. (1962a, frog and toad), Brooks (1970b, frog), and Berezin and da Silva Sasso (1974, toad).

Pattle et al. (1977) found the surfactant of newt lung to be less stable than that of frogs and of *Ichthyophis;* it often formed non-spherical bubbles. Osmiophilic bodies were present but rare. No TLS was found in the newt lung by Pattle et al. (1977) or by Okada et al. (1962a). In the epithelial cells of *Necturus*, the status of whose surfactant is unknown, Brooks (1970b) found pairs of lines with a spacing of about 50 nm - many times that of the TLS - and also a 12-nm spacing.

4. Surfactant in Fishes

The surfactant in lungfishes resembles that of mammals (Hughes, 1967, 1973). LOPBs and osmiophilic wisps have been found. In the lungs of the primitive bony fish *Amia*, an imperfect surfactant and copious LOPBs are present (Hughes, 1970). LOPBs are also present in the swimbladders of other fish (Copeland, 1969, cod; Brooks, 1970a, trout), in which the surfactant has not been tested. Their function is unknown; Brooks (1970b) thinks they may be antioxidant. Surfactant has not been found in the swimbladder of the goldfish or of a dace-roach hybrid (Hughes, 1967). No sign of TLS has been found in lungfishes or bony fishes.

5. Surfactant, Osmiophilic Bodies, and Trilaminar Substance in Bird Lungs

a) Historical

Surfactant was detected in bird lungs shortly after, and by the same methods as, it was found in mammalian lungs; Pattle (1958, p. 230) found that bubbles squeezed from pigeon lung were as stable as those from rat lung. This finding seems to have been missed by later workers. Miller and Bondurant (1961), using the Whilhelmy trough, found less surface hysteresis in extracts of bird, turtle, and frog lungs than with extracts of mammalian lungs; as, however, they found a minimum surface tension of 19 mN/m with mammalian lung extracts, their results cannot be regarded as demonstrating a highly surpellic substance therein, and so cannot give any information about surfactant in lungs of birds etc. Although Pattle and Hopkinson (1963) demonstrated surfactant in bird, reptile, and amphibian lungs, later workers have continued to follow Miller and Bondurant (1961); thus Harlan et al. (1966) classify bird lung as "without surfactant" and Jones and Radnor (1972b) describe the surfactant status of birds as "controversial". The present author considers it proved beyond doubt (see also Fujiwara et al., 1970) that bird lung has a surfactant lining similar to that of mammalian lung.

b) Lamellated Osmiophilic Bodies and Trilaminar Substance in Birds

There can be little doubt that the surfactant of the lungs of mammals, reptiles, amphibians, and lungfishes is represented in E.M.s by the lamellated osmiophilic bodies in the lining cells and by osmiophilic remnants in the air spaces. In bird lungs, LOPBs very similar to those of mammals are found in the atrial cells (Jones and Radnor, 1972a, b). These resemble the mammalian bodies in being easily eroded in conventional E.M. embedding, and in being better preserved by the lead ferricyanide staining methods of Schock et al. (1973; Figs. 1, 2) than by osmium staining alone. In bird lung, however, there is also a substance, the TLS, which, as has been seen above, resembles nothing found in non-avian lungs; its connection with surfactant is obscure, and its function, if it is not a form of surfactant, is unknown.

c) The Trilaminar Substance: A Review

The TLS appears to have first been recorded by Bargmann and Knoop (1961) in pigeon and canary lungs. They found sheets of TLS (their "membrankonvolute") both in the air capillaries and in the niches between capillaries, where they often penetrated below the surface of the cells. Other authors who have described this substance are Tyler and Pangborn (1964, chicken), Lambson and Cohn (1968, goose), Petrik and Riedel (1968a, chicken; 1968b, pigeon, sparrow, and chicken), Akester (1970, chicken), Fujiwara et al. (1970, turkey), and Jones and Radnor (1972b, chicken). Okada et al. (1965), however, did not notice the TLS in their E.M.'s. It seems likely that TLS is to be found in all bird lungs.

The descriptions, and particularly the measurements, given by various authors differ somewhat; this may be due to different methods of staining, or to measurements being made on different structures. For instance, a pair of sheets each with an osmiophilic line down the middle may appear as two osmiophilic regions separated by a nonosmiophilic gap. Rather than review each worker's contribution, I shall attempt a synthetic view.

The basic unit of TLS appears to be a sheet with two osmiophilic layers, which when stained by the method of Schock et al. (1973) (P.U.) are about 6 nm apart, flanked by two nonosmiophilic layers about 4.5 nm thick, giving a total thickness of 15 nm. These sheets may occur singly or in packs; sometimes they seem to occur in pairs. The TLS is distinguished by its resistance to erosion and by the huge quantities (in relation to extracellular surfactant) in which it is to be found. It can even be seen with the optical microscope (Lambson and Cohn, 1968). In the atria, dense masses of TLS, of dimensions up to 2 μm, can be seen (Petrik and Riedel, 1968a; Jones and Radnor, 1972b). Within some of the atrial lining cells are skeins or masses of this material (avian inclusion bodies, Jones and Radnor, 1972b). The TLS is secreted into the atrium, and probably also into the air capillaries, mainly by breaking away of part of the cell which is underlain by sheets and rolls of TLS. This seems to be an example of true "apocrine" secretion, i.e., secretion with breaking away of part of the cell cytoplasm, a phenomenon first postulated before the advent of electron microscopy and whose very existence Ham (1970) regards as doubtful. Ordinary LOPBs, by contrast, are extruded, as in mammals, by merocrine secretion (Petrik, 1967; Akester, 1970, Fig. B). The air capillaries may be free from TLS, or may be lined by one or more 15 nm sheets. In the epithelial cells may be found clefts, lined by cell plasma membrane, which contain a single 15 nm sheet of TLS which extends from the bottom of the cleft to the cell surface (Petrik and Riedel, 1968b). Structures similar to folded clefts also exist deep in the lining cells of the air capillaries, but whether the bounding membrane of these structures is continuous with the plasma membrane, or is part of the endoplasmic reticulum, is not known.

The TLS and the "avian inclusion bodies" composed of it are morphologically quite different from ordinary surfactant and ordinary LOPBs. Both are present in bird lung, but many authors have failed to see the difference between the LOPBs, which are easily aroded and show a spacing of about 5 nm, and the masses of TLS, which resist erosion and show a much larger spacing. Jones and Radnor (1972b) make the distinction clear and tabulate some of the contrasting characteristics. They state that the two kinds of material arise in different types of atrial cell ("granular" and "non-granular" pneumocytes); this is demonstrated in Figure 1. The broken cell in Figure 2, however, seems to contain both LOPBs and masses of TLS.

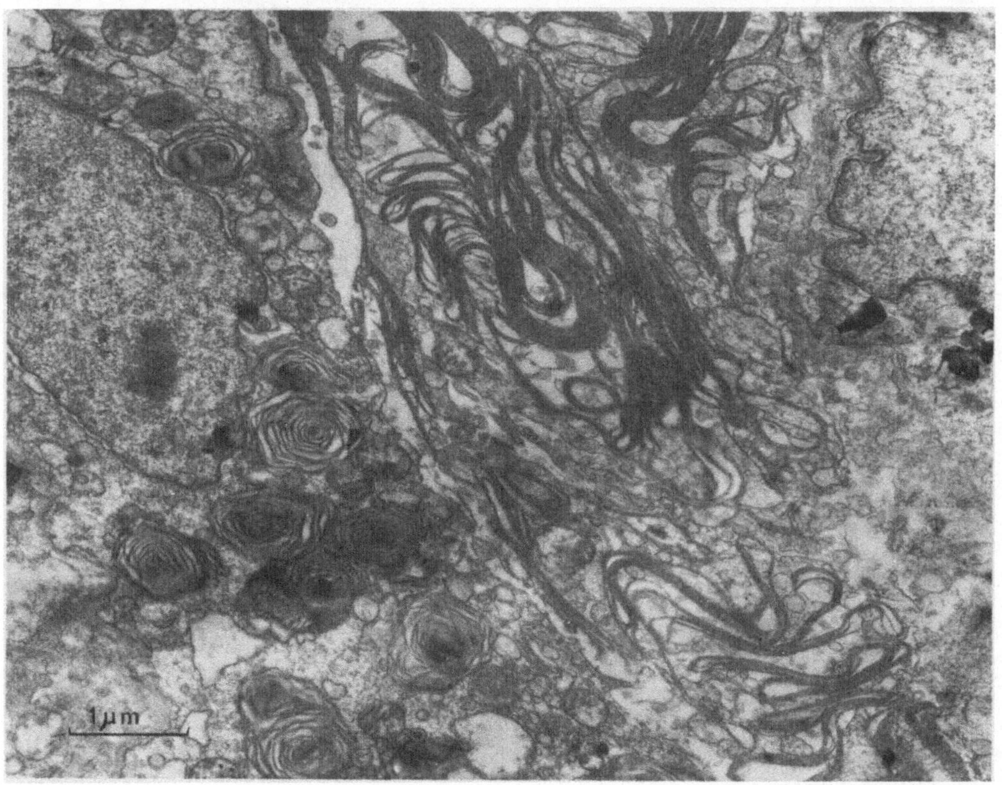

Fig. 1. Lung of chick, one day after hatching. Two cells, one with ordinary lamellated osmiophilic bodies and one with trilaminar substance

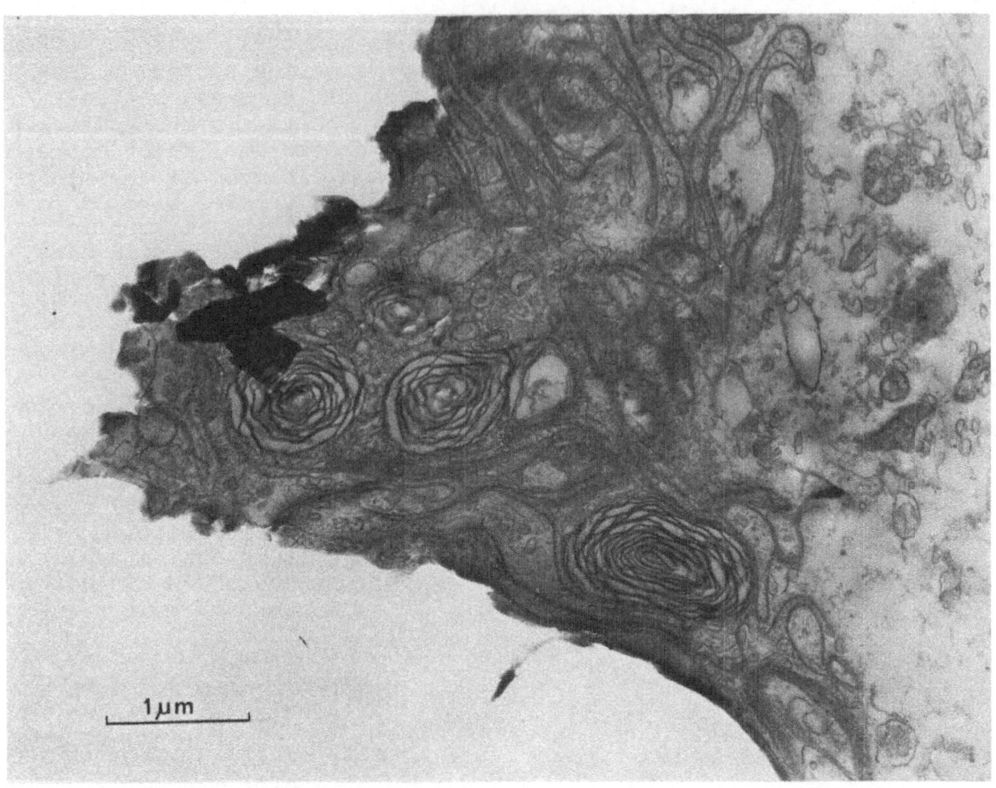

Fig. 2. Broken cell from chick lung, apparently containing both ordinary laminated osmiophilic bodies and trilaminar substance

d) Possible Functions of the Trilaminar Substance

We now have to consider what function the TLS may serve. There are various possibilities.

An Additional Surfactant. Some workers, especially those who have not recognised the morphological distinction between the LOPBs and the TLS, considered that the latter was composed of surfactant. The following arguments, however, suggest that surfactant in bird lung is represented by LOPBs.

Surfactant very similar to that of bird lung is found in mammals, reptiles, anuran and caecilian amphibians, and lungfishes. LOPBs are found in all of these, and the connection of LOPBs with surfactant has been proved in mammals (Gil and Reiss, 1973). Merocrine secretion of LOPBs has been found in mammals, (Bensch et al., 1964) Caecilians, frog, and bird. TLS, however, has been found nowhere but in birds and, perhaps, in *Necturus*. Bird lung surfactant was found by Fujiwara et al. (1970) to have a lipid: protein ratio of 10:1, which is not compatible with the value of about 3:5 for the ratio of osmiophilic to non-osmiophilic matter found in TLS.

In the chick the TLS appears only at the time of hatching (20-21 days), according to Jones and Radnor (1972b). LOPBs however appear after 17 days (see also Petrik, 1967). I have found that chick lung can form stable bubbles after 18 days. Confirmation of these times with bubble tests and micrographs from the same specimens would be desirable.

Jones and Radnor (1972b) have found that TLS is produced in the lining cells of the capillaries only in the first week after hatching, and at later times only in the atria. I have found only accasional traces of TLS in the air capillaries of the adult chicken. These are at least as sharply curved as mammalian alveoli, and the blood-air barrier is just as thin as in the mammal, and so reduction in surface tension is just as necessary. The rigid structure of bird lung makes transudation from the blood the main risk if lowering of surface tension is inadequate, while in the mammal the first effect of absence of surfactant is alveolar collapse. We must therefore suppose that the air capillaries are lined mainly with ordinary surfactant which, like that of mammals, does not show up in E.M.s.

The TLS, however, might possibly act as a second kind of surfactant. A unit membrane-like bilayer, flanked by two sheets of protein, could split in two, leaving lipid layers facing the air and the protein facing the wall of the air capillary, and this might constitute a very stable surpellic film about 6 nm thick. It is possible, but not necessary, to interpret the micrographs of Petrik and Riedel (1968b, Figs. 6 and 7) as showing such a process.

A Hydration Product of Surfactant. The TLS might be a hydration product of surfactant, like the myelin figures often found in mammalian alveoli. The secretion of TLS within cells, its very regular geometry, and the absence of rectangular patterns, negate this hypothesis.

An Absorbent for Liquid. Alternatively, the TLS might be concerned with keeping the air capillaries free from liquid. In mammalian lungs, any free liquid present probably concentrates in the niches, away from the blood-air barrier (see Gil and Weibel, 1969, 1970). In birds the blood-air barrier is in the regions of greatest curvature, i.e., in the air capillaries, where liquid will, if present, tend to gather, blocking flow through them. A liquid film lining a cylindrical space is mechanically unstable. It seems likely that in the bird there is no liquid lining layer, and that there must be a special mechanism for keeping the air capillaries dry. The TLS may perhaps be capable of absorbing water without liquefying. The large scale on which it is secreted would be compatible with such a function. So would the variations in the unit dimensions reported in the literature.

A Haemostatic Substance. The TLS might be a coagulant or other haemostatic product. In bird lung, as in other lungs, the blood-air barrier is very thin, and much blood might be shed without much "tissue juice" coagulant (tissue thromboplastin) being released. The TLS might represent a second line of defence. There is some evidence that a special haemostatic mechanism exists in mammalian alveoli (P.U.).

Of the above four possibilities, the writer considers the most likely is that the TLS has something to do with the absorption of surplus liquid; but the question of the function of the TLS remains completely open.

6. Mechanism of Renewal of Surface Film

It is not clear how lung surfactant is adsorbed to the surface in birds. For a film of very low surface tension to be formed, it is necessary first for an equilibrium film (of tension about 30 mN/m) to be formed, and then for the area to be reduced to lower the surface tension. In the mouse lung the reduction in area necessary is 20%. During a deep inspiration fresh surfactant is adsorbed, and the film is compressed on expiration (Pattle, 1977). It is by no means clear what is the corresponding process in bird lung.

References

Akester, A.R.: Osmiophilic inclusion bodies as the source of the laminated membrane in the epithelial lining of avian lung. J. Anat. 107, 189-190 (1970)

Avery, M.E., Mead, J.: Surface properties in relation to stelectasis and hyaline membrane disease. A.M.A.J. Dis. Children 97, 517-523 (1959)

Bargmann, W., Knoop, A.: Vergleichende elektronenmikroskopische Untersuchungen der Lungenkapillaren. Z. Zellforsch. mikrosk. Anat. 44, 263-281 (1956)

Bargmann, W., Knoop, A.: Elektronenmikroskopische Untersuchungen an der Reptilien- und Vogellunge. Z. Zellforsch. mikrosk. Anat. 54, 541-548 (1961)

Bensch, K., Schaefer, K., Avery, M.E.: Granular pneumocytes: electron microscopic evidence of their exocrine function. Science 145, 1318-1319 (1964)

Berezin, A., da Silva Sasso, W.: Electron microscopy of the pulmonary alveolar cells (granular pneumocytes) of normal and vagotomised amphibian (*Bufo icterus icterus*). Experientia 30, 1074 (1974)

Brooks, R.E.: Ultrastructure of the physostomatous swimbladder of the rainbow trout (*Salmo gairdneri*). Z. Zellforsch. mikrosk. Anat. 106, 473-483 (1970a)

Brooks, R.E.: Lung alveolar cell cytosomes: a consideration of their significance. Z. Zellforsch. mikrosk. Anat. 106, 484-497 (1970b)

Clements, J.A.: (Chairman) Biochemical aspects of pulmonary function. Federation Proc. 33, 2231-2262 (1974)

Clements, J.A., Brown, E.S., Johnson, R.P.: Pulmonary surface tension and the mucus lining of the lungs: some theoretical considerations. J. Appl. Physiol. 12, 262-268 (1958)

Copeland, D.E.: Fine structural study of gas secretion in the physoclistous swim bladder of *Fundulus heteroclitus* and *Gadus calliaris*, and in the euphysoclistous swimbladder of *Opsonus tau*. Z. Zellforsch. mikrosk. Anat. 93, 305-331 (1969)

Creasey, J.M., Pattle, R.E., Schock, C.: Ultrastructure of inclusion bodies of type II cells of lung, human and sub-simian. J. Physiol. 237, 35P-37P (1974)

Fujiwara, T., Adams, F.H., Nozaki, M., Dermer, G.B.: Pulmonary surfactant phospholipids from turkey lung: comparison with rabbit lung. Am. J. Physiol. 218, 218-225 (1970)

Gil, J., Reiss, O.K.: Isolation and characterisation of lamellar bodies and tubular myelin from rat lung homogenates. J. Cell Biol. 58, 152-171 (1973)

Gil, J., Weibel, E.R.: Improvements in demonstration of lining layer of lung alveoli by electron microscopy. Resp. Physiol. 8, 13-36 (1969/1970)

Goerke, J.: Lung surfactant. Biochimica biophysica acta 344, 241-261 (1974)

Ham, A.W.: Histology, 7th. ed. Philadelphia: J.B. Lipincott, 1970, pp. 199-201

Harlan, W.R., Jr., Margraf, J.H., Said, S.I.: Pulmonary lipid composition of species with and without surfactant. Am. J. Physiol. 211, 855-861 (1966)

Heppleston, A.G., Young, A.E.: Alveolar lipo-proteinosis: an ultrastructural comparison of the experimental and human forms. J. Pathol. 107, 107-117 (1972)

Hughes, G.M.: Evolution between air and water. In: Development of the Lung. Ciba Foundation Symp. De Reuck, A.V.S., Porter, R. (eds.). London: Churchill, 1967, pp. 64-80

Hughes, G.M.: Ultrastructure of the air-breathing organs of some lower vertebrates. Proc. 7th. Int. Cong. Electron Microsc., Paris, Soc. Franc. Microsc. 1970, pp. 599-600

Hughes, G.M.: Ultrastructure of the lung of *Neoceratodus* and *Lepidosiren* in relation to the lung of other vertebrates. Folia Morphologica, Prague, 21, 155-161 (1973)
Jones, A.W., Radnor, C.: The development of the chick tertiary bronchus. I. General development and the mode of production of the osmiophilic inclusion body. J. Anat. 113, 303-324 (1972a)
Jones, A.W., Radnor, C.: The development of the chick tertiary bronchus. II. The origin of the surface lining system. J. Anat. 113, 325-340 (1972b)
King, R.J., Clements, J.A.: Surface active materials from dog lung. II. Composition and physiological correlations. Am. J. Physiol. 223, 715-726 (1972)
King, A.S., Molony, V.: The anatomy of respiration. In: Physiology and Biochemistry of the Domestic Fowl. Bell, D.J., Freeman, B.M. (eds.). London: Academic Press, 1971, pp. 93-169
Lambson, R.O., Cohn, J.E.: Ultrastructure of the lung of the goose and its lining of surface material. Am. J. Anat. 122, 631-649 (1968)
Miller, D.A., Bondurant, S.: Surface characteristics of vertebrate lung extracts. J. Appl. Physiol. 16, 1075-1077 (1961)
Okada, Y., Ishiko, S., Daido, S., Kim, J., Ikeda, S.: Comparative morphology of the lung with special reference to the alveolar epithelial cells. 1. Lung of the amphibia. Acta tuberc. Jpn. 11, 63-72 (1962a)
Okada, Y., Ishiko, S., Daido, S., Kim, J., Ikeda, S.: Comparative morphology of the lung with special reference to the alveolar epithelial cells. 2. Lung of the reptilia. Acta tuberc. Jpn. 12, 1-10 (1962b)
Okada, Y., Ishiko, S., Daido, S., Ikeda, S., Genka, K., Morihisha, K.: Comparative morphology of the lung with special reference to alveolar epithelial cells. 3. Lung of the bird. Acta tuberc. Jpn. 14, 35-41 (1965)
Pattle, R.E.: Properties, function, and origin of the alveolar lining layer. Nature (London) 175, 1125 (1955)
Pattle, R.E.: Properties, function and origin of the alveolar lining layer. Proc. Roy. Soc. Lond. ser. B 148, 217-240 (1958)
Pattle, R.E.: Surface lining of lung alveoli. Physiol. Rev. 45, 48-79 (1965)
Pattle, R.E.: Lung surfactant in the evolutionary tree. In: Respiration of Amphibious Vertebrates. Hughes, G.M. (ed.). London: Academic Press, 1976, pp. 233-255
Pattle, R.E.: The relation between surface tension and area in the alveolar lining film. J. Physiol. 269, 591-604 (1977)
Pattle, R.E., Hopkinson, D.A.W.: Lung lining in bird, reptile, and amphibian. Nature (London) 200, 894 (1963)
Pattle, R.E., Robards, G.J., Sutherland, P.D.: Method for demonstrating difference between surface properties of sheep and human amniotic fluids, and attempting to predict human respiratory distress syndrome. J. Physiol. 263, 110P-111P (1976)
Pattle, R.E., Rossdale, P.D., Schock, C., Creasey, J.M.: The development of the lung and its surfactant in the foal and other species. J. Reprod. Fert. Supp. 23, 651-657 (1975)
Pattle, R.E., Schock, C., Creasey, J.M., Hughes, G.M.: Surpellic films, lung surfactant, and their cellular origin in newt, caecilian, and frog. J. Zool. Lond. 182, 125-136 (1977)
Pattle, R.E.: unpublished
Petrik, P.: The ultrastructure of the chicken lung in the final stages of embryonic development. Folia Morphologica, Prague 15, 176-186 (1967)
Petrik, P., Riedel, B.: A continuous osmiophilic non-cellular membrane at the respiratory surface of the lungs of fetal chickens and young chicks. Lab. Invest. 18, 54-62 (1968a)
Petrik, P., Riedel, B.: An osmiophilic bilaminar lining film at the respiratory surface of avian lungs. Z. Zellforsch. mikrosk. Anat. 88, 204-219 (1968b)
Roth, J., Winkelmann, A., Meyer, H.W.: Electron microscopic studies in mammalian lungs by freeze-etching. IV. Formation of the superficial layer of the surfactant system by lamellar bodies. Exp. Path., Jena 8, 354-362 (1973)
Scarpelli, E.M.: The Surfactant System of the Lung. Philadelphia: Lea and Fibiger, 1968
Schock, C., Pattle, R.E., Creasey, J.M.: Methods for electron microscopy of the lamellated osmiophilic bodies of the lung. J. Microsc. Oxford 97, 321-330 (1973)
Schürch, S., Goerke, J., Clements, J.A.: Direct determination of surface tension in the lung. Proc. Nat. Acad. Sci. US 73, 4698-4702 (1976)

Schulz, H.: The Submicroscopic Anatomy and Pathology of the Lung. Berlin-Göttingen-Heidelberg: Springer, 1959, p. 21
Smith, D.S., Smith, U., Ryan, J.W.: Freeze-fractured lamellar body membranes of the rat lung great alveolar cell. Tissue and Cell 4, 457-468 (1972)
Tierney, D.F.: Lung metabolism and biochemistry. Ann. Rev. Physiol. 36, 209-231 (1974)
Tyler, W.S., Pangborn, J.: Laminated membrane surface and osmiophilic inclusions in avian lung epithelium. J. Cell Biol. 20, 157-164 (1964)
Villee, C.A., Villee, D.B., Zuckerman, J. (eds.): Respiratory Distress Syndrome. New York: Academic Press, 1973
Weibel, E.R.: Morphological basis of alveolar-capillary gas exchange. Physiol. Rev. 53, 419-495 (1973)

Intrapulmonary Carbon Dioxide Sensitive Receptors: Amphibians to Mammals

M. R. Fedde and W. D. Kuhlmann

Summary

Direct neural recordings have been made from intrapulmonary CO_2-sensitive receptors in all classes of tetrapods. Specific characteristics of these receptors, such as their relative sensitivity to chemical (CO_2, O_2, and several drugs) and mechanical stimuli, the responsiveness to static and dynamic CO_2 concentrations in their microenvironment, their location in the lung, and the influence of intracellular H^+ in controlling their discharge have been studied in detail in some animals.

Amphibian lungs, as exemplified by the bullfrog, possess both rapidly and slowly adapting CO_2-sensitive mechanoreceptors, but no receptors whose sole physiological stimulus is CO_2 concentration. Reptilian lungs, at least in turtles and Tegu lizards, possess CO_2-sensitive mechanoreceptors as well as receptors strikingly sensitive to CO_2 but not to stretch of the lung. The CO_2 receptors respond to static concentration of CO_2 in the lungs as well as to rapid changes in intrapulmonary CO_2 concentration. Avian lungs apparently possess only CO_2 receptors without mechanical sensitivity. Mammalian lungs seemingly have only CO_2-sensitive mechanoreceptors (slowly adapting pulmonary stretch receptors).

In all vertebrate classes studied, the discharge from both CO_2 receptors and CO_2-sensitive mechanoreceptors inhibits respiratory neuronal output from the brain. Intensive investigations in many laboratories are currently in progress to determine the importance of these receptors in controlling breathing.

1. Introduction

Many features about the control of breathing remain to be discovered. The mechanisms responsible for maintaining constant partial pressure of carbon dioxide (P_{CO_2}) in the arterial blood while alveolar ventilation changes with exercise are currently being investigated in many laboratories. Despite extensive study of pulmonary mechano- and irritant receptors and the known chemoreceptors which sample arterial blood, it has not been possible thus far adequately to explain these mechanisms.

Recently, interest has been renewed in investigations of possible chemoreceptors located in the lungs. Such receptors might be in an ideal location to monitor the load of CO_2 returning to the lung from the

---

Department of Anatomy and Physiology, Kansas State University, Manhattan, Kansas 66506, USA.

mixed venous blood, either by changes in P_{CO_2} in that blood or by changes in blood flow. Investigating intrapulmonary chemoreceptors is difficult in mammals but easier in the lower vertebrate classes. The lungs of these animals can be unidirectionally ventilated by passing gas into the trachea, through the lungs, and into the atmosphere through a hole in the airsacs, in the case of birds, or in the apex of the lung itself, in the case of reptiles and amphibians.

Unidirectional ventilation of the lung: (1) allows independent controlling of chemical and mechanical stimuli to intrapulmonary receptors and, (2) allows the stimulus to be changed quickly at the receptive site so both the dynamic and static response of any chemoreceptor to a given P_{CO_2} change may be studied. Furthermore, a high gas flow through the lung causes the gas concentration at the receptive site to remain relatively unaffected by the mixed venous blood and therefore to approximate to that in the tracheal gas where it can be readily measured. The additional feature of a bag-like lung in some reptiles and amphibia, which can be exteriorized from the thorax, offers a possibility for direct receptor site localization.

This review is designed to summarize current evidence for the presence of CO_2-sensitive receptors in the lungs of various classes of vertebrates and to discuss their possible role in controlling breathing.

2. Amphibia

a) Carbon Dioxide-Sensitive Mechanoreceptors

Frogs respond to inhaled CO_2 by an initial inhibition followed by a general stimulation of breathing (Smyth, 1939). The response is present after cutaneous and palate denervation and after removing the carotid body. The lungs of the frog also contain mechanoreceptors responsive to lung deformation (Neil et al., 1950; Bonhoeffer and Kolatat, 1958; Downing and Torrance, 1961; Taglietti and Casella, 1966, 1968; McKean, 1969), but pulmonary chemoreceptors sensitive to CO_2 have not been reported.

We recently attempted to determine if intrapulmonary CO_2 receptors are present in the bullfrog (*Rana catesbeiana*). The CO_2 sensitivity of pulmonary mechanoreceptors was also studied. The animals were anesthetized, paralyzed, and the left lung was unidirectionally ventilated. Intrapulmonary pressure could be changed independent of the CO_2 concentration and vice versa. Single-unit afferent activity was recorded from the left cervical vagus in 15 bullfrogs. In six of the frogs, the entire vagus was fasciculated into fine strands and searched for receptors that were solely stimulated by changes in intrapulmonary CO_2 concentration. None were found. However, 247 mechanoreceptors were found and 55 of these were studied in detail to determine their sensitivity to changes in CO_2 concentration. The discharge of both rapidly adapting and slowly adapting (Fig. 1) receptors increased by lowering intrapulmonary CO_2 concentration. The CO_2 sensitivity differed among the various receptors and was not apparently influenced by distension of the lung.

Thus, lungs of the frog possess CO_2-sensitive mechanoreceptors, but not pure chemoreceptors whose adequate stimulus is CO_2.

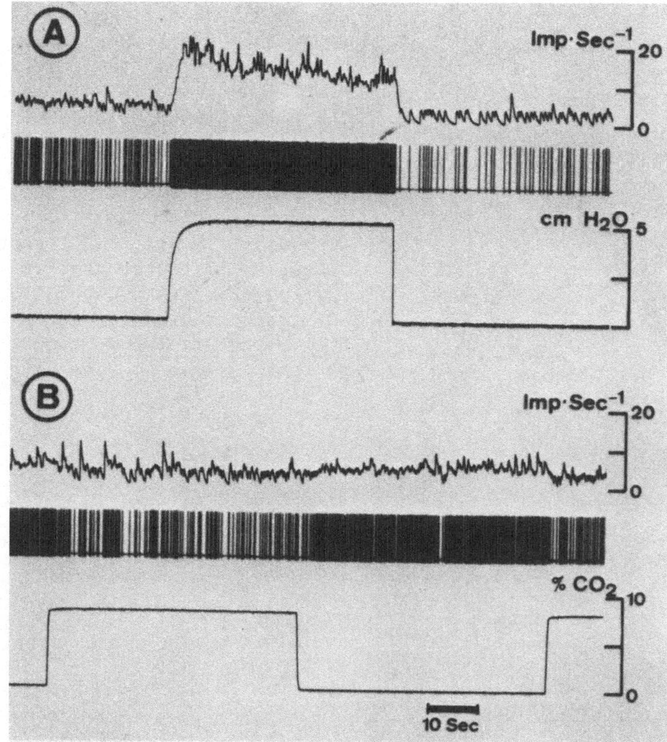

Fig. 1. Sensitivity of a slowly adapting intrapulmonary mechanoreceptor in the frog to stretch and carbon dioxide. (A) Increase in intrapulmonary pressure (*bottom tracing*) evoked an increase in discharge which exhibited a small dynamic overshoot when the stretch was applied and an undershoot when the stretch was removed. *Upper tracing* is the instantaneous discharge frequency of the receptor; *middle tracing*, normalized nerve impulses. Intrapulmonary CO_2 concentration was near 0%. (B) Reduced intrapulmonary CO_2 concentration (*bottom tracing*) increased the discharge of the receptor maintained at a constant degree of stretch. *Upper and middle tracings*, as in (A). Intrapulmonary pressure was 1 cm H_2O

3. Reptilia

The first breath response to changes in inspired CO_2 concentration in lizards may be due to pulmonary CO_2 receptors (Saalfeld, 1934; Boelaert, 1941; Nielsen, 1961; Templeton and Dawson, 1963). More recently, the apneic response to reduced intrapulmonary CO_2 concentration in the Tegu lizard (*Tupinambis nigropunctatus*) has been shown to result from stimulating receptors in the lungs (Gatz et al., 1975).

Single-unit recordings from vagal afferent fibers in both the Tegu lizard (Fedde et al., 1977; Scheid et al., 1977) and the turtle (Milsom and Jones, 1976) have illustrated that receptors in the lungs of these animals are CO_2-sensitive. Two basic kinds of receptor have been found.

a) Carbon Dioxide Receptors

One kind of intrapulmonary CO_2-sensitive receptor changes its discharge frequency when intrapulmonary CO_2 concentration is changed, but is not sensitive to mechanical distortion of the lung. The response of this

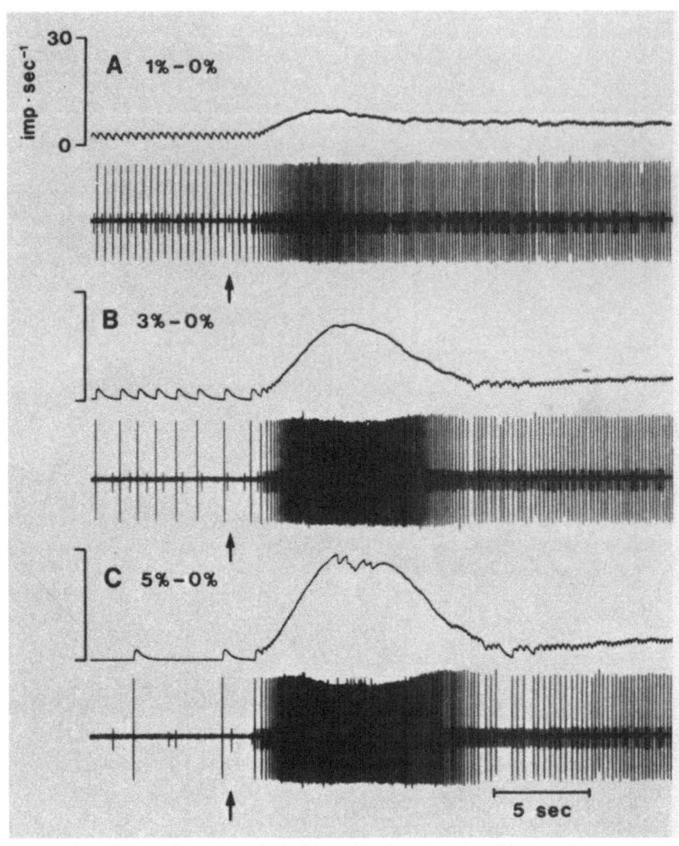

Fig. 2. Response of an intrapulmonary CO_2 receptor in the Tegu lizard to sudden reduced intrapulmonary CO_2 concentration (*at the arrows*). The lung was unidirectionally ventilated at a flow of 1 l/min. *Top tracing* in each record is the instantaneous discharge frequency; *bottom tracing*, impulses in a single afferent fiber. (A) CO_2 concentration lowered from 1% to 0%. (B) CO_2 concentration lowered from 3% to 0%. (C) CO_2 concentration lowered from 5% to 0%

type of receptor to a sudden reduction in intrapulmonary CO_2 concentration in a unidirectionally ventilated lung of the Tegu lizard is shown in Figure 2. The receptors exhibited a stable discharge frequency dictated by the static CO_2 concentration in the ventilating gas stream (left of the arrows in Fig. 2). A change in CO_2 concentration in the ventilating gas stream resulted in increased discharge frequency, which had an overshoot proportional to the magnitude of the change in CO_2 concentration. At high discharge frequencies, some receptors became irregular and the impulse amplitude was diminished (Fig. 2C). Following the overshoot, the discharge adapted in a few seconds to a static frequency characteristic of that for an input CO_2 concentration of 0%. When the input CO_2 concentration was returned to its prior level, the receptor discharge was quickly reduced and exhibited an undershoot whose length was proportional to the magnitude of the change in CO_2 concentration (Fig. 3). The discharge gradually returned to a static level characteristic of that for the given input CO_2 concentration.

This kind of receptor is totally unaffected by inflation or deflation of the lung (Fig. 4). Furthermore, neither discharge nor the CO_2 sensitivity of the receptor changes when exposed to either a hypoxic (7% O_2) or a hyperoxic gas mixture (96% O_2). Some receptors can respond to changes in intrapulmonary CO_2 concentration at a frequency of 1.28 Hz and, hence inform the brain about rapid changes in CO_2 concentration in their microenvironment, as might occur in polypnea. Thus, chemoreceptors, whose adequate stimulus is CO_2 concentration, are present in the lung of the Tegu lizard and they can convey information to the

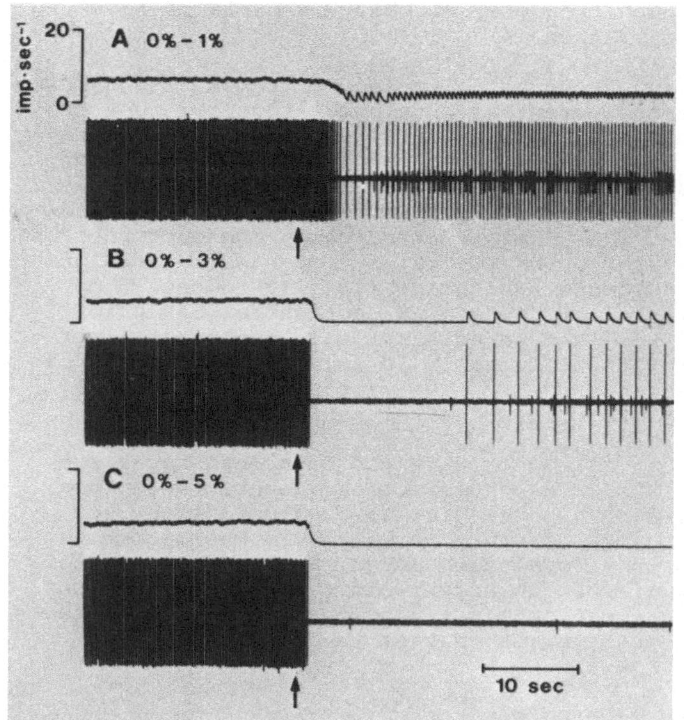

Fig. 3. Response of an intrapulmonary CO_2 receptor (same receptor as in Fig. 2) in the Tegu lizard to a sudden increased intrapulmonary CO_2 concentration (at the *arrows*). Tracings as in Figure 2. Changes in CO_2 concentration are above each record. Note that these records were taken at half the chart speed as those in Figure 2

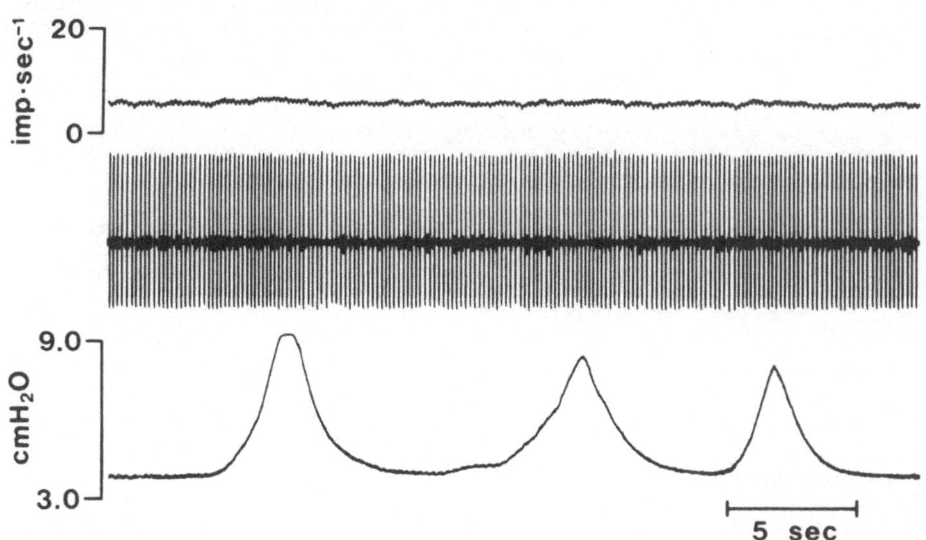

Fig. 4. Lack of influence of changes in intrapulmonary pressure (causing stretch of the lung) on the discharge of an intrapulmonary CO_2 receptor (same receptor as Fig. 2) in the Tegu lizard. *Upper tracing*, instantaneous discharge frequency; *middle tracing*, nerve impulses; *bottom tracing*, changes in intrapulmonary pressure resulting in stretch of the lung. Intrapulmonary CO_2 concentration near 0% during the test

brain about both magnitude and frequency of change and static level of intrapulmonary CO_2 concentration.

There have been questions about the actual molecule or ion responsible for changing the transmembrane potential of the CO_2 receptor when the CO_2 concentration in its microenvironment is altered. Because the receptor loses much of its sensitivity to changes in CO_2 concentration when the carbonic anhydrase inhibitor, acetazolamide, is injected into the lizard, apparently the CO_2 molecule per se does not directly control the receptor discharge. That experiment suggests one or more of the ions produced by the hydration of CO_2 may be important. Infusion of H^+ following carbonic anhydrase inhibition markedly reduces the discharge of the receptor, similar to high CO_2 concentrations prior to inhibition; infusion of HCO_3^- has no effect. Thus, changes of intracellular H^+, associated with the changes in CO_2 concentration, may be crucial to controlling the discharge of these receptors. Additional studies must be conducted before the mechanism is completely clear.

Study of the location of the CO_2 receptor in the Tegu lung has also provided insight about their spatial organization. Several receptors (up to 5 have been found), separated by distances of more than 2 cm, apparently are innervated by one afferent nerve fiber. These chemoreceptors may be organized in a complex receptor field in the wall of the lung. The implications of such an organization are not known, nor is it clear how such an organization might provide information to the brain about regional CO_2 concentration changes in the lung.

b) Carbon Dioxide-Sensitive Mechanoreceptors

A second kind of receptor in the reptilian lung alters its discharge frequency when the lung is inflated or deflated and, thus, is definitely mechanosensitive (Fig. 5). Most of these receptors also are sensitive to the intrapulmonary CO_2 concentration because their discharge frequency decreases as the CO_2 concentration increases (Fig. 5A). Some receptors are more influenced by changes in CO_2 concentration than others; some may be almost totally inhibited from discharging when the lung is stretched by an inflating gas containing a CO_2 concentration of 10% although others are only slightly inhibited.

Clearly, the lungs of at least some reptiles contain both chemoreceptors and mechanoreceptors which are modulated by CO_2. Both kinds of receptors may play a role in controlling breathing in this vertebrate class.

4. Aves

a) Carbon Dioxide Receptors

Many years ago scientists (Bordoni, 1888; Foa, 1911; Dooley and Koppanyi, 1929) demonstrated that apnea could be produced quickly in the turkey, pigeon, and chicken by passing air, O_2, or H_2 into the trachea through the lungs and out the opened caudal airsacs; and in the duck by blowing air or O_2 into the pneumatic humerous through the lung and out the trachea. Hyperventilation of the lungs with these gases lowers the intrapulmonary CO_2 concentration below normal. The rapidity of the apneic response suggested that the receptor system responding to the change in intrapulmonary gas concentration may be in or near the lungs (Foa, 1911) and was conclusively demonstrated recently (Peterson and

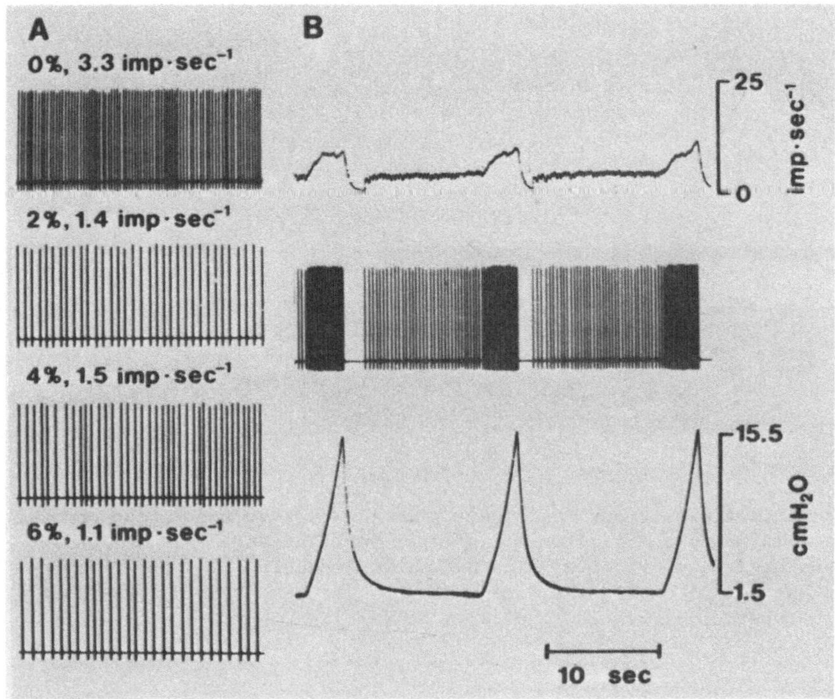

Fig. 5. Response of a slowly adapting intrapulmonary mechanoreceptor to changes in intrapulmonary CO_2 concentration and pressure in the Tegu lizard. (A) Change in static discharge frequency at intrapulmonary CO_2 concentrations ranging from 0% to 6%; low intrapulmonary pressure (approximately 1.5 cm H_2O). (B) Response of this mechanoreceptor to changes in intrapulmonary pressure. *Upper tracing*, instantaneous discharge frequency; *middle tracing*, normalized nerve impulses; *lower tracing*, intrapulmonary pressure. Pressure changes conducted when intrapulmonary CO_2 concentration was near 0%

Fedde, 1968). When unidirectionally ventilated chickens with transiently occluded blood flow to the lungs were rapidly induced into apnea by lowering the intrapulmonary CO_2 concentration, it was clear that the receptors sensing the stimulus were not receiving a blood-born signal and must therefore reside in the lung. The rapid apneic response was prevented by vagotomy. This provided evidence for the afferent pathway to the brain.

Direct neural recordings from intrapulmonary CO_2 receptors in many species of birds (chickens, ducks, emus, geese) have now been made, and their responses to static intrapulmonary CO_2 concentrations and to rapid changes in concentrations of that gas are well established (Fedde and Peterson, 1970; Osborne and Burger, 1971; Osborne and Burger, 1974; Fedde et al., 1974a; Burger et al., 1976a, b; Barnas, 1977). These receptors respond to suddenly reduced intrapulmonary CO_2 concentration with increased discharge frequency, generally exhibiting an overshoot of varying degree, then producing a static discharge frequency related to the CO_2 concentration in the microenvironment of the receptor (Fig. 6). The static discharge frequency at various CO_2 concentrations differs considerably from receptor to receptor (Fig. 7). The static stimulus-response relation is best described by a logarithmic regression above an F_{CO_2} of approximately 1.5%, although hyperbolic and

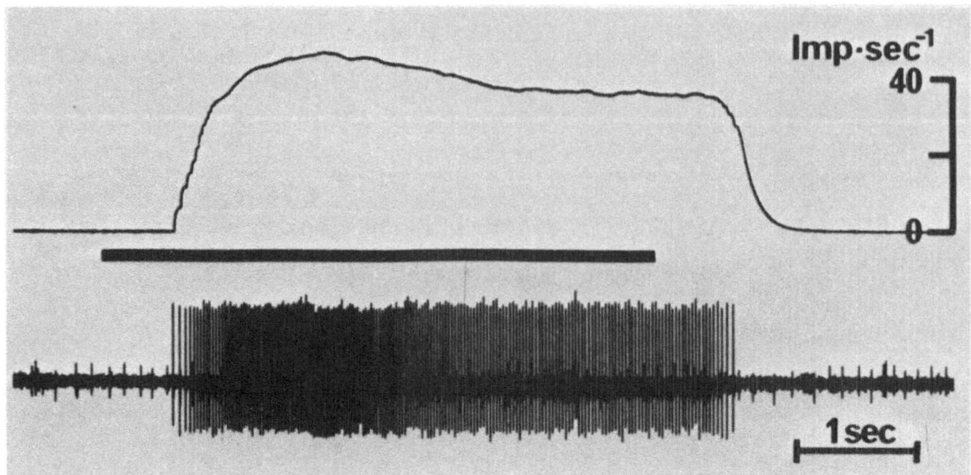

Fig. 6. Response of an intrapulmonary CO_2 receptor in the duck to the sudden removal of CO_2 from the ventilating gas during unidirectional ventilation. *Upper tracing*, instantaneous discharge frequency; *lower tracing*, nerve impulses. CO_2 concentration in the ventilating gas was 0% during the time marked by the black bar and was 5.2% before and after that period. (From Fedde et al., 1974a)

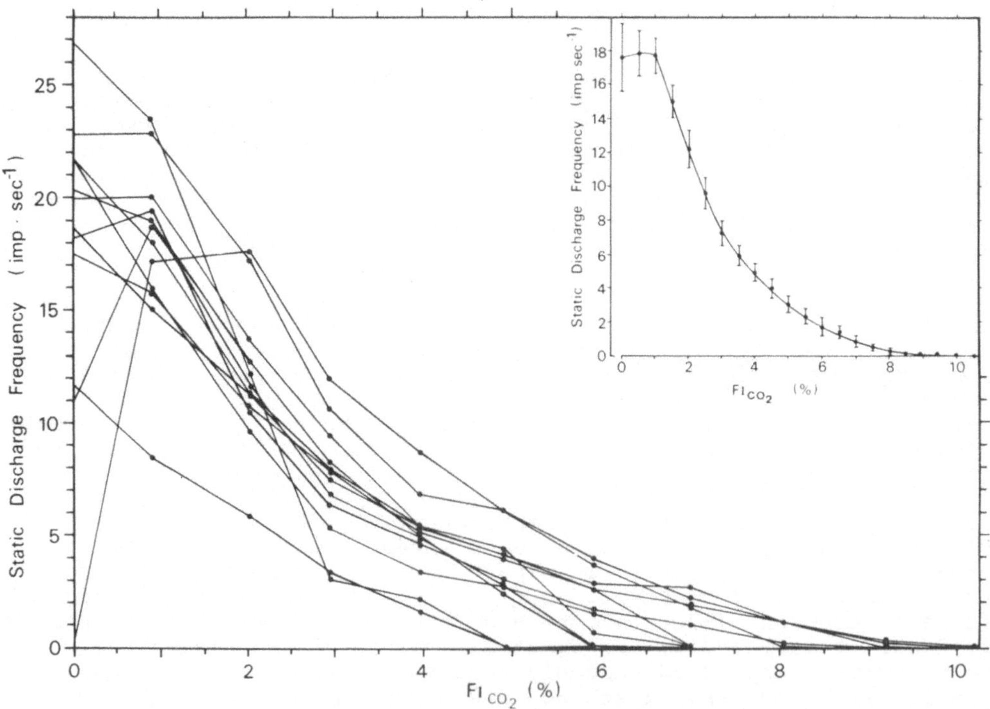

Fig. 7. Discharge frequencies of 12 intrapulmonary CO_2 receptors in an emu at various static CO_2 concentrations in the unidirectional gas stream ventilating the lungs. Insert shows the mean (± SE) response of these receptors. (From Burger et al., 1976a)

quadratic regressions also may be used to characterize the relation (Nye, 1977). Below a maintained F_{CO_2} of about 1.5%, some receptors begin to discharge irregularly and may even cease, as shown in Figure 7. The reason is not known, but may relate to severe alterations in intracellular acid-base balance in such an alkalotic situation.

Intrapulmonary CO_2 receptors apparently are true chemoreceptors whose adequate stimulus is CO_2. They are not sensitive to hypoxia or hyperoxia (Fedde and Peterson, 1970; Leitner, 1972; Molony, 1972; Tschorn and Fedde, 1974), to intravenous injections of lactic acid, sodium cyanide, lobeline, veratridine, or acetylcholine (Fedde and Peterson, 1970; Burger et al., 1974), and do not respond consistently to changes in extracellular H^+ concentration (Burger et al., 1974). Furthermore, they are not influenced by elevated intrapulmonary pressure, which would markedly stretch the respiratory system, when the intrapulmonary CO_2 concentration remains constant (Fig. 8) (Fedde et al., 1974a). They are sensitive to changes in CO_2 delivery to the lungs from the mixed venous blood (Banzett and Burger, 1977; Nye, 1977), as demonstrated by either changing the CO_2 load returning to the lung or stopping the pulmonary blood flow by occlusion of the pulmonary artery.

Several attempts have been made to determine the location of the CO_2 receptors within the lung (Burger et al., 1974; Scheid et al., 1974; Nye, 1977). The receptors are located in the gas exchange regions of the lung and most apparently are in paleopulmonic parabronchi near their origin from the mediodorsal secondary bronchi. Some may be in the middle of the paleopulmonic parabronchi but none appear in the trachea or extrapulmonary primary bronchus. Perhaps a few are in the ends of the paleopulmonic parabronchi near the ventromedial secondary bronchi. Most of the receptors are in that part of the lung that undergoes marked fluctuations in CO_2 concentration with each breath and is influenced by CO_2 evolving from the mixed venous blood. They, thus, appear capable of monitoring the balance between ventilation and CO_2 load returning to the lung.

Intrapulmonary CO_2 receptors may have an important role in controlling avian respiration. When their activity is removed from the brain by cutting the pulmonary branches of the vagus nerve, respiratory rate is markedly slowed, and respiratory amplitude is greatly increased (Fedde et al., 1963). Electrical stimulation of the pulmonary branches of the vagus produces apnea in a manner similar to sudden reduced intrapulmonary CO_2 concentration (Peterson and Nightingale, 1976). Furthermore, respiration can be paced by intrapulmonary CO_2 oscillations in unidirectionally ventilated, unanesthetized chickens (Kunz and Miller, 1974a, b). Ventilation is influenced by both the mean CO_2 concentration in the lung and the amplitude of the CO_2 oscillations (Miller and Kunz, 1975). The discharge from these receptors may adjust ventilation to the rate of metabolic CO_2 production and control individual breath size.

b) Mechanoreceptors

Afferent fibers from mechanoreceptors also run centrally in the vagus (Fig. 9). These receptors are sensitive to inflation of the respiratory system (Molony, 1972, 1974; Leitner and Roumy, 1974; Fedde et al., 1974b) but possess little or no sensitivity to changes in CO_2 concentration (Molony, 1974; Fedde et al., 1974b). These receptors may be located in the lungs or in other parts of the respiratory system; or they may be in almost any other visceral organ in the thoracic or abdominal cavity (Duke et al., 1977). When the respiratory system is in-

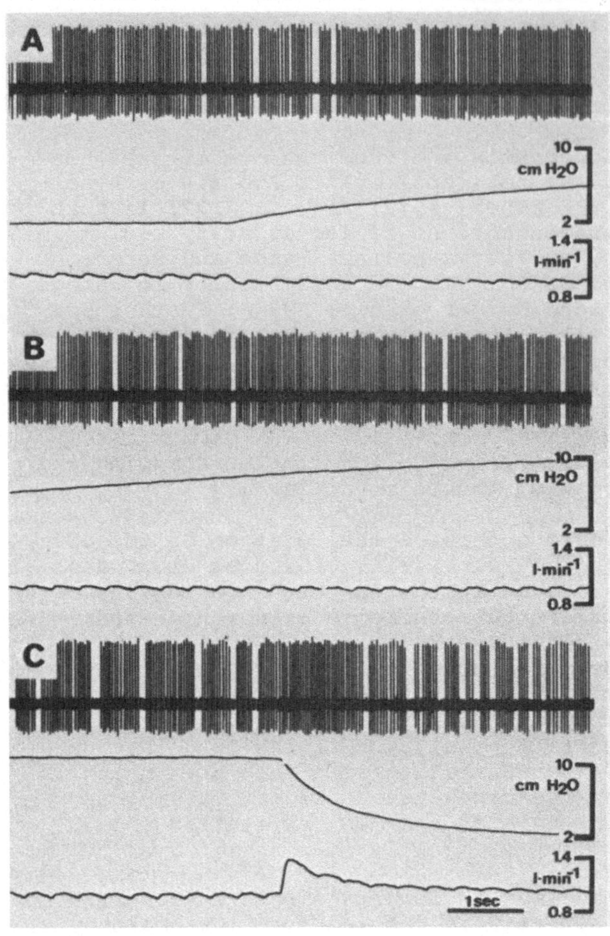

Fig. 8. Response of an intrapulmonary CO_2 receptor in the duck to changes in intrapulmonary pressure. Unidirectionally ventilated bird. *Upper tracing* in each record, nerve impulses; *middle tracing*, intrapulmonary pressure; *lower tracing*, flow of gas through the lungs. (A) and (B) (continuous). Intrapulmonary pressure was increased while gas flow was kept constant. (C) Intrapulmonary pressure was suddenly decreased which caused a transient increase in gas flow. Discharge frequency increased as gas flow increased and as pressure decreased. (From Fedde et al., 1974a)

flated, stresses are placed on all of these organs; this prevents precise localization of the receptors by inflating the respiratory system.

The role of these mechanoreceptors in controlling body function is not currently known. However, if the respiratory system is inflated with a gas containing similar CO_2 concentrations as end-expired gas (5% or higher), thereby preventing increased discharge from the CO_2 receptors, apnea is prevented and respiratory frequency is only slightly decreased (Eaton et al., 1971). That would suggest that the mechanoreceptor discharge which would increase during the inflation may not be significant in controlling breathing.

5. Mammalia

a) Carbon Dioxide-Sensitive Mechanoreceptors

As early as 1847, suggestions were made that CO_2, evolved into the lungs from the venous blood, excited the vagus nerve, and thereby stimulated breathing (Cf. Perkins, 1964 for review). Adrian (1933) indicated

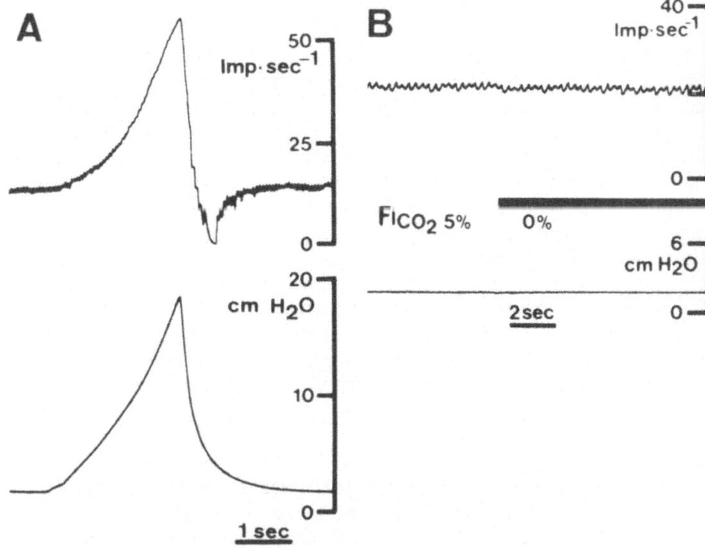

Fig. 9. Discharge characteristics of a mechanoreceptor with a vagal afferent pathway in the duck. (A) Inflation of the respiratory system increased discharge proportional to the intrapulmonary pressure. *Upper tracing*, instantaneous discharge frequency; *lower tracing*, intrapulmonary pressure. (B) Response of this kind of receptor to changes in intrapulmonary CO_2 concentration during unidirectional ventilation. *Upper tracing*, instantaneous discharge frequency; *lower tracing*, intrapulmonary pressure. CO_2 concentration in the unidirectional gas stream was suddenly reduced to 0% at the beginning of the bar. (Modified from Fedde et al., 1974b)

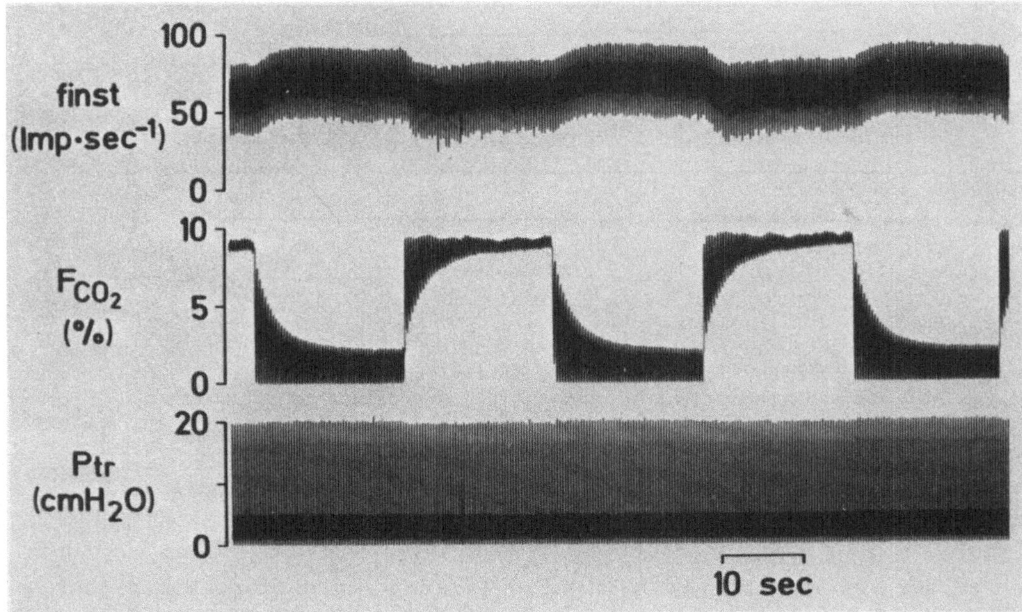

Fig. 10. Sensitivity of a slowly adapting pulmonary stretch receptor in the cat to changes in intrapulmonary CO_2 concentration. *Upper tracing*, instantaneous discharge frequency; *middle tracing*, CO_2 concentration in the ventilating gas; *lower tracing*, tracheal pressure. (From Kunz et al., 1976)

that CO_2 inhaled in concentrations greater than 10% produced a slightly diminished discharge of slowly adapting pulmonary stretch receptors in the cat. The reduced discharge, he indicated, was not beyond the range of experimental error. Little attention was given to the action of that gas on intrapulmonary receptors until Mustafa and Purves (1972) and Schoener and Frankel (1972) studied this in detail. They reported an increased discharge when these receptors were stretched during hypocapnia (18-27 torr) and a decreased discharge during hypercapnia (58-67 torr). An example of CO_2 sensitivity exhibited by these receptor in the cat is shown in Figure 10. Both the inspiratory and expiratory discharge frequency were elevated when the intrapulmonary CO_2 concentration was lowered although the inflation pressure remained constant. Slowly adapting pulmonary stretch receptors also communicate with the central nervous system through the thoracic rami communicants in dogs (Kostreva et al., 1975); hence a non-vagal pathway is also present whereby changes in pulmonary CO_2 concentration may influence respiration.

When blood flows through the lungs, it is difficult to lower the partial pressure of end-tidal CO_2 below about 20 torr, even with severe hyperventilation. Hence, to compare the behavior of slowly adapting pulmonary stretch receptors in mammals with the behavior of CO_2 receptors in birds during exposure to low CO_2 concentrations, the techniques of cardio-pulmonary bypass (Bradley et al., 1976) and pulmonary vascular occlusion (Banzett et al., 1976) had to be used. With these techniques applied to dogs, it is clear that the discharge frequency markedly increases as intrapulmonary CO_2 concentration decreases (Fig. 11). Al-

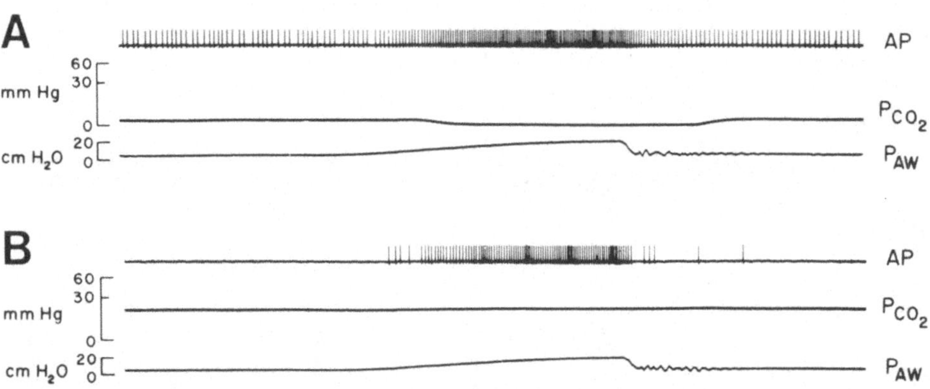

Fig. 11. Influence of low intrapulmonary CO_2 concentrations on the discharge of a slowly adapting pulmonary stretch receptor in the dog. Pulmonary artery was occluded and the lung was ventilated with O_2 only (A) and approximately 3% CO_2 in O_2 (B). AP, action potentials recorded from a left vagal filament; P_{CO_2}, tidal CO_2 from the left lung; P_{AW}, airway pressure in the left lung. Average discharge frequency throughout the ventilatory cycle in *A* is 35.2 imp/s; in *B*, 17.9 imp/s. Each tracing is 6.5 s long. Courtesy of H.M. Coleridge, J.C.G. Coleridge and R.B. Banzett

though the CO_2 sensitivity of the slowly adapting pulmonary stretch receptors is qualitatively similar to that of the CO_2 receptors in birds, the stretch receptors are generally less inhibited by high CO_2 concentrations and display less sensitivity to changes in CO_2 concentration in the range of normal alveolar concentrations (P_{ACO_2} = 40 torr; Fig. 12); indeed, most receptors do not completely cease discharging

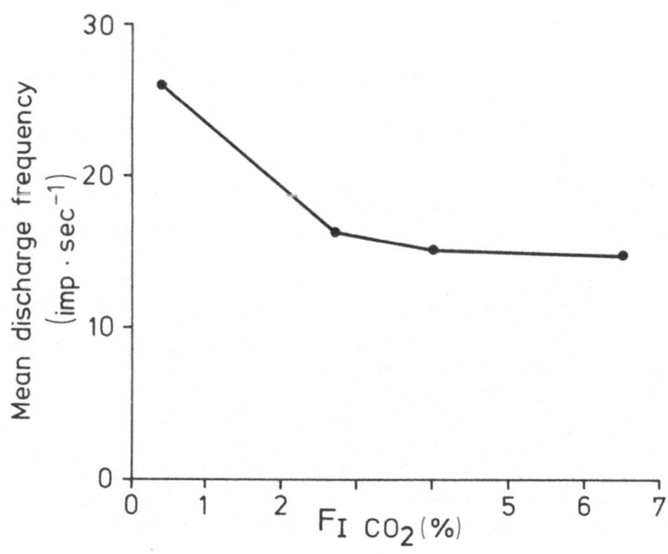

Fig. 12. Response of slowly adapting pulmonary stretch receptors to inspired CO_2 and O_2 during constant volume ventilation. Vascularly occluded left lung of the dog. Mean discharge frequency throughout the ventilatory cycle from 17 receptors is shown. Courtesy of H.M. Coleridge, J.C.G. Coleridge and R.B. Banzett

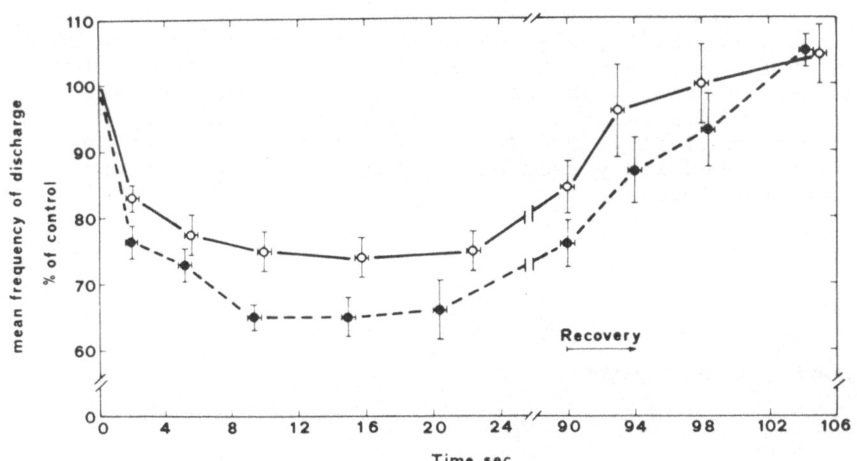

Fig. 13. Time course of CO_2 inhibition on the average rate of discharge of 12 slowly adapting pulmonary stretch receptors located in the intrapulmonary airways of the dog. *Continuous line* before, and *broken line* after ligature of the branch of the pulmonary artery supplying the lobe where the receptors were localized. The average frequency of discharge for the entire respiratory cycle is represented as percentage of the control breaths before administering 8% CO_2 in O_2. *Bars* indicate one standard error. (From Sant'Ambrogio et al., 1974)

during inspiration even though the CO_2 concentration in the inflating gas is high. This supports the early observations (Partridge, 1933; Adrian, 1933) of lack of appreciable CO_2 sensitivity of the receptors above normal alveolar concentrations.

The initial reduction in pulmonary stretch receptor discharge when CO_2 is inhaled is very rapid, sometimes appearing during the first inspiration (Fig. 13) (Sant'Ambrogio et al., 1974; Bradley et al., 1976). Those receptors located in the trachea are not affected by changes in CO_2 concentrations; those in the extrapulmonary bronchi are affected to only a small degree; and those located in the intrapulmonary airways are most affected (Sant'Ambrogio et al., 1974; Bartlett and Sant'Ambrogio, 1976). Furthermore, systemic hypercapnia is generally ineffective in decreasing their discharge as compared with changes in airway gas concentrations (Bartlett and Sant'Ambrogio, 1976). These results coincide with a location of these receptors just beneath the basement membrane of the epithelium in these airways (Bitensky et al., 1975), which would make them accessible to changes in airway CO_2 concentrations.

Reduced partial pressure of CO_2 in the alveoli (P_{ACO_2}) also apparently increases the discharge frequency of rapidly adapting pulmonary stretch (irritant) receptors, but only if the CO_2 concentration is very low (below 19 torr) (Sampson and Vidruk, 1975; Banzett et al., 1976). Furthermore although extremely high concentrations of CO_2 (20-100%) stimulates J-receptors in cats (Dickinson and Paintal, 1970), that gas seemingly does not alter the discharge of "pulmonary" or "bronchial" C-fibers (Banzett et al., 1976) over a physiological range of 20 to 50 torr. These receptors may not be significant in the CO_2-induced changes in ventilation.

It now appears that the reduced respiratory frequency observed when P_{ACO_2} is reduced during cardiopulmonary bypass in dogs (Bartoli et al., 1974) results from increased slowly adapting pulmonary stretch receptor discharge (Bradley et al., 1976). Elevation of the CO_2 concentration in the inspired gas of these dogs results in an immediate increase in respiratory frequency because of a shortening of the expiratory duration with no change in tidal volume (Fig. 14). These reflex changes are associated with immediately reduced discharge frequency of the pulmonary stretch receptors.

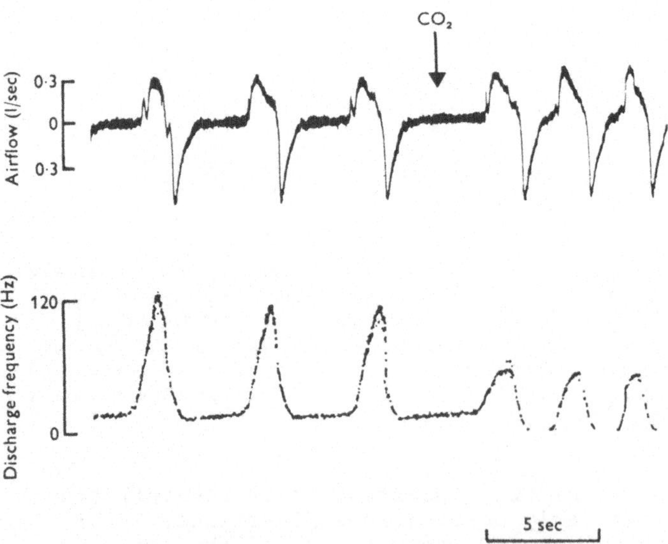

Fig. 14. Relationship between the reflex response to changes in inhaled CO_2 concentration and the effect on slowly adapting pulmonary stretch receptor discharge in the spontaneously breathing dog. The *upper tracing* indicates the airflow signal (inspiration upwards); *Lower tracing*, the instantaneous discharge frequency of a single receptor. CO_2 was added to the inspired gas at the *arrow*. (From Bradley et al., 1976)

As first suggested by Burger (1968), the CO_2-induced depression of the pulmonary mechanoreceptors, which reduces central respiratory inhibition, may be important in controlling respiratory frequency in mammals. However, the significance of this reflex under normal P_{ACO_2} is not clear. Because the slowly adapting pulmonary stretch receptors do not change their discharge markedly when P_{ACO_2} is elevated, it is difficult to conceive how they could contribute to the hyperpnea of exercise (H.M. Coleridge, J.C.G. Coleridge, and R.B. Banzett, personal communication). These investigators suggest that the reflex may be important in limiting the degree of hyperventilation (and thus the degree of hypocapnia) induced by heat stress. The increased discharge from the stretch receptors as P_{ACO_2} is reduced would inhibit the central respiratory neuronal pool and thereby prevent further CO_2 depletion.

6. Conclusions

It presently appears that a dichotomy has occurred in the evolutionary development of intrapulmonary CO_2-sensitive receptors (Fig. 15). Carbon dioxide receptors in the constant-volume avian lung provide an inhib-

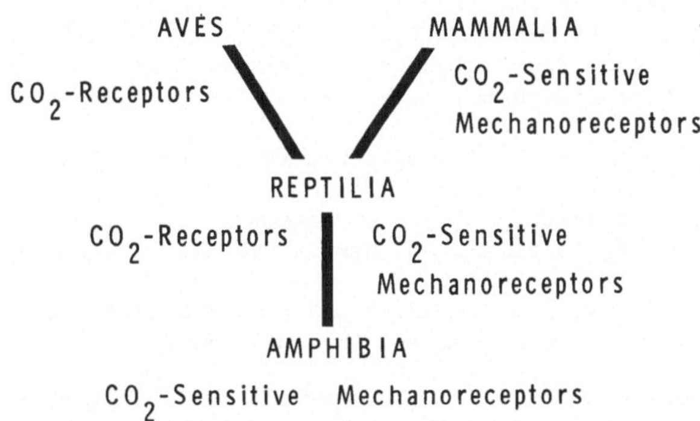

Fig. 15. Summarization of the evolutionary development of known intrapulmonary CO_2-sensitive receptors in the various classes of tetrapods

itory input to the brain in response to overventilation of the lung, similar to the slowly adapting stretch receptors in the mammalian lung, which respond to both changes in CO_2 concentration and mechanical stimuli. These classes of vertebrates apparently have developed sensing systems most suitable to the important types of stimuli in their lungs during each breath. Reptiles, at least as represented by the Tegu lizard, have both receptor types. Hence, the foundation for the separate development in birds and mammals is present.

Many studies must still be conducted before the importance of intrapulmonary CO_2-sensitive receptors in controlling breathing is fully understood. Their discovery, however, allows attention to be focused on an additional sensing system that may lead us closer to discovering exactly which variable is controlled during both normal and abnormal breathing.

Acknowledgments. This study was supported, in part, by a grant-in-aid from the American Heart Association, Kansas Affiliate, Inc. Contribution No. 78-53a, Department of Anatomy and Physiology, KAES, Kansas State University, Manhattan, Kansas 66506, USA.

References

Adrian, E.D.: Afferent impulses in the vagus and their effect on respiration. J. Physiol. (London) 79, 332-357 (1933)

Banzett, R.B., Burger, R.E.: Response of avian intrapulmonary chemoreceptors to venous CO_2 and ventilatory gas flow. Respir. Physiol. 29, 63-72 (1977)

Banzett, R.B., Coleridge, H.M., Coleridge, J.C.G.: Effect of CO_2 on pulmonary vagal afferents in dogs. Physiologist 19, 115 (1976)

Barnas, G.M.: Relationship between the discharge of avian intrapulmonary CO_2 receptors and bronchial smooth muscle contraction. M.S. thesis, Univ. Nebraska, Lincoln, Nebraska 1977

Bartlett, D., Jr., Sant'Ambrogio, G.: Effect of local and systemic hypercapnia on the discharge of stretch receptors in the airways of the dog. Respir. Physiol. 26, 91-99 (1976)

Bartoli, A., Cross, B.A., Guz, A., Jain, S.K., Noble, M.I.M., Trenchard, D.W.: The effect of carbon dioxide in the airways and alveoli on ventilation; a vagal reflex studied in the dog. J. Physiol. (London) 240, 91-109 (1974)

Bitensky, L., Chambers, D.J., Chayen, J., Cross, B.A., Guz, A., Jain, S.K., Johnston, J.J.: Evidence concerning the site of receptors mediating the Hering-Breuer inflation reflex. J. Physiol. (London) 249, 30-31 (1975)

Boelaert, R.B.: Sur la physiologie de la respiration de lacentiens. Arch. Int. pharmacodyn. Ther. 51, 379-436 (1941)

Bonhoeffer, K., Kolatat, T.: Druckvolumendiagramm und Dehnungsrezeptoren der Froschlunge. Pflügers Arch. 265, 477-484 (1958)

Bordoni, L.: Sull'apnea spermentale. Sperimentale 61, 113-132 (1888)

Bradley, G.W., Noble, M.I.M., Trenchard, D.: The direct effect of pulmonary stretch receptor discharge produced by changing lung carbon dioxide concentration in dogs on cardiopulmonary bypass and its action on breathing. J. Physiol. (London) 261, 359-373 (1976)

Burger, R.E.: Pulmonary chemosensitivity in the domestic fowl. Federation Proc. 27, 328 (1968)

Burger, R.E., Osborne, J.L., Banzett, R.B.: Intrapulmonary chemoreceptors in *Gallus domesticus:* Adequate stimulus and functional localization. Respir. Physiol. 22, 87-97 (1974)

Burger, R.E., Nye, P.C.G., Powell, F.L., Ehlers, C., Barker, M., Fedde, M.R.: Response to CO_2 of intrapulmonary chemoreceptors in the emu. Respir. Physiol. 28, 315-324 (1976a)

Burger, R.E., Coleridge, J.C.G., Coleridge, H.M., Nye, P.C.G., Powell, F.L., Ehlers, C., Banzett, R.B.: Chemoreceptors in the paleopulmonic lung of the emu: Discharge patterns during cyclic ventilation. Respir. Physiol. 28, 249-259 (1976b)

Dickinson, C.J., Paintal, A.S.: Stimulation of type-J pulmonary receptors in the cat by carbon dioxide. Clin. Sci. 38, 33P (1970)

Dooley, M.S., Koppanyi, T.: The control of respiration in the domestic duck (*Anas. boscos*). J. Pharmac. Exp. Ther. 36, 507-518 (1929)

Downing, S.E., Torrance, R.W.: Vagal baroreceptors of the bull-frog. J. Physiol. (London) 156, 13P (1961)

Duke, G.E., Kuhlmann, W.D., Fedde, M.R.: Evidence for mechanoreceptors in the muscular stomach of the chicken. Poultry Sci. 56, 297-299 (1977)

Eaton, J.A., Jr., Fedde, M.R., Burger, R.E.: Sensitivity to inflation of the respiratory system in the chicken. Respir. Physiol. 11, 167-177 (1971)

Fedde, M.R., Burger, R.E., Kitchell, R.L.: Localization of vagal afferents involved in the maintenance of normal avian respiration. Poultry Sci. 42, 1224-1236 (1963)

Fedde, M.R., Gatz, R.N., Slama, H., Scheid, P.: Intrapulmonary CO_2 receptors in the duck: I. Stimulus specificity. Respir. Physiol. 22, 99-114 (1974a)

Fedde, M.R., Gatz, R.N., Slama, H., Scheid, P.: Intrapulmonary CO_2 receptors in the duck. II. Comparison with mechanoreceptors. Respir. Physiol. 22, 115-121 (1974b)

Fedde, M.R., Kuhlmann, W.D., Scheid, P.: Intrapulmonary receptors in the tegu lizard: I. Sensitivity to CO_2. Respir. Physiol. 29, 35-48 (1977)

Fedde, M.R., Peterson, D.F.: Intrapulmonary receptor response to changes in airway-gas composition in *Gallus domesticus*. J. Physiol. (London) 209, 609-625 (1970)

Foa, C.: Recherches sur l'apnée des oiseaux. Archs. ital. Biol. 55, 412-422 (1911)
Gatz, R.N., Fedde, M.R., Crawford, E.C., Jr.: Lizard lungs: CO_2-sensitive receptors in *Tupinambis nigropunctatus*. Experentia 31, 455 (1975)
Kostreva, D.R., Zuperka, E.J., Hess, G.L., Coon, R.L., Kampine, J.P.: Pulmonary afferent activity recorded from sympathetic nerves. J. Appl. Physiol. 39, 37-40 (1975)
Kunz, A.L., Kawashiro, T., Scheid, P.: Study of CO_2-sensitive vagal afferents in the cat lung. Respir. Physiol. 27, 347-355 (1976)
Kunz, A.L., Miller, D.A.: Pacing of avian respiration with CO_2 oscillation. Respir. Physiol. 22, 167-177 (1974a)
Kunz, A.L., Miller, D.A.: Effects of feedback delay upon the apparent damping ratio of the avian respiratory control system. Respir. Physiol. 22, 179-189 (1974b)
Leitner, L.-M.: Pulmonary mechanoreceptor fibres in the vagus of the domestic fowl. Respir. Physiol. 16, 232-244 (1972)
Leitner, L.-M., Roumy, M.: Vagal afferent activities related to the respiratory cycle in the duck: Sensitivity to mechanical, chemical and electrical stimuli. Respir. Physiol. 22, 41-56 (1974)
McKean, T.A.: A linear approximation of the transfer function of pulmonary mechanoreceptors of the frog. J. Appl. Physiol. 27, 775-781 (1969)
Miller, D.A., Kunz, A.L.: Avian ventilatory response to dynamic CO_2 signals. J. Appl. Physiol. 39, 129-134 (1975)
Milsom, W.K., Jones, D.R.: Are reptilian pulmonary receptors mechano- or chemosensitive. Nature (London) 261, 327-328 (1976)
Molony, V.: A study of vagal afferent activity in phase with breathing and its role in the control of breathing in *Gallus domesticus*. Ph.D. thesis, Univ. Liverpool, Liverpool, England 1972
Molony, V.: Classification of vagal afferents firing in phase with breathing in *Gallus domesticus*. Respir. Physiol. 22, 57-76 (1974)
Mustafa, M.E.K.Y., Purves, M.J.: The effect of CO_2 upon discharge from slowly adapting stretch receptors in the lungs of rabbits. Respir. Physiol. 16, 197-212 (1972)
Neil, E., Ström, L., Zotterman, Y.: Action potential studies of afferent fibres in the IXth and Xth cranial nerves of the frog. Acta physiol. Scand. 20, 338-350 (1950)
Nielsen, B.: On the regulation of the respiration in reptiles. I. The effect of temperature and CO_2 on the respiration of lizards (*Lacerta*). J. exp. Biol. 38, 301-314 (1961)
Nye, P.C.G.: Stimulus-response relations and location of intrapulmonary chemoreceptors in the lung of *Gallus domesticus*. Ph.D. thesis, Univ. California, Davis, California 1977
Osborne, J.L., Burger, R.E.: Static and dynamic response characteristics of CO_2-sensitive chemoreceptors in the avian lung. Physiologist 14, 205 (1971)
Osborne, J.L., Burger, R.E.: Intrapulmonary chemoreceptors in *Gallus domesticus*. Respir. Physiol. 22, 77-85 (1974)
Partridge, R.C.: Afferent impulses in the vagus nerve. J. Cell. Comp. Physiol. 2, 367-380 (1933)
Perkins, J.F., Jr.: Historical development of respiratory physiology. In: Handbook of Physiology, Section 3, Respiration, Vol. I. Fenn, W.A., Rahn, H. (eds.). Washington, D.C.: Am. Physiol. Soc., 1964, pp. 1-62
Peterson, D.F., Fedde, M.R.: Receptors sensitive to carbon dioxide in lungs of chicken. Science 162, 1499-1501 (1968)
Peterson, D.F., Nightingale, T.E.: Functional significance of thoracic vagal branches in the chicken. Respir. Physiol. 27, 267-275 (1976)
Saalfeld, E. von: Die nervöse Regulierung der Atembewegungen bei *Uromastix*. Pflügers Arch. ges. Physiol. 233, 449-468 (1934)
Sampson, S.R., Vidruk, E.H.: Properties of irritant receptors in canine lung. Respir. Physiol. 25, 9-22 (1975)
Sant'Ambrogio, G., Miserocchi, G., Mortola, J.: Transient response of pulmonary stretch receptors in the dog to inhalation of carbon dioxide. Respir. Physiol. 22, 191-197 (1974)
Scheid, P., Kuhlmann, W.D., Fedde, M.R.: Intrapulmonary receptors in the tegu lizard: II. Functional characteristics and localization. Respir. Physiol. 29, 49-62 (1977)

Scheid, P., Slama, H., Gatz, R.N., Fedde, M.R.: Intrapulmonary CO_2 receptors in the duck: III. Functional localization. Respir. Physiol. 22, 123-136 (1974)

Schoener, E.P., Frankel, H.M.: Effect of hyperthermia and $PaCO_2$ on the slowly adapting pulmonary stretch receptor. Am. J. Physiol. 222, 68-72 (1972)

Smyth, D.H.: The central and reflex control of respiration in the frog. J. Physiol. (London) 95, 305-327 (1939)

Taglietti, V., Casella, C.: Stretch receptors stimulation in frog's lungs. Pflügers Arch. 292, 297-308 (1966)

Taglietti, V., Casella, C.: Deflation receptors in frog's lung. Pflügers Arch. 304, 81-89 (1968)

Templeton, J.R., Dawson, W.R.: Respiration in the lizard *Crolophytus collaris*. Physiol. Zool. 36, 104-121 (1963)

Tschorn, R.R., Fedde, M.R.: Effects of carbon monoxide on avian intrapulmonary carbon dioxide-sensitive receptors. Respir. Physiol. 20, 313-324 (1974)

Circulation: A Comparison of Reptiles, Mammals, and Birds

F. N. White

Summary

The evolution of endothermy has required structural elaboration of the airways and respiratory exchange surfaces. The presence of a completely septated heart in endotherms precludes the use of predominantly neurogenic control of pulmonary vascular resistance, as seen in noncrocodilian reptiles. The occurrence of ventilation-perfusion inequities in the finely subdivided airways of endotherms has required local vascular control in response to prevailing alveolar P_{O_2} with potential for intrapulmonary shunting from poorly ventilated to more adequately ventilated respiratory units. No evidence for this mechanism is seen in the pulmonary circulation of turtles. Although mean air convection requirements are similar in reptiles and endotherms at similar body temperatures and mean blood flow requirements are similar when adjusted for differences in the O_2 capacity of blood, many noncrocodilian reptiles exhibit temporal variation in pulmonary blood flow during a given respiratory cycle. Right-to-left intracardiac shunting is intensified with the length of the apneic period. Such shunts are not possible in endotherms because of complete cardiac septation. An hypothesis is presented to the effect that intracardiac shunting may aid in depletion of available O_2 stores during protracted apnea. The shunt may profitably be viewed as a CO_2 shunt past the alveolar gas phase. The resulting influence on alveolar and systemic P_{CO_2} may augment O_2 binding to Hb in the lungs while intensifying unloading through the Bohr effect at the tissue level. From this perspective, the reptilian blood flow distribution patterns may be viewed as adaptive to the protracted apnea and lower respiratory frequency characterizing the group. These differences may be related to differences in metabolic intensity, respiratory frequency, and the consequent adaptive value of differing control mechanisms for pulmonary vascular resistance in ectotherms and endotherms.

1. Introduction

Comparisons of the circulatory physiology of contemporary reptiles and endotherms may profitably begin from the perspective of the phylogenetic affinities between these groups as indicated by the fossil record. While birds and mammals are reptilian derivatives, the affinities between extant reptiles and mammals are rather distant. Mammals arose from synapsid stock: the mammal-like reptiles of the Permian period. Contemporary noncrocodilian reptiles developed along lines which di-

---

Physiological Research Laboratory, Scripps Institution of Oceanography, University of California, San Diego, La Jolla, California 92093/USA.

verged from synapsids. Turtles were osteologically quite similar to present day species as early as the Triassic (15). Crocodilian affinities with birds appear to be somewhat closer, both groups having been derived from archosaurian stocks of the Triassic. Nonetheless, major structural and metabolic modifications were required in the avian line while crocodilians remained conservative. The divergences leading to contemporary reptiles, birds, and mammals took place perhaps 2×10^8 years ago, and the student of transitional relationships among extant groups must take cognizance of this long period of evolutionary time.

A characteristic of endotherms is high metabolic intensity, and a consequent greater demand on the cardiorespiratory system for gas transport. At similar body mass and body temperature ($\dot{V}_{O_2}$) the O_2 consumption of reptiles is 0.14-0.1 that of birds and mammals at rest (19). These differences are reflected in proportionally lower reptilian values for alveolar surface area, respiratory frequency, cardiac output, hematocrit, and maximum oxygen content of blood (20, 21, 31).

2. Discussion

a) Structural and Functional Comparisons

Distinctive differences exist in cardiac structure between reptiles and endotherms. Among reptiles we may conveniently recognize two general heart types - crocodilian and noncrocodilian (23). The crocodilian heart resembles the avian heart in two ways: complete septation between atria and ventricles; and the persistence of the right aortic arch as the parent vessel of the carotid arteries. However, like other reptiles, both left and right aortic arches persist in the adult. While the right arch originates from the left ventricle as in birds, the left arch originates alongside the pulonary artery from the right ventricle. Both aortic arches communicate just distal to the aortic valves via the formaen of Panizza. Crocodilians are unique in this feature. During respiration the distribution of the cardiac output is much as in birds and mammals, there being no demonstrable intracardiac shunting. During the protracted apnea of diving, a large pressure difference may develop between right ventricle and pulmonary artery owing to increased impedence of the pulmonary outflow tract of the heart. Right ventricular pressure may become suffiently high to cause ejection into the left aortic arch and a right-to-left intracardiac shunt may develop (24). Owing to the circuitry of the crocodilian heart, left-to-right shunts are not possible. No measure of the magnitude of cardiac output or shunts exist for crocodilians.

Noncrocodilian reptiles exhibit an incompletely divided ventricle (23, 25). As a result, mandatory separation of the pulmonary and systemic circuits, as in mammals, does not exist. The degree of separation of these circuits depends on the prevailing resistances in the two circuits. Thus, when pulmonary vascular resistance is high, as in the diving turtle, the ratio of pulmonary to systemic arterial blood flow may fall to low levels (29). Millen et al. (13) have reported complete pulmonary bypass in diving *Pseudemys scripta*. The intensity of right-to-left shunting appears to depend on apneic time and such shunts characterize normally breathing turtles (28). While mean apneic time for resting *P. floridana* at 25°C is 1.4 min, apnea of up to 12 min duration is not uncommon (9). Left-to-right shunts may occur during the ventilation phase of the respiratory cycle (18, 29). Just prior to ventilation, heart rate increases, and with it cardiac output (Fig. 1). Blood flow

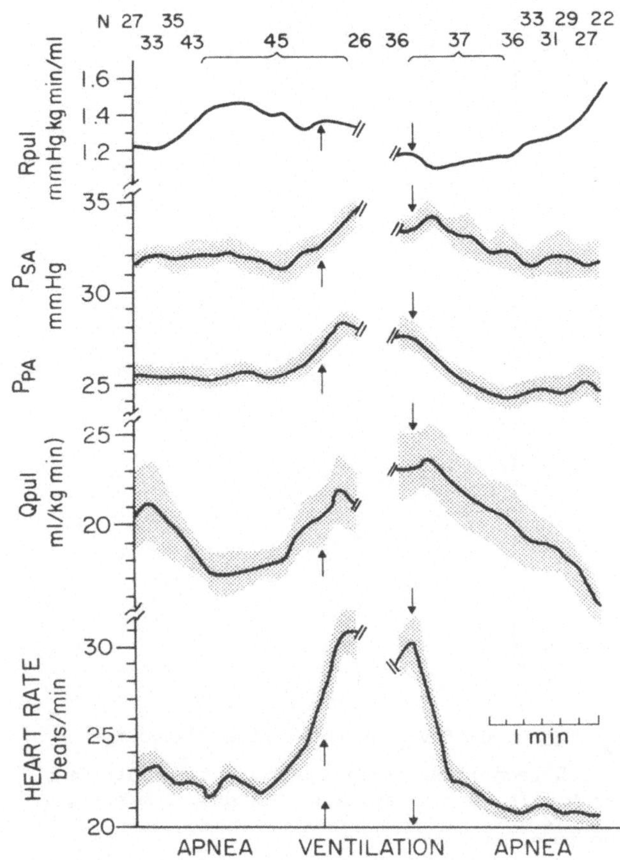

Fig. 1. Variations in mean pulmonary blood flow and heart rate in the turtle, *Pseudemys floridana*, in relation to the respiratory cycle. *Shaded zones* represent mean ± SE (standard error of the mean). N, number of observations. Pulmonary blood flow determined by electromagnetic flow meter

to the lungs reaches maximum values during ventilation and early apnea, declining as apnea progresses. Thus, maximum flow is associated with the phase of maximum P_{AO_2} and minimum P_{ACO_2} while pulmonary bypass becomes progressively greater as P_{AO_2} declines and P_{ACO_2} increases.

While variation in pulmonary blood flow characterizes the respiratory cycle, the mean blood flow ($\dot{V}_b$), determined from $\dot{V}_{O_2}$ and pulmonary arterial and venous O_2 content by the Fick method, exhibits a direct proportionality to oxidative metabolism ($\dot{V}_{O_2}$) at different body temperatures (Fig. 2). This constancy in $\dot{V}_b/\dot{V}_{O_2}$ characterizes mammals undergoing nonfatiguing exercise. The blood convection requirement for *P. floridana* at rest and 37°C is 1.1 l/mmol O_2 (8, 27) while that of man is 0.52 (4) and hen, 0.4 (14). These values are in rough inverse proportion to the O_2 capacity of blood ($C^{max}_{HbO_2}$). Ventilation requirements for resting turtle and man are quite similar at the same body temperature: turtle, 0.5 l/mmol O_2 (8); man, 0.56 l/mmol O_2 (4). However,

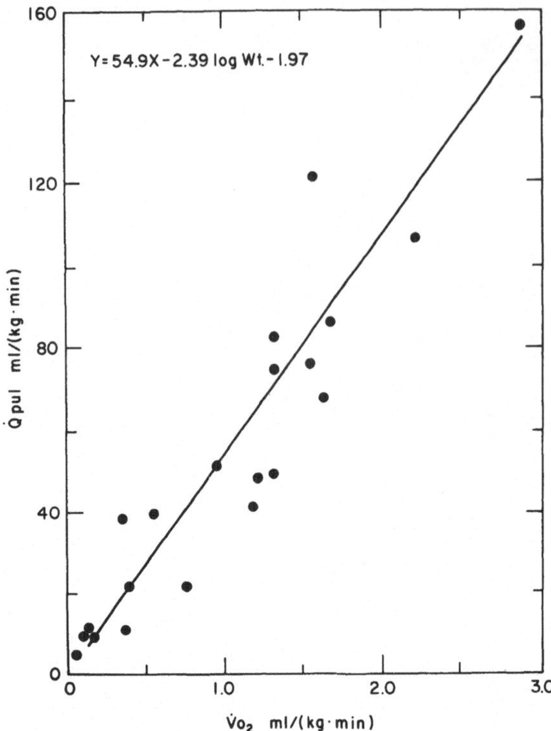

Fig. 2. Mean pulmonary blood flow as a function of $\dot{V}_{O_2}$ in the turtle, *Pseudemys floridana*. Blood flows determined by Fick technique. A range of body temperatures were used to achieve various levels of $\dot{V}_{O_2}$ in resting animals. The body temperature for a given $\dot{V}_{O_2}$ and body weight can be found from the equation:
Log $\dot{V}_{O_2}$, ml/(kg • min) = 0.047 BT, °C - 0.24 log wt, kg - 1.32 (8, 27)

when body temperature declines, the air convection requirement of turtles increases, an adjustment which has been attributed to regulation of the arterial pH parallel to the neutral pH of water (5, 7). This results in an inverse relationship between ventilation ($\dot{V}_E$) and pulmonary blood flow ($\dot{Q}_{pul}$) and body temperature or $\dot{V}_{O_2}$. It is only when the data are compared at similar body temperature that the close similarity in overall ventilation and perfusion requirements between turtle and man become apparent. Fixing of $\dot{V}E/\dot{V}_{O_2}$ at similar levels would appear to reflect similar requirements for the regulation of P_{aCO_2} and pH in these distantly related lung breathers. O_2 extraction, at similar body temperature, is of the same order of magnitude in turtle, man, and bird (8).

The regulatory mechanisms that govern mean blood flow requirements through the gas exchanger of both resting turtles and endotherms achieve similar results when allowances are made for differences in $C^{max}_{HbO_2}$. However, the nature of the effectors governing blood flow through the lungs appears to be different. While variations in pulmonary vascular resistance are associated with distributional inequities between systemic and pulmonary circuits in noncrocodilian reptiles, such variations in distribution are not possible in adult birds and mammals because of complete cardiac septation. In mammals, the autonomic innervation of the pulmonary vasculature appears to influence vascular resistance only slightly. However, exposure to transient or longer term hypoxia elevates pulmonary resistance in both birds (3) and mammals (4, 22). The hypoxic response is attributed to a local effect and has the advantage of inducing blood shunts from poorly ventilated gas exchange units to those which are more adequately ventilated. This redistribution of intrapulmonary blood flow has the consequence of insuring better blood gas and acid-base composition of the arterial blood.

White and Kinney (27) have directly observed the microcirculation of the alveoli of the turtle through glass plates mounted in the carapace. Through cannulae which enter the alveolar spaces, gases of varying composition can be introduced. Under conditions of local hypoxia, hypercapnea, or mixed hypoxia and hypercapnea, no visual changes in microcirculatory flow patterns have been observed. We are led to doubt the presence of local vasoconstrictor mechanisms of the mammalian type in the turtle lung.

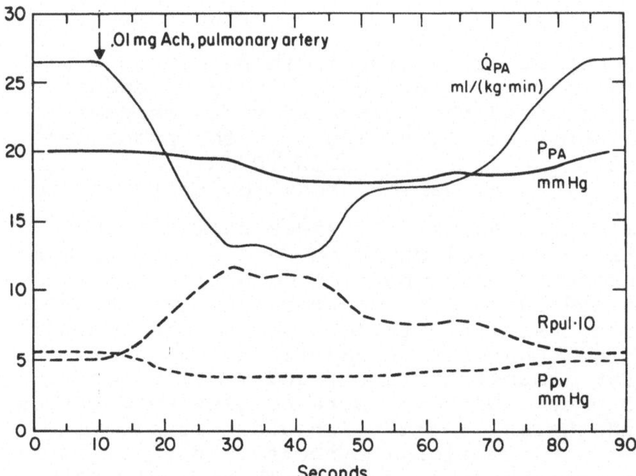

Fig. 3. Effect of acetylcholine on pulmonary hemodynamics in the turtle, *Pseudemys floridana*. Resistance ($R_{pul} \cdot 10$) calculated from the ratio of arterio-venous pressure difference ($P_A - P_V$) to flow ($\dot{Q}_{PA}$). PA, pulmonary artery; PV, pulmonary vein. Flow determined by electromagnetic flow meter

The anatomy of the pulmonary vessels of noncrocodilian reptiles offers a further contrast with mammals and birds. The basal portion of the pulmonary arch is largely devoid of smooth muscle, while each pulmonary artery becomes abruptly muscular near the lung hilus. Krogh (9) and Luckhardt and Carlson (11) made observations on frog and turtle that demonstrated the presence of cholinergic vasoconstrictor fibers in the pulmonary vasculature. Figure 3 demonstrates the influence of acetylcholine on blood flow and vascular resistance through the lung of the turtle. Atropine abolishes this response. Stimulation of the pulmonary vagus nerve produces a similar effect (Fig. 4). Cholinergic vasoconstrictor effects on the gas exchanger have been reported for fishes and amphibians as well as reptiles, while catecholamines produce

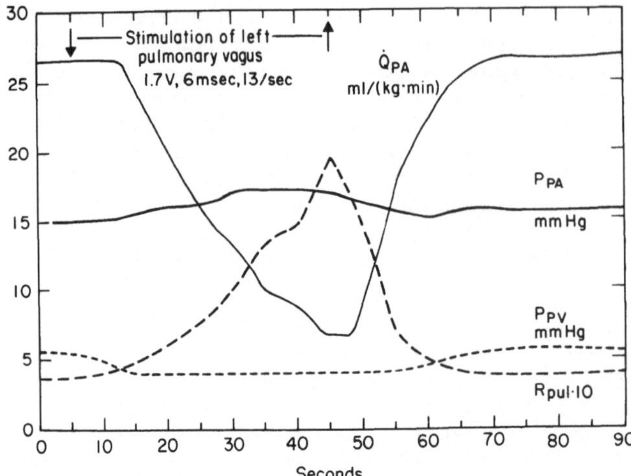

Fig. 4. Effect of stimulation of pulmonary branch of the vagus nerve on pulmonary hemodynamics in the turtle, *Pseudemys floridana*. Resistance ($R_{pul} \cdot 10$) calculated from the ratio of arterio-venous pressure difference (PA-PV) to flow ($\dot{Q}_{PA}$). PA, pulmonary artery; PV, pulmonary vein. Flow determined by electromagnetic flow meter

dilator responses through β-receptor activation in these groups (26). The pulmonary vascular resistance of the turtle is highly dependent upon vagal influences since section of the pulmonary vagus induces vasodilation as does atropine administration. Thus, in the probable absence of local hypoxic effects the adjustments in $\dot{V}E/\dot{Q}_{pul}$ and maintenance of constant overall $\dot{Q}_{pul}/\dot{V}_{O_2}$ at various body temperatures appear to be under reflex cholinergic control. It seems unlikely that this type of pulmonary resistance control would be well suited to mammals and birds where such neurogenic vasoconstriction would impose large right ventricular loads and pulmonary hypertension. The presence of local vascular control through sensitivity to alveolar gas composition may be less critical in the amphibian or reptilian lung where airway resistance is low, alveolar dimensions large, and ventilation inequalities are less likely to occur. With the more complex subdivisions of the airways required by endotherms, ventilation and perfusion inequality is more likely, due to gravitational effects or partial obstruction of finer airways by lung secretions (22). These contrasts suggest that with the development of endothermy and the consequent demand for greater respiratory surface area, a predominantly neural control of pulmonary vascular resistance became inappropriate for regulation of blood gas composition. Instead, because of the more uneven distribution of airway resistance and gas flow, local vacular resistance control as a means of insuring a relatively constant arterial gas composition came to be important. Unfortunately, the mammal-like reptiles which may have exhibited this transition disappeared in the Permian past.

b) Functional Significance of Reptilian Intracardiac Shunts

The data required for a quantitative evaluation of the effect of reptilian blood-shunting patterns on gas exchange are not available. A knowledge of alveolar gas composition, blood gases and hemoglobin saturation in multiple loci, and systemic and pulmonary blood flows, all on a chronological basis through varying apneic periods, will be required for such analysis. We can, nonetheless, deduce from existing data a probable significance of intracardiac shunts.

Reptiles exhibit protracted apnea during the normal respiratory cycle, and numerous species of all reptilian groups are either totally or semi-aquatic, and submergence may last from minutes to hours. Arterial P_{CO_2} values exceeding 100 torr have been observed after 2 h submergence in *Pseudemys scripta*, while P_{O_2} values fall to a few torr (6). Utilization of virtually all lung and blood O_2 stores occurs during protracted breath-holding and as hypoxia progresses, anaerobic glycolysis is utilized as an energy source. As apnea progresses, P_{AO_2} progressively falls while P_{ACO_2} rises - both conditions unfavorable to oxygenation of blood owing to the Bohr shift. However, the respiratory exchange ratio (R_E) in the lungs becomes progressively lower with apneic time in *Pseudemys* as demonstrated by Wilson (30). This observation is suggestive of right-to-left intracardiac shunting. A similar progressive reduction in CO_2 elimination from blood into the lung was apparent from the decreasing slope of the line relating P_{ACO_2} to P_{AO_2} during 20 min breath-holding in *Chelys* (10). In this study, the P_{O_2} gradient between alveolar gas and arterial blood decreased with time, a condition which was interpreted as due more to the presence of right-to-left shunting than to changes in conditions affecting diffusion from gas phase into the blood. The P_{50} (lungs) will inexorably become greater as apnea progresses, P_{ACO_2} increases, and P_{AO_2} declines. Right-to-left shunts may be viewed as a means of reducing the rate at which P_{50} increases owing to the reduction in CO_2 acquisition rate by lung gas. Consequently, a progressively lower lung blood flow may become as effective in O_2 extraction from lung gas as higher flows.

Right-to-left shunts will have the converse effect on arterial blood due to mixing of venous blood with that exposed to the lung gases. Intensification of the Bohr effect at the tissue level through this mechanism will augment tissue uptake of O_2.

A peculiarity of the CO_2 dissociation curve of turtle blood must also be considered (Fig. 5). Unlike mammalian blood, as P_{CO_2} is elevated, the slope of the dissociation curve becomes decidedly flatter. As a consequence, the combination of a given quantity of CO_2 with blood has an inordinately strong effect on P_{CO_2} and pH. This difference in turtle and mammalian blood has been related to differences in Hb concentration, since elevating the hematocrit of turtle blood to mammalian levels increases the slope of the curve to a similar level as for human blood (17, 30). Lenfant et al. (10) found that as breathholding by *Chelys* progressed and P_{aCO_2} increased, there was an initial moderate change in pH and increment in HCO_3^-; then, as P_{CO_2} increased further, the HCO_3^- rise tapered off with a fall in pH. This is in contrast to the behavior of human blood under conditions of respiratory acidosis.

The acidification effect at higher CO_2 levels should be expected to augment tissue O_2 transport further via the Bohr shift. At elevated P_{CO_2} in mixed venous blood, the impact on CO_2 transport to lung gas under nonshunt conditions would propel P_{ACO_2} upward at an accelerated rate. Progressive right-to-left shunting may become increasingly important in blunting this trend. Although crocodilians apparently do

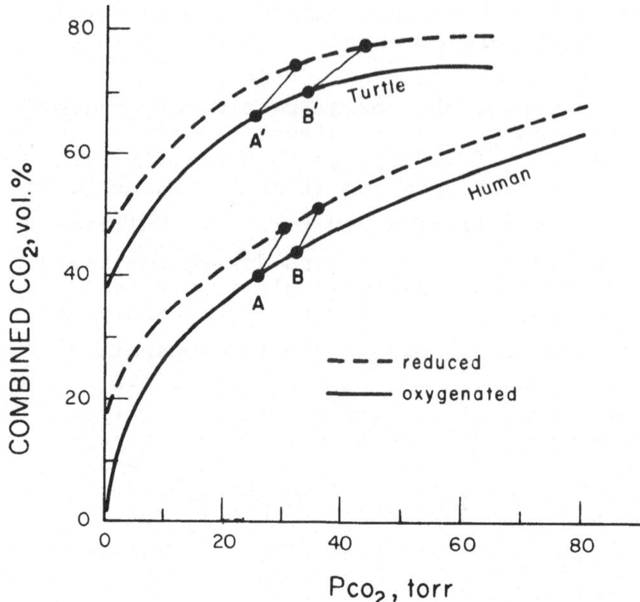

Fig. 5. Comparison of CO_2 dissociation curves for man and turtle (based on 17, 30). As CO_2 content increases in the turtle blood the slope of the curve decreases. At higher CO_2 contents, a given increment (A´ to B´) in CO_2 at the tissue level has a greater influence on P_{CO_2} than for human blood (A to B). Increasing hematocrit above normal in turtle blood increases the slope of the curve (17)

not exhibit right-to-left shunting while on land, the observed shunt during submersion (24) may be interpreted in a similar manner.

Intracardiac shunting of resting reptiles during normal ventilatory activity may not reach levels which materially affect O_2 transport. These patterns may be a part of a continuum of responses associated with ventilatory state, and which have contributed to the exploitation of both terrestrial and aquatic habitats. Left-to-right intracardiac shunts have been observed only in turtles. The larger cardiac output and pulmonary blood flow associated with ventilation may be viewed as a mechanism for accelerating CO_2 elimination and O_2 uptake following protracted apnea or exercise.

A high degree of separation between systemic and pulmonary circuits has been reported for *Varanus niloticus* with little apparent tendency to develop shunts (12). Varanid lizards exhibit an unusually high aerobic capacity and are highly active predators. While resting oxidative metabolism is similar to other reptiles of similar size, the ability for sustained oxidative metabolism is remarkably developed. In searching for correlates of this feature, Bennett (1) found no unusual comparative differences between the iguanid lizard, *Sauromalus*, and *V. gouldii* among resting metabolic rate, resting or active ventilation rate, active heart rate, hematocrit, blood oxygen capacity and affinity, resting pH and lactate content, blood becarbonate and phosphate concentrations, and aerobic enzymatic activities in liver or skeletal muscle. The major contrasts with other lizards are: a complex lung structure with greater surface area, high myoglobin levels, and excellent noncarbonic blood buffers. For *Sauromalus*, the noncarbonic buffering strength is -2.26 log P_{CO_2}/pH unit; for *V. gouldii*, -8.55 log P_{CO_2}/pH unit. This higher buffering capacity should ameliorate the impact of blood acidification and its influence on P_{50} (lungs) which appears to characterize turtles under conditions of elevated blood CO_2. The apparent lack of intracardiac shunting in varanids, higher ratio of systemic to pulmonary resistance, capacity for sustained aerobic activity, lungs of large surface area and superior noncarbonic buffering capacity places this group in a functionally intermediate position between other reptiles and birds and mammals.

Acknowledgment. Supported by Grant No. PCM-12754 from the National Science Foundation.

References

1. Bennett, A.F.: Blood physiology and oxygen transport during activity in two lizards, *Varanus gouldii* and *Sauromalus* hispidus. Comp. Biochem. Physiol. 46A, 673-690 (1973)
2. Berger, P.J.: The vagal and sympathetic innervation of the isolated pulmonary artery of a lizard and a tortoise. Comp. Gen. Pharmac. 3, 113-124 (1972)
3. Burton, R.R., Besch, E.L., Smith, A.H.: Effect of chronic hypoxia on the pulmonary arterial blood pressure of the chicken. Am. J. Physiol. 214, 1438-1442 (1968)
4. Dejours, P.: Principles of Comparative Respiratory Physiology. Amsterdam, Oxford: North Holland, 1975, p. 253
5. Howell, B.J., Baumgardner, F.W., Bondi, K., Rahn, H.: Acid-base balance in cold-blooded vertebrates as a function of body temperature. Am. J. Physiol. 218, 600-606 (1970)
6. Jackson, D.C.: Metabolic depression and oxygen depletion in the diving turtle. J. Appl. Physiol. 24, 503-509 (1968)
7. Jackson, D.C.: The effect of temperature on ventilation in the turtle, *Pseudemys scripta elegans*. Respir. Physiol. 12, 131-140 (1971)
8. Kinney, J.L., Matsuura, D.T., White, F.N.: Cardiorespiratory effects of temperature in the turtle, *Pseudemys floridana*. Respir. Physiol. (in press)
9. Krogh, A.: Über vasomotorische Nerven zu den Lungen. Centralbl. Physiol. 20, 802-806 (1906)
10. Lenfant, C., Johansen, K., Petersen, J.A., Schmidt-Nielsen, K.: Respiration in the fresh water turtle, *Chelys fimbriata*. Respir. Physiol. 8, 261-275 (1970)
11. Luckhardt, A.B., Carlson, A.J.: Studies on the visceral sensory nervous system. VII. On the presence of vasomotor fibers in the vagus nerve to the pulmonary vessels of the amphibian and the reptilian lung. Am. J. Physiol. 56, 72-112 (1921)
12. Millard, R.W., Johansen, K.: Ventricular outflow dynamics in the lizard, *Varanus niloticus:* Responses to hypoxia, hypercarbia and diving. J. Exp. Biol. 60, 871-880 (1974)
13. Millen, J.E., Murdaugh, H.V., Bauer, C.B., Robin, E.D.: Circulatory adaptations to diving in the fresh water turtle. Science 145, 591-593 (1964)
14. Piiper, J., Drees, F., Scheid, P.: Gas exchange in the domestic fowl during spontaneous breathing and artigicial ventilation. Respir. Physiol. 9, 234-245 (1970)
15. Romer, A.S.: Vertebrate Paleontology. 3rd ed. Chicago: Chicago Univ. Press, 1966, p. 468
16. Shelton, G., Burggren, W.: Cardiovascular dynamics of the Chelonia during apnoea and lung ventilation. J. Exp. Biol. 64, 323-343 (1976)
17. Southworth, F.C., Redfield, A.C.: The transport of gas by the blood of the turtle. J. Gen. Physiol. 9, 387-403 (1926)
18. Steggerda, F.R., Essex, H.E.: Circulation and blood pressure in the great vessels and heart of the turtle (*Chelydra serpentina*). Am. J. Physiol. 190, 310-326 (1957)

19. Templeton, J.R.: Reptiles. In: Comparative Physiology of Thermoregulation, Vol. I. Whittow, G.C. (ed.). New York-London: Academic Press, 1970, pp. 167-221
20. Tenney, S.M., Remmers, J.E.: Comparative quantitative morphology of the mammalian lung: Diffusing area. Nature (London) 197, 54-56 (1963)
21. Tenney, S.M., Tenney, J.B.: Quantitative morphology of cold-blooded lungs: Amphibia and Reptilia. Respir. Physiol. 9, 197-215 (1970)
22. West, J.B.: Ventilation/blood flow and gas exchange. Oxford: Blackwell Scientific Publications, Ltd., 1970, p. 117
23. White, F.N.: Functional anatomy of the heart of reptiles. Am. Zool. 8, 211-219 (1968)
24. White, F.N.: Redistribution of cardiac output in the diving alligator. Copeia 3, 567-570 (1969)
25. White, F.N.: Circulation. In: Biology of the Reptilia, Vol. V. Gans, C., Dawson, W.R. (eds.). London, New York, San Francisco: Academic Press, 1976, pp. 276-334
26. White, F.N.: Comparative aspects of vertebrate cardiorespiratory physiology. Ann. Rev. Physiol. (in press)
27. White, F.N., Kinney, J.: Ventilation-perfusion relationships in the turtle. Physiologist 19, 409 (1976)
28. White, F.N., Ross, G.: Blood flow in turtles. Nature (London) 208, 759 (1965)
29. White, F.N., Ross, G.: Circulatory changes during experimental diving in the turtle. Am. J. Physiol. 211, 15-18 (1966)
30. Wilson, J.W.: Some physiological properties of reptilian blood. J. Cell. Comp. Physiol. 13, 315-326 (1939)
31. Wood, S.C., Lenfant, C.J.M.: Respiration: Mechanics, control and gas exchange. In: Biology of the Reptilia, Vol. V. Gans, C., Dawson, W.R. (eds.). London, New York, San Francisco: Academic Press, 1976, pp. 225-274

Mechanisms Controlling the Oxygen Affinity of Bird and Reptile Blood: A Comparison Between the Funtional Properties of Chicken and Crocodile Hemoglobin

C. Bauer, W. Jelkmann, and H. Rollema

Summary

We have compared the mechanisms which determine the O_2 affinity of bird (chicken) and crocodile (*Crocodylus porosus*) blood. It was established that the two species, even though being distant relatives, have evolved entirely different strategies for decreasing the O_2 affinity of the respective hemoglobins: in the bird it is the O_2-linked binding of myo-inositolpentaphosphate and in the crocodile the formation of O_2-linked carbamino compounds which largely account for the O_2 affinity of whole blood.

1. Introduction - Summary of Evolution

The nearest reptilian relatives of birds are members of the order Crocodilia. As this Symposium is devoted to the respiratory function in birds, it appears appropriate to compare more carefully the mechanisms controlling the O_2 affinity of these distant relatives rather than to enumerate indiscriminately the respiratory pattern of blood from a large variety of reptiles and birds. To start with, we want to trace briefly the genealogical tree of crocodiles and birds and then to consider the characteristics of O_2 binding of blood and hemoglobin solutions.

Reptiles have evolved from stem reptiles: the ancestral Cotylosaurs, which appeared during the Pennsylvanian (Carboniferous), some 300 million years ago. The Chelonians (turtles), the Mesosaurs, and the Ichthyosaurs (fish reptiles) seem to have descended directly from the Cotylosaurs. In the upper Permian and lower Triassic (230 million years ago), the Eosuchians appeared which are the direct ancestors of the lizards and snakes of the order Squamata. Other descendents of the primitive Eosuchians are the Rynchocephalians from which only one species, *Sphenodon*, has survived. The ancestors of the subclass Archosauria, the so-called ruling reptiles (Dinosaurs, Pterosaurs, Crocodiles and primitive birds) are represented by the order Thecodontia, which made their first appearance during the Triassic period, about 220 million years ago. It is likely that the Thecodonts rose directly from the primitive Cotylosaurs rather than by the way of Eosuchians. This order Thecodontia is, in all likelihood, the stock from which crocodiles and birds arose, the former during the Triassic period and the latter during the jurassic period about 150 million years ago. The common features of crocodiles and birds are the diapsid condition, i.e., two temporal openings in the skull (which have fused into one in birds) and the

Physiologisches Institut der Universität, D-8400 Regensburg, Federal Republic of Germany.

presence of a four-chambered heart. The combination of these two characteristics is not shared by any other reptiles (Romer, 1966).

2. Methods and Results

Let us now consider the O_2-binding properties of blood and hemoglobin (Hb) solutions and the influence on them of allosteric cofactors. The O_2 pressure at which the Hb is half saturated with O_2 (P_{50}) was 30.7 torr in the crocodile and 27.8 torr in the chicken (20°C, extracellular pH 7.47, P_{CO_2} = 40 torr). In order to facilitate comparison at temperatures other than 20°C we have determined the temperature dependency of P_{50} in the range between 20°C and 37°C and found $\Delta \log P_{50}/\Delta T = 0.014$ and 0.015 in crocodile and chicken blood, respectively.

If the Hb of each species was freed of all intracellular phosphate compounds by gel chromatography and its O_2 affinity examined in the absence of CO_2, P_{50} (pH 7.3 and 20°C) was 2.6 torr in chicken and 3.8

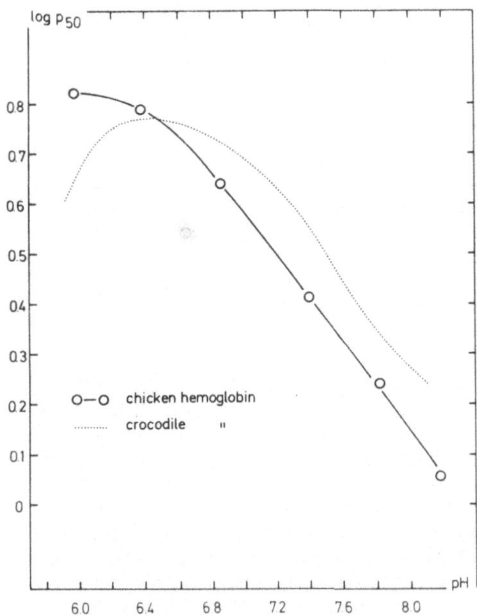

Fig. 1. Plot of log P_{50} against pH in chicken (o-o) and crocodile (---) hemoglobin. The Bohr coefficient $\Delta \log P_{50}/\Delta pH$ is -0.40 for chicken and -0.43 for crocodile Hb in the range between pH 7.2 and pH 8.0. [Hb] = 120 µM Fe, T = 20°C

torr in crocodile Hb (Fig. 1). Thus, it is apparent that the intrinsic O_2 affinity of bird and crocodile Hb is high and therefore not responsible for the relatively low O_2 affinity of whole blood. The obvious thing to do then, is to investigate the effect of the various phosphate compounds which were shown to be of importance for the control of O_2 affinity. In the erythrocytes of major birds, myoinositolpentaphosphate (IP_5) and adenosine triphosphate (ATP) constitute over 70% of the total phosphate content; whilst in the developing chick 2,3-diphosphoglycerate (DPG) and ATP are the most important phosphate compounds (Isaacks et al., 1976a, b). We have prepared IP_5 from bird blood (Bartlett and Borgese, 1976) and compared its effect on P_{50} of chicken Hb with that of ATP

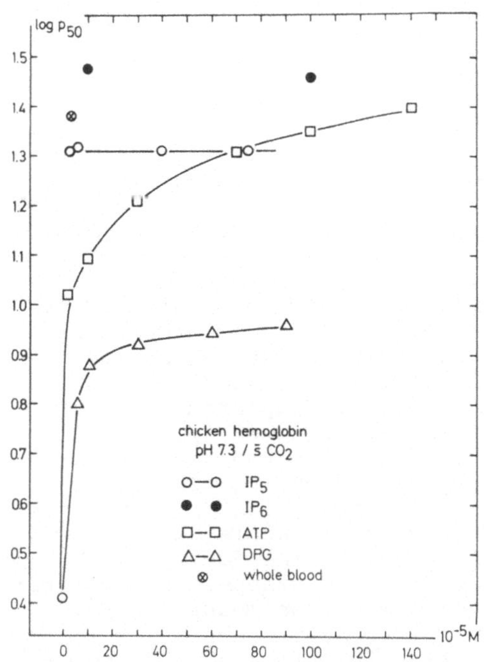

Fig. 2. Plot of log P_{50} against the concentration of various phosphates in absence of CO_2 ($\bar{s}$ CO_2). Hb = 120 µM Fe, T = 20°C. IP_5: myo-inositolpentaphosphate, IP_6: myo-inositolhexaphosphate, ATP: adenosine triphosphate, DPG: 2,3-diphosphoglycerate. The sign ⊗ indicates log P_{50} of whole blood at extracellular pH 7.47, P_{CO_2} = 40 torr, 20°C. Intracellular pH is assumed to be 7.3

and DPG (Fig. 2). It turned out that, not unexpectedly, IP_5 has the strongest effect of all, at low concentrations.

Addition of 1 mol IP_5 to 1 mol Hb decreased the O_2 affinity by a factor of 8 (P_{50} = 21 mmHg) but did not further change with increasing concentrations of IP_5. It thus turned out that the P_{50} of the Hb solution (pH 7.3, 20°C) in the presence of IP_5 is quite close to that of whole blood (extracellular pH 7.47, 20°C). This result suggests that the interaction between IP_5 and chicken Hb largely determines the O_2 affinity of whole blood. The slightly higher P_{50} of the blood is probably due to the uncertainty in estimating intracellular pH as addition of either CO_2 (at constant pH) or ATP, or both, to Hb solutions containing IP_5 did not alter P_{50} in comparison to the value found with IP_5 alone. A specific action of CO_2 via formation of O_2-linked carbamate can therefore be excluded in chicken blood which confirms results obtained with IP_6 in other bird species (Baumann and Baumann, in press; Petschow et al., 1977). Note, that at equal concentrations, IP_6 increases P_{50} by a factor of almost 12, indicating that the difference in binding between oxyhemoglobin and deoxyhemoglobin is substantially larger for IP_6 than for IP_5.

ATP at high concentrations (ATP/Hb_4 > 30) was even more effective than IP_5 in increasing P_{50}. Of the phosphates DPG proved to be the least effective, increasing P_{50} (at maximal concentrations) by a factor of 3.5. In order to establish the binding constants which would describe the influence of phosphates on the O_2 affinity of bird Hb, we have used the approach of Szabo and Karplus (1976). We found following association constants:

	ATP	DPG
Oxyhemoglobin	5.5×10^2 M^{-1}	1.2×10^4 M^{-1}
Deoxyhemoglobin	9×10^6 M^{-1}	1.9×10^6 M^{-1}

As the maximal effect of IP_5 on P_{50} was already reached at equimolar concentrations of the phosphate it is only possible to obtain numbers on the *difference* between the association constants for oxy- and deoxyhemoglobin. Using again the formula given by Szabo and Karplus (1976) this difference turned out to be 4.4×10^3.

We next wish to consider the relatively low temperature dependency of the O_2 affinity in chicken blood. As DPG decreases the apparent heat of oxygenation in human Hb (Benesch et al., 1969) we suspected that the low temperature effect of chicken blood may be related to the strong binding of IP_5 to chicken Hb. At pH 7.3, the apparent heat of oxygenation (ΔH) obtained from a plot of log P_{50} against 1/T (K) in the range between 20 and 37°C was -7.8 kcal/mol O_2 bound, in the absence and -2.4 kcal/mol O_2 bound, in the presence of IP_5 (IP_5/Hb_4 = 1.5). This latter figure is in reasonable agreement with an apparent ΔH of -3.6 kcal/mol O_2 bound, which we estimated in whole blood. The higher value found in whole blood is likely due to the fact that the slope of the line connecting pH and temperature is steeper in the intra-erythrocytic compartment than in the plasma, where we measured pH.

The major phosphate which has been identified in alligator and crocodile erythrocytes is ATP (Rapoport and Guest, 1941; Bartlett, 1976 and our own observations) which, however, is present only in low concentrations, i.e., about 1 mmol/l RBC. Nevertheless, it still could be that ATP interacts as strongly with crocodile hemoglobin as IP_5 does with chicken hemoglobin. This, however, proved not to be the case. Neither ATP nor DPG nor IP_5 each used in a 10-20 fold excess over Hb_4 had any effect on the O_2 affinity of crocodile hemoglobin. Even an increase in ionic strength produced by inorganic anions (0.1 M PO_4) or use of more remote organic phosphate compounds like fructose 1,6-diphosphate, cyclic 3',5'-AMP, or thimidine triphosphate did not shift P_{50} to any significant extent. The only allosteric cofactor which considerably increased P_{50} was CO_2. At pH 7.3 addition of CO_2 equivalent to a P_{CO_2} of 40 torr shifted P_{50} from 3.9 to 28.8 torr (Fig. 3). Note that P_{50} of whole blood at 20°C, P_{CO_2} = 40 torr, extracellular pH 7.47, was almost identical to the one found in dilute Hb solutions in the presence of CO_2. This large increase in P_{50} is most likely due to an extensive formation of O_2-linked carbamino compounds in crocodile hemoglobin. As some tetrameric reptile hemoglobins associate to form higher aggregates like octamers (Sullivan and Riggs, 1967) the possibility arises that CO_2 simply increases the formation of higher aggregates which could have a lower O_2 affinity than the parent molecules, similar to what is observed in lamprey hemoglobin (Briehl, 1963). This possibility can be safely excluded for the following reasons. Firstly, the high molecular weight component (apparent molecular weight 110,000) and the low molecular weight component (apparent molecular weight 43,000) of crocodile hemoglobin which were isolated by gel chromatography showed the very same effect of CO_2 on P_{50} as the whole hemolysate (Bauer and Jelkmann, 1977). Secondly, gel chromatography in the absence and presence of CO_2 did not give any indication that the proportion of the high and the low molecular weight component changed if CO_2 was present. We thus come to the conclusion that it is most likely the marked formation of O_2-linked carbamate compounds which is responsible for the observed effect. The lack of knowledge on the primary structure of reptile Hbs prevents us from assigning the O_2-linked car-

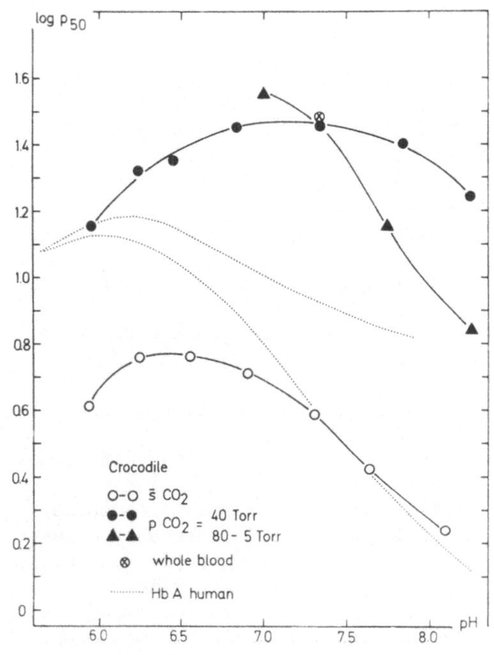

Fig. 3. Plot of log P_{50} against pH of unfractionated crocodile hemoglobin in the absence ($\overline{s}$ CO_2) and presence of CO_2 at 20°C. [Hb] = 120 μM Fe. pH was also varied by changing P_{CO_2} at constant [HCO_3^-] yielding a Bohr factor $\Delta\log P_{50}/\Delta pH = -0.66$. The sign ⊗ indicates log P_{50} of whole blood at 20°C, extracellular pH 7.47 and P_{CO_2} = 40 torr (from Bauer and Jelkmann, 1977). *Dotted lines* were obtained from similar experiments on human Hb at P_{CO_2} < 1 torr (*lower curve*) and P_{CO_2} = 40 torr (*upper curve*) of P_{CO_2} = 40 torr. (From Bauer et al., 1975)

bamate to possible binding places in crocodile Hb. In case they are represented by the N-terminal α-amino groups, as in mammalian Hb, it is likely that the ionization constant is quite high and that the carbamate equilibrium constant changes much more upon oxygenation than even at the β-subunits of human Hb where most of the O_2-linked carbamate is being formed (Bauer et al., 1975).

3. Conclusions

It becomes clear that the intrinsic O_2 affinity of both chicken and crocodile Hb is much higher than it is in its natural environment, the erythrocytes. However, the ways in which the affinity is decreased in the two species differ completely. In the bird, it is a fairly complicated phosphate compound, IP_5, and possibly ATP which serve to decrease the O_2 affinity of Hb by a preferential binding to the deoxygenated compared to the oxygenated structure. The same preferential binding prevails in crocodile hemoglobin, only the allosteric cofactor is merely CO_2, an ever present product of the intermediary metabolism (e.g., citric acid cycle). Clearly, both the interaction of phosphates and of CO_2 with the respective Hb requires a coordinated structural tuning of the pigment, which is hard to explain if selectively neutral mutations of Hb were fixed at random (Kimura and Ohta, 1971). It is much more likely that Darwinian evolution has acted to optimize the interactions between Hb and those cofactors which decrease its O_2 affinity. This statement is substantiated by the work of Goodman et al. (1975) who have shown that amino acid replacements which improved Hb function were accepted with a high rate at the beginning of Hb evolution and conserved once an optimal combination was present.

References

Bartlett, G.R.: Phosphate compounds in red cells of reptiles, amphibians and fish. Comp. Biochem. Physiol. 55 A, 211-214 (1976)

Bartlett, G.R., Borgese, T.A.: Phosphate compounds in red cells of the chicken and duck embryo and hatchling. Comp. Biochem. Physiol. 55 A, 207-210 (1976)

Bauer, C., Baumann, R., Engels, U., Pacyna, B.: The carbon dioxide affinity of various human hemoglobins. J. Biol. Chem. 250, 2173-2176 (1975)

Bauer, C., Jelkmann, W.: CO_2 governs the oxygen affinity of crocodile blood. Nature (London) 269, 825-827 (1977)

Baumann, F., Baumann, R.: A comparative study of the respiratory properties of bird blood. Respir. Physiol. (in press)

Benesch, R.E., Benesch, R., Yu, C.I.: The oxygenation of hemoglobin in the presence of 2,3-diphosphoglycerate. Effect of temperature, pH, ionic strength, and hemoglobin concentration. Biochemistry 8, 2567-2571 (1969)

Briehl, R.W.: The relation between the oxygen equilibrium and aggregation of subunits in lamprey hemoglobin. J. Biol. Chem. 238, 2361-2366 (1963)

Goodman, M., Moore, G.W., Matsuda, G.: Darwinian evolution in the geneology of haemoglobin. Nature (London) 253, 603-608 (1975)

Isaacks, R.E., Harkness, D.R., Froemann, G.A., Goldman, P.H., Adler, J.L., Sussman, S.A., Roth, S.: Studies on avian erythrocyte metabolism. - II. Relationship between the major phosphorylated metabolic intermediates and oxygen affinity of whole blood in chick embryos and chicks. Comp. Biochem. Physiol. 53 A, 151-156 (1976a)

Isaacks, R.E., Harkness, D.R., Sampsell, R.N., Adler, J.L., Kim, C.Y., Goldman, P.H.: Studies on avian erythrocyte metabolism. - IV. Relationship between the major phosphorylated metabolic intermediates and oxygen affinity of whole blood in adults and embryos in several galliformes. Comp. Biochem. Physiol. 55 A, 29-33 (1976b)

Kimura, M., Ohta, T.: On the rate of molecular evolution. J. Molec. Evolution 1, 1-17 (1971)

Petschow, D., Würdinger, I., Baumann, R., Duhm, J., Braunitzer, G., Bauer, C.: Causes of high blood O_2 affinity of animals living at high altitude. J. Appl. Physiol. 42, 139-143 (1977)

Rapoport, S., Guest, G.M.: Distribution of acid-soluble phosphorus in the blood cells of various vertebrates. J. Biol. Chem. 138, 269-282 (1941)

Romer, A.S.: Vertebrate Paleontology. Chicago, London: University of Chicago Press, 1966

Sullivan, B., Riggs, A.: The subunit dissociation properties of turtle hemoglobins. Biochim. Biophys. Acta 140, 274-283 (1967)

Szabo, A., Karplus, M.: Analysis of the interaction of organic phosphates with hemoglobin. Biochemistry 15, 2869-2878 (1976)

Session II

Adjustments to Changes in Oxygen Availability and Requirement: Hypoxia, Diving, and Flight

Chairman: K. Schmidt-Nielsen

Effects of Hypoxia on Heart Activity in Diving, Flying, and Land Birds

D. Sturkie and A. Abati

Summary

Isolated hearts of chickens, diving ducks, and pigeons were made hypoxic by substituting 95% N_2 for O_2 for as long as 8 min.

Hypoxia increased coronary flow and vasodilatation; it also increased diastolic pressure, attributable to increased resting tension of heart muscle.

Hypoxia decreased left ventricular pressure from a control level of 80-105 mmHg to 12% of control in chickens and about 45% in diving ducks.

Hypoxia decreased dP/dt most in chickens (15% of control) and in pigeons (31% of control), and 36 and 52% in the two breeds of diving ducks. The changes in $\frac{dP/dt}{IP}$ were almost identical to dP/dt.

Before hypoxia, the levels of glycogen in the heart were highest in the pigeon and of the similar magnitude in chickens and ducks. After hypoxia the depletion of glycogen was 52% in pigeon, 34% in chickens, and 19% in ducks.

1. Introduction

This report is based mainly on a thesis by A. Abati (Rutgers University), much of which has not been previously reported. A short abstract of some of the work appeared in Federation Proc. 33, 308 (1974).

It is well documented that hypoxia produces profound cardiovascular effects such as bradycardia, decreased cardiac output, and changes in distribution of blood flow and also increased peripheral resistance; these effects are most pronounced in diving species. In the intact animal it is difficult if not impossible to determine the extent to which such changes are caused by decreased contractile force of heart muscle per se, because of the intervention of several variables such as heart rate and changes in muscle fiber length occasioned by changes in end diastolic pressure, volume (preload or Frank-Starling mechanism), and increased peripheral resistance (afterload).

Preload, afterload, and heart rate can be kept constant in the isovolumic, isolated heart; true contractility can therefore be accurately assessed. The demand for and utilization of O_2 and the tolerance of O_2 deficiency (hypoxia) differ greatly in diving, flying, and land birds. The main objective of this study was to determine whether such differences may have had an adaptive influence on the contractility of

Dept. Environmental Physiology, Rutgers University, New Brunswick, N.J., U.S.A.

heart muscle of these species as revealed by the effects of hypoxia on isolated hearts.

The main objectives of this study were to determine the effects of hypoxia on: (1) isovolumic ventricular pressure and its rate of rise to peak pressure (dP/dt); (2) end diastolic pressure; (3) coronary flow; and (4) glycogen stores in heart muscle.

2. Methods

Adult chickens (White Leghorns), pigeons (*Columbia livia*), Mallard duck (*Anas platyrhynchos*), and two related species of deep-diving ducks (*Athya americana* and *Athya fuligula*) were used. They were anesthetized, heparinized, killed, and their hearts removed, and the brachiocephalic arteries were tied off. The aorta was cannulated and attached to the nozzle of a Langendorff apparatus for retrograde perfusion with Hank's solution at 38°C and 60 mmHg. A thin-walled balloon was inserted through the left atrium and thence to left ventricle and enough Hank's fluid injected (10-15 ml) to fill the ventricle and produce an end diastolic pressure of 2 mmHg (isovolumic preparation). The output of the Statham pressure transducer from the balloon catheter was fed to appropriate recorders and an oscilloscope. Photographs were taken of the pressure pulse and rate of rise to peak pressure (dP/dt) was determined manually. IP, the integrated pressure was determined by measuring the area under half of the pressure curve.

Hearts were paced electronically at a rate of 270/min. After aerobic perfusion with 95% O_2 and 5% CO_2 for 20 min, 95% N_2 was substituted for the O_2 for 8 min and the hearts were then reoxygenated.

3. Results

Ventricular pressures were normally from 80-105 mmHg, and after N_2 perfusion (hypoxia) pressures began to drop within 1 min and continued to do so for 8 min. The drop ranged from 12% of normal for chickens to 45% of normal for deep-diving ducks. The changes in ventricular pressure were much like the changes in dP/dt shown in Figure 1. The decrease in contractility after 8 min of hypoxia was greatest in chickens (15% of control values), intermediate in pigeons (31%), and 36 and 52% respectively in Mallards and deep-diving ducks. After reoxygenation, the return toward normal occurred in inverse order (deep-diving ducks first). The $\frac{dP/dt}{IP}$ curves were almost the same as dP/dt; thus ventricular pressure, dP/dt, and $\frac{dP/dt}{IP}$ were all good measures of contractility under these conditions.

Coronary flows were increased after hypoxia (a measure of vasodilatation) 26% in pigeon, 19.5% in the duck, and 17% in the chicken.

End diastolic pressure which was 2 mmHg under aerobic conditions increased after hypoxia in all species to a level of 10-11 mmHg, but there were no significant species differences. This increased pressure is believed to be due to increased resting tension of heart muscle and not to increased ventricular volume.

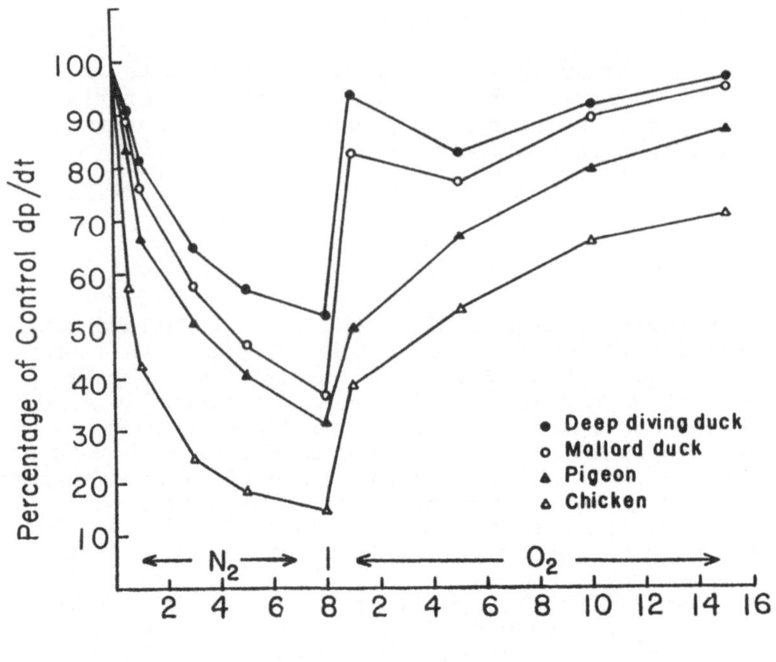

Fig. 1. Effects of anoxia (N_2) and reoxygenation (O_2) on dP/dt of isolated avian hearts

Table 1. Glycogen content of hearts before and during hypoxia, expressed as µg glycogen/mg protein

Hypoxia	Chicken	Pigeon	Mallard
Before	22.9	39.8	22.6
After	16.0	18.4	18.2
Percent depletion	34.4	52.2	19.3
Protein content mg/g	1.07	1.27	1.01

Glycogen content and rate of glycolysis of heart muscle have been used as indicators of anaerobic energy sources, which are called upon in times of hypoxic stress. The results of this study are presented in Table 1. Before hypoxia the levels of glycogen were highest in the pigeon and of the same magnitude in chickens and Mallard ducks. After hypoxia (8 min) the percentage depletions of glycogen were 34.4% in chicken, 52.2% in pigeon, and 19.3% in the duck. These data do not indicate that glycogen content or depletion is a measure of protection from hypoxia, but do suggest that the habitats of these avian species, particularly the divers, have had an adaptive influence on the contractility of heart muscle.

Cardiopulmonary Responses to Acute Hypoxia in Domestic Fowl

E. L. Besch and H. Kadono

Summary

Adult, male, single-comb, white leghorn-type fowls weighing between 1600 and 3000 g were exposed, in series, in an environmental chamber to acute hypoxia. The chamber was ventilated with room air in which the O_2 concentration was progressively reduced from 21% (control) to 10% (hypoxia) by dilution with tanked N_2. Pulmonary arterial blood pressure, mean systemic (femoral artery) blood pressure, and respiratory frequency all remained unchanged until the inspired O_2 concentration dropped below 15%. Between 15% and 10% O_2, the pulmonary artery blood pressure increased about 31% and the respiratory frequency increased about 44% compared to control values. Heart rate increased progressively with hypoxia to the 12% O_2 level and then decreased to a value not significantly different from controls. The pulmonary arterial blood pressure was negatively correlated (r = -0.94) whereas, the femoral artery pressure was positively correlated (r = 0.96) with O_2 concentration in the inspired air. Peripheral resistance and arterial oxygen tension $P_{a\,O_2}$ also were decreased ($P < 0.05$).

1. Introduction

Physiological adjustments by the respiratory and cardiovascular systems are necessary if body tissues are to receive an adequate supply of O_2 during exposure to acute or chronic hypoxia (Van Liere and Stickney, 1963). Physiological adaptation of domestic fowl to chronic hypoxia, e.g., to high altitudes, apparently consists of several adaptates including: (a) cardiac hypertrophy (Burton and Smith, 1967); (b) erythrocytic polycythemia (Burton and Smith, 1967; Smith and Abbott, 1961); (c) pulmonary arterial hypertension (Burton et al., 1968); (d) reduced buffering capacity of the blood (Besch et al., 1971); and (e) increased tolerance to acute hypoxia (Burton et al., 1969).

On the other hand, responses to acute hypoxia of fowl reared at sea-level are not as well documented. Compared with mammals and other species of birds, the sea-level chicken reportedly has little tolerance to acute hypoxia (Altland, 1961). Of the reports available, effects of acute hypoxia in the chicken include: decreased body temperature (Kadono and Besch, 1974) and O_2 consumption (Boyd and McDaniel, 1972; Butler, 1967); increased respiratory frequency (Butler, 1967; Ray and Fedde, 1969), plasma volume (Maddison and Besch, 1972), and heart rate (HR) (Butler, 1967; Durfee and Sturkie, 1963; Kadono and Besch, 1972). Although these reported decreases in SBP, and the increases in HR and

---

Department of Metabolism, College of Veterinary Medicine, J. Hillis Miller Health Center, University of Florida, Gainesville, Florida 32610/U.S.A.

plasma volume suggest a systemic vasodilation, virtually no information is available on changes in cardiac output (CO) or peripheral resistance (PR) under hypoxic conditions. Moreover, little is known about the relationship between pulmonary arterial blood pressure (P_ABP) and systemic blood pressure during acute hypoxia. Thus, the object of the present work was to examine cardiopulmonary responses of the fowl to acute hypoxia.

2. Materials and Methods

Animals. Unanesthetized, adult, male, single-comb, white leghorn-type domestic fowls, weighing between 1600 and 3000 g were used. These animals were housed in outdoor pens and exposed to seasonal temperature changes. Because data from both Kansas and Florida are reported, summer and winter animals were categorized primarily by ambient temperature conditions. Generally, summer animals were exposed to an average daily temperature of 24.1 ± 0.4°C in Kansas (1 June - 30 September) and 20.6 ± 0.1°C in Florida (2 Feb. - 6 May). Winter animals were exposed to average daily temperatures of 5.7 ± 0.8°C in Kansas (1 Oct. - 31 Jan.); no equivalent conditions were used in Florida. Age, photoperiods, feed, housing, and husbandry practices were essentially the same for all animals. Each animal was placed in an indoor cage and fasted for 12 to 24 h prior to being used in an experiment. Cannulation surgery was performed using a local anesthetic.

a) Chamber. Lightly restrained animals on small boards were placed in a 210-l, airtight, plexiglas exposure chamber similar to the one described elsewhere (Maddison and Besch, 1972). After placing the animal in this chamber, tanked air entered the chamber via a gas-mixing manifold for the normoxic (21% O_2) condition; this air was diluted with tanked N_2 for the hypoxic condition. For the progressive hypoxia experiments, the O_2 concentration of inlet air was decreased 1% every 10 min until an ambient atmosphere of 10% O_2 was produced, usually within 110-120 min. In the other experiments, O_2 concentration was decreased to 13% usually within 80 min followed by exposure to the hypoxic condition for a minimum of 2 h. In all cases, ventilation rate of the chamber was adjusted to give approximately two air changes per hour. The O_2 concentration of the inlet gas was monitored by an O_2 analyzer (Beckman, Model E-2). CO_2, monitored similarly with a medical gas analyzer (Beckman, LB-1), never rose above 0.2%. The intrachamber temperature, monitored regularly, increased 0.8°C (24.1-24.9°C) between the beginning and the end of each experiment. All procedures or manipulations on the animal within the chamber were accomplished through glove ports on the front side of the chamber.

b) Measurements. Systemic blood pressure was usually measured from the femoral artery using a blood pressure transducer (Statham, Model p23); in cardiac output experiments, SBP was measured from the external ischiatic artery. Measurement of P_ABP was according to the method described by Burton et al. (1968). Heart rate and respiratory frequency (RF) values were obtained from blood pressure recordings or from electrocardiogram waveforms using a portable recorder (Beckman Dynograph, Type RS). Rectal temperatures were monitored with a tele-thermometer (Yellow Springs, Model 43T).

Arterial blood was collected from the external ischiatic artery in airtight syringes and analyzed in vitro at 37°C using suitable electrodes (Instrument Laboratories blood gas analyzer, Model 113). Arterial gas tensions (P_{aO_2} and P_{aCO_2}) and arterial pH (pH_a) were corrected for temperature to 40°C (Kelman and Nunn, 1966). To prevent blood from clotting, each animal received an intravenous injection of heparin (200 units/kg) at the beginning of each trial. Upon collection, blood samples were rapidly chilled to 2°C until determinations were made, usually within 45 min. Changes in P_{O_2} resulting from O_2 uptake by blood cells are negligible (Besch, 1966).

CO was measured by dye dilution (Sturkie and Vogel, 1959) with Evans Blue dye (T-1824) injected as a bolus into the right atrium via a cannula (0.058 cm inner diameter) in the right cutaneous ulnar vein. Cannula placement was verified at the end of each experiment. Stroke volumes (SV) (ml/beat) were calculated by dividing CO by HR. Total peripheral resistant (TPR), in arbitrary units, was calculated as the quotient of mean arterial blood pressure ($\overline{SBP}$) (mmHg) and CO (ml/min). This value was multiplied by 100 to obtain a convenient unit (Nightingale, 1977).

3. Results

No changes in SBP or P_ABP, HR, or RF were observed in animals in the chamber ventilated with air (control condition) for 40 min prior to the exposure to hypoxia. $\overline{SBP}$ parallels the decrease in the O_2 concentration of the exposure chamber (Fig. 1). Between concentrations of 21% and 15% O_2 in the chamber, there is relatively little change ($P > 0.05$) in mean SBP. For a 30-min period at 15% O_2, the SBP remained unchanged, but abruptly decreased at O_2 concentrations below 15%.

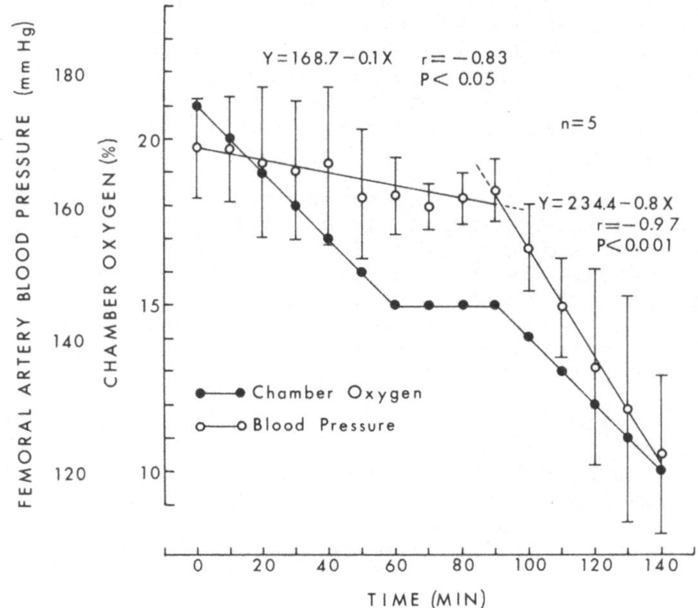

Fig. 1. Effect of progressive hypoxia on femoral arterial blood pressure in winter male fowl (see Materials and Methods). Chamber oxygen was decreased from 21% to 15% within 60 min; held constant at 15% for 30 min; then decreased to 10% within 50 min. Data are from five animals expressed as mean ± *SE* and *N* equals the number of animals

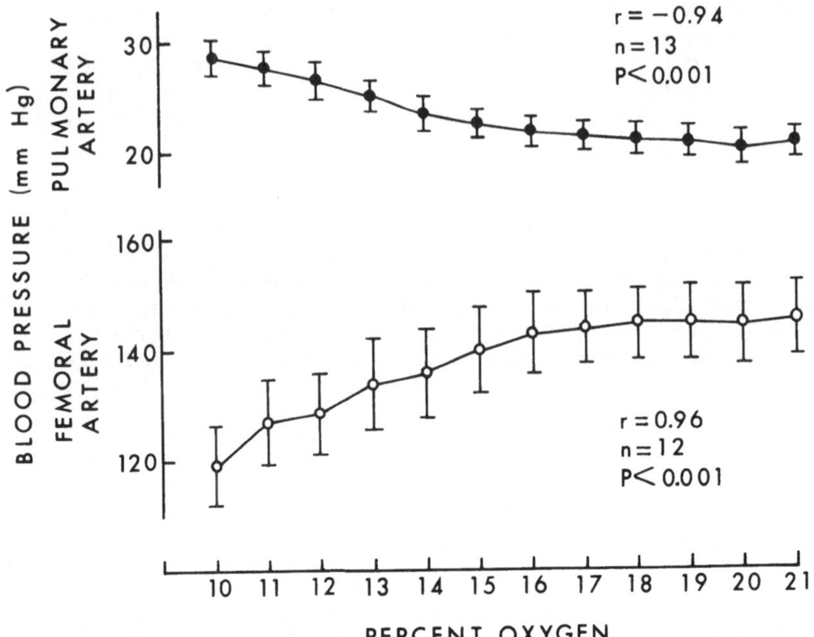

Fig. 2. Relationship between ambient O_2 concentrations and pulmonary arterial and femoral arterial blood pressure. Each data point represents the mean ± SE and *N* equals the number of animals

Progressive hypoxia consistently results in increased P_ABP and decreased $\overline{SBP}$ (Fig. 2). However, P_ABP and $\overline{SBP}$ remained relatively unchanged until the inspired O_2 dropped below about 15%. Between 15% and 10% O_2, P_ABP increased about 31% ($P < 0.01$); the $\overline{SBP}$ decreased about 16% ($P < 0.05$). P_ABP was negatively correlated ($r = -0.94$; $P < 0.001$) and $\overline{SBP}$ positively correlated ($r = 0.96$; $P < 0.001$) with O_2 concentration in the inspired air.

During progressive hypoxia HR and RF both increased (Fig. 3). Relatively little change (8%) occurred in RF between 21% and 15% O_2; but between 15% and 10%, the RF increased about 44% ($P < 0.001$). HR increased progressively with hypoxia to the 12% O_2 level and then decreased to a value not significantly different from the controls. Slight cyanosis of the comb was apparent at about 15% O_2; the cyanosis became more pronounced at concentrations between 15% and 11%. At about 12% O_2, the animals usually became restless.

It is apparent that the $\overline{SBP}$ values of Figure 1 are significantly greater than those of Figure 2 especially at O_2 levels above 15%. The basis for this difference appears to be related to the ambient temperatures at which the animals are housed prior to the BP measurements. The hyperbolic relationship between $\overline{SBP}$ and ambient O_2 for winter and summer birds (Fig. 4) is best described by the equation:

$$A - \overline{SBP} = B\,e^{-k\,O_2}$$

where: A is the maximum arterial blood pressure (mmHg);
$\overline{SBP}$ is the mean arterial blood pressure (mmHg) at any O_2 concentration (%);
B is an integration constant; and,
k is a constant.

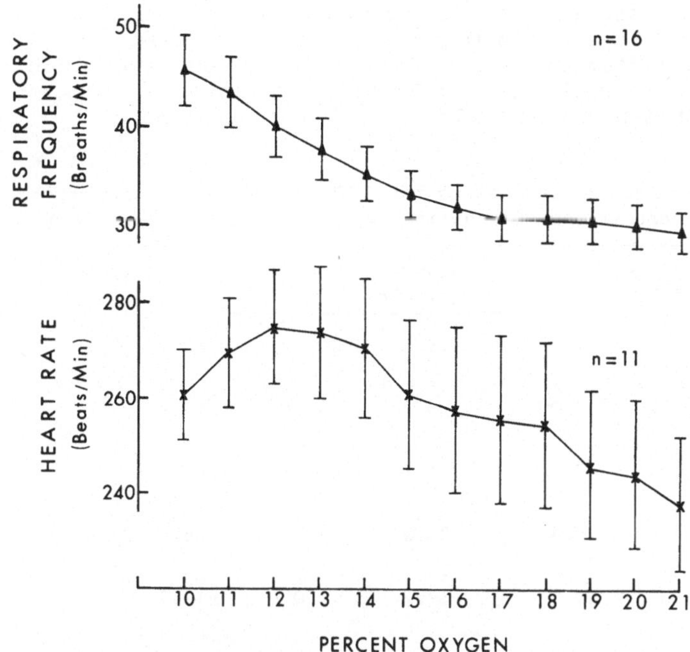

Fig. 3. Respiratory frequency and heart rate changes resulting from changes in ambient oxygen concentration. Each data point represents the mean ± SE and N equals the number of animals

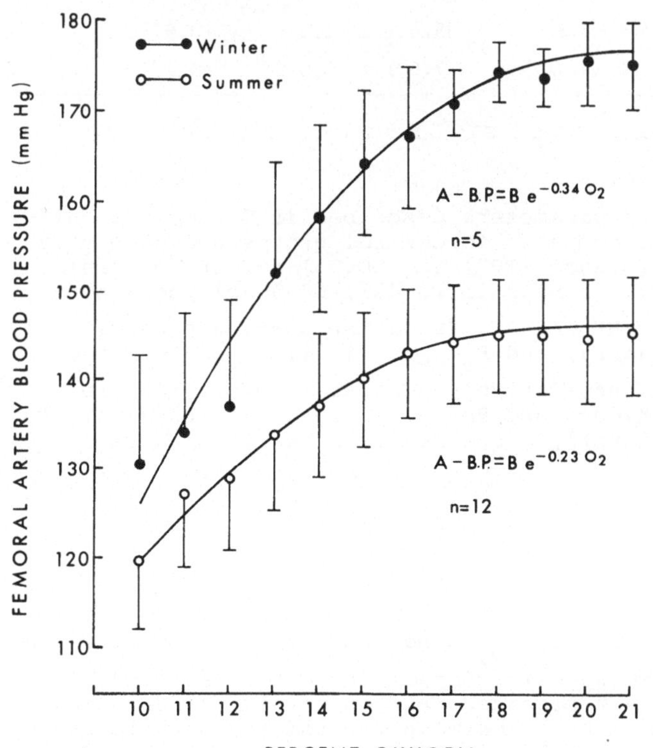

Fig. 4. Femoral arterial blood pressure changes in winter and summer fowl at different ambient oxygen concentrations. Each data point represents the mean ± SE and N equals the number of animals

Below 15% O_2, there are no significant differences in $\overline{SBP}$ for winter and summer birds; between 15% and 21% O_2, $\overline{SBP}$ values for these two groups are significantly different ($P < 0.05$). However, no significant differences were apparent in P_ABP, HR, or RF between winter and summer birds at any O_2 level between 10% and 21%.

Table 1. Comparison of cardiopulmonary parameters for SCWL-type males at normoxic (21% O_2) and hypoxic (13% O_2) conditions. Six animals were used at each condition

	Air (21% O_2)	Hypoxia (13% O_2)	% Difference[a]
Cardiac output (CO) (ml/min/kg)	123.0 ± 7.8[b]	133.1 ± 10.8	+8.2
Stroke volume (SV) (ml)	0.97 ± 0.09	0.98 ± 0.18	+1.0
Heart Rate (HR) (beats/min)	253.0 ± 8.4	280.7 ± 15.6	+11.0
Respiratory frequency (RF) (respirations/min)	17.8 ± 1.4	21.0 ± 1.5	+18.0
Mean blood pressure ($\overline{SBP}$) (mmHg)	130.5 ± 10.1	107.2 ± 15.0	-17.8
Total peripheral resistance (TPR)	62.5 ± 7.7	40.5 ± 4.6	-35.2[c]
Body temperature (^{O}C)	39.6 ± 0.1	38.6 ± 0.5	-2.5
$P_{a\ O_2}$ (mmHg)	89.8 ± 3.3	63.7 ± 5.0	-29.1[d]
$P_{a\ CO_2}$ (mmHg)	27.8 ± 1.3	25.4 ± 1.5	-8.6
pH	7.570 ± 0.017	7.601 ± 0.018	+0.4

[a] % Difference = $\frac{\text{hypoxia-air}}{\text{air}} \times 100$; [b] Mean ± SE; [c] $P < 0.05$; [d] $P < 0.01$

Of all of the cardiopulmonary parameters described in Table 1, significant differences in hypoxic compared to control groups are seen only in decreased peripheral resistance (TPR) ($P < 0.05$) and in decreased $P_{a\ O_2}$ ($P < 0.01$). As a result of hypoxia to 13%, CO displayed a nonsignificant increase, as did HR and RF. SV of the heart was essentially unchanged. SBP, body temperature, and $P_{a\ CO_2}$ all displayed decreases that are quantitatively and qualitatively similar to previously reported data (Butler, 1967; Kadono and Besch, 1972, 1974). Body weights were slightly but nonsignificantly decreased as a result of acute hypoxia.

4. Discussion

Arterial gas tensions ($P_{a\ O_2}$ and $P_{a\ CO_2}$) and pHa (Table 1) confirm that animals breathing 13% O_2 were hypoxic compared to controls. These data are similar to those previously reported (Burton et al., 1968; Besch et al., 1971). The decrease in $P_{a\ O_2}$ between control and hypoxic birds represents a decrease in O_2 saturation of about 30% (from 95% to 65%).

The decrease in pHa is consistent with the change in $P_{a\ CO_2}$ and probably resulted from the increased RF in hypoxic birds.

During the initial stages of acute hypoxia, the cardiovascular responses of the intact fowl are similar to those of dogs (Nahas et al., 1954) and humans (Doyle et al., 1952). In the later stages, however, there are dissimilarities: the fowl displays a tachycardia, decreased systemic arterial blood pressure (SBP) and little increase in CO; unanesthetized man, displays tachycardia, and increased CO, but little significant change in SBP (Doyle et al., 1952). In conscious dogs acutely exposed to 8% O_2, there is an increased CO and elevated SBP (Nahas et al., 1954); in general, hypoxic rabbits responded similarly (Vogel et al., 1969).

In fowl, acute hypoxia responses in SBP, P_ABP, HR, and RF (Figs. 2 and 3) suggest different physiological regulation. Butler (1967) reported that hypotension resulting from progressive hypoxia may be due to hypoxia itself; whereas, the autonomic nervous system appears to be involved in HR and RF changes. Durfee and Sturkie (1963) found that, during hypoxia, resting HR and reflex changes in HR are mediated by the vagus nerve. Burton and Smith (1967) and Burton et al. (1968) have shown that increased P_ABP in chickens exposed to chronic hypoxia is due partly to increased viscosity of the blood and partly to hypoxia per se. This also may be true for acute hypoxia.

The decreased SBP and relatively unchanged CO are consistent with the observed decreased total peripheral resistance (TPR). This, together with the increased P_ABP suggests that marked changes may have occurred in the vascular bed. Although not measured, it is reasonable to assume that during hypoxia there were changes in regional blood flow because of the differences in SBP and P_ABP and CO compared to normoxic birds.

Seasonal changes in avian BP reported here (Fig. 4) are similar to those reported by Weiss et al. (1961) between 1.6^o and 21.1^{o}C. They found a 1.5 mmHg decrease per ^{o}C between winter and summer conditions for males compared to 1.6 mmHg in the present study. Moreover, they reported a 10% decrease in BP between winter and summer and a 20% increase between summer and winter compared to 17% and 21%, respectively, reported here. The differences in SBP between winter and summer normoxic birds become nonsignificant as ambient O_2 concentrations decrease below 15% O_2, suggesting that differences between 15% and 21% may be related to peripheral vasoconstriction.

Relatively little change in SBP, P_ABP, HR, or RF occurred at chamber O_2 concentrations above about 15% (Figs. 1, 2, 3) or about 110 mmHg. That approximates the P_{O_2} obtained in a normobaric (20.9% O_2) environment at a barometric pressure of about 530 mmHg - the ambient pressure encountered at about 3000 m (9840 ft) altitude. These findings are consistent with reports (Van Liere and Stickney, 1963) that the first subjective symptoms of altitude sickness in humans are usually noted at 3050 m (10,000 ft) altitude.

Because of the length of the experiments (3.5 ± 0.5 h) reported here, comparisons were made in animals exposed to different periods of light restraint at normoxic (20.9%) conditions. The results are similar to those reported elsewhere for restrained chickens (Whittow et al., 1965): increased HR and SV; decreased BP and TPR. Although the data of Table 1 were adjusted for the period of restraint, a significant decrease in total peripheral resistance was observed in hypoxic fowl.

Acknowledgments. Part of this research was conducted at the Department of Physiological Sciences, College of Veterinary Medicine, and Institute for Environmental Research, College of Engineering, Kansas State University. Work supported by USAFOSR Contract F 44620-68-C-0020; NIH Contract 71-2511; NIH Grant RR00753; and NIH Grant RR00942.

References

Altland, P.D.: Altitude tolerance of chickens and pigeons. J. Appl. Physiol. 16, 141-143 (1961)

Besch, E.L.: Respiratory activity of avian blood cells. J. Cell. Physiol. 67, 301-306 (1966)

Besch, E.L., Burton, R.R., Smith, A.H.: Influence of chronic hypoxia on blood gas tensions and pH in domestic fowl. Am. J. Physiol. 220, 1379-1382 (1971)

Boyd, R.L. III, McDaniel, G.R.: A chamber for exposure to hypoxia and measurement of oxygen consumption. Poultry Sci. 51, 1785-1786 (1972)

Burton, R.R., Smith, A.H.: The effects of polycythemia and chronic hypoxia on heart mass in the chicken. J. Appl. Physiol. 22, 782-785 (1967)

Burton, R.R., Besch, E.L., Smith, A.H.: Effects of chronic hypoxia on the pulmonary arterial blood pressure of the chicken. Am. J. Physiol. 214, 1438-1442 (1968)

Burton, R.R., Smith, A.H., Carlisle, J.C., Sluka, S.J.: Role of hematocrit, heart mass and high altitude in acute hypoxia tolerance. J. Appl. Physiol. 27, 49-52 (1969)

Butler, P.J.: The effect of progressive hypoxia on the respiratory and cardiovascular systems of the chicken. J. Physiol. 191, 309-324 (1967)

Doyle, J.T., Wilson, J.S., Warren, J.V.: The pulmonary vascular responses to short term hypoxia in human subjects. Circulation 5, 263-270 (1952)

Durfee, W.K., Sturkie, P.D.: Some cardiovascular responses to anoxia in the fowl. Federation Proc. 22, 182 (1963)

Kadono, H., Besch, E.L.: Effect of progressive hypoxia on blood pressure in domestic fowl. Federation Proc. 31, 3386 (1972)

Kadono, H., Besch, E.L.: The effect of acute hypoxia on body temperature in domestic fowl. Environ. Physiol. Biochem. 4, 1-6 (1974)

Kelman, G.R., Nunn, J.F.: Nomograms for correction of blood P_{O_2}, P_{CO_2}, pH and base excess for time and temperature. J. Appl Physiol. 21, 1484-1490 (1966)

Liere, E.J. Van, Stickney, J.C.: Hypoxia. Chicago: University Chicago Press, 1963, 381 p

Maddison, W.H., Besch, E.L.: The effects of acute hypoxia on the body fluids of domestic fowl. Environ. Physiol. Biochem. 2, 154-161 (1972)

Nahas, G.G., Josse, J.W., Muchow, G.C.: Influence of acute hypoxia on peripheral and central venous pressure in the non-narcotized dog. Am. J. Physiol. 177, 315-318 (1954)

Nightingale, T.E.: Comparison of cardiopulmonary parameters in awake and anesthetized chickens. Poultry Sci. 56, 147-153 (1977)

Ray, P.J., Fedde, M.R.: Responses to alterations in respiratory P_{O_2} and P_{CO_2} in the chicken. Resp. Physio. 6, 135-143 (1969)

Smith, A.H., Abbott, U.K.: Adaptation of the domestic fowl to high altitude. Poultry Sci. 40, 1459 (1961)

Sturkie, P.D., Vogel, J.A.: Cardiac output, central blood volume and peripheral resistance in chickens. Am. J. Physiol. 197, 1165-1166 (1959)

Vogel, J.A., Pulver, R.I., Burton, T.M.: Regional blood flow distribution during simulated high altitude. Federation Proc. 28, 1155-1159 (1969)

Weiss, H.S., Fisher, H., Giminger, P.: Seasonal changes in avian blood pressure related to age, sex, diet, confinement and breed. Am. J. Physiol. 201, 655-659 (1961)

Whittow, G.C., Sturkie, P.D., Stein, G., Jr.: Cardiovascular changes in restrained chickens. Poultry Sci. 44, 1452-1459 (1965)

Oxygen Transport During Progressive Hypoxia in Bar-headed Geese (Anser indicus) Acclimatized to Sea Level and 5600 Meters

C. P. Black, S. M. Tenney, and M. van Kroonenburg

Summary

The Bar-headed Goose (*Anser indicus*) has been observed flying at altitudes up to 9200 m during its semi-annual migration across the Himalayas, suggesting that this species may possess unique adaptations to hypoxia. Therefore, Bar-headed geese and individuals of a comparison species (Pekin Duck - *Anas platyrhynchos* forma domestica) were subjected to progressive hypoxia corresponding to altitudes up to 11,580 m. After 15 min exposure to each level of hypoxia, tidal volume, respiratory frequency, and levels of O_2 and CO_2 in arterial and venous blood were determined. In separate experiments the dissociation curve and Bohr effect factor were determined for each species also. The Bar-headed Goose showed a lower resting normoxic level of resting ventilation than the Pekin Duck, as well as a threshold of ventilatory increase in response to hypoxia and maximum ventilation at lower levels of inspired oxygen. After five weeks of acclimatization at 5640 m, both species showed hyperventilation under normoxic conditions, threshold of ventilatory increase, and level of maximum ventilation at lower levels of inspired oxygen than prior to acclimatization. Both species show about the same partial pressures of O_2 and CO_2 in arterial and venous blood under normoxic conditions. During progressive hypoxia, the Bar-headed Goose is able to maintain only marginally higher partial pressures of O_2 in arterial and venous blood, but O_2 content is much greater than that observed in the Pekin Duck. This is primarily due to the relatively greater affinity of Bar-headed Goose hemoglobin for O_2. Acclimatization produces polycythemia, increased hemoglobin levels, and a large rightward shift in the dissociation curve for the Pekin Duck, while none of these changes were observed in the Bar-headed Goose.

1. Introduction

The Bar-headed Goose (*Anser indicus*) winters throughout the Indian subcontinent and breeds mostly on large lakes throughout the south-central portion of Asia including Tibet, Mongolia, and parts of the Soviet Union (Ali and Ripley, 1968; Delacour, 1959). It has been recorded breeding in large numbers at altitudes as high as 5500 m (Würdinger, 1973), and it migrates directly across the Himalayan Mountains, apparently without more than brief rest stops en route (Swan, 1961). During migration, reliable observers have repeatedly seen flocks crossing directly over summits of the highest peaks (Swan, 1970) including at least one instance of a flock crossing directly over the summit of Mount Everest (Swan, 1961). The Bar-headed Goose is thus unique, since

Department of Physiology, Dartmouth Medical School, Hanover, New Hampshire 03755/ USA.

no other known animal species routinely goes from near sea-level conditions to elevations as high as 9200 m in such a short period of time. This investigation was therefore undertaken to obtain information on O_2 transport during exposure to progressive hypoxia in both acclimatized and unacclimatized individuals of this species. All experiments were also performed on the Pekin Duck (*Anas platyrhynchos* forma domestica) for comparative purposes.

2. Methods

Experiments were performed on birds obtained from commercial breeders and kept in aviaries with mixed grain and water for bathing and drinking available ad lib. Prior to an experiment involving blood gas determinations, catheters of PE 90 tubing were placed in the right wing vein and artery under local anesthesia. The venous catheter was advanced until its tip was in the right ventricle (about 13 cm).

The bird was then placed completely unrestrained in a whole-body plethysmograph (Drorbaugh and Fenn, 1955) large enough (30 × 50 × 70 cm) so that it could stand and turn around, and blood was withdrawn for hematocrit and hemoglobin determinations. The plethysmograph was sequentially ventilated with gas mixtures equivalent to sea level, 3050, 6100, 7620, 9250, 10,670, 11,580 (Pekin and Bar-Head), and 12,190 (Bar-head only) m elevation [fractional concentration of oxygen in inspired gas (F_{IO_2}) = 0.209, 0.14, 0.088, 0.070, 0.053, 0.038, 0.032, and 0.028 respectively]. P_{aO_2}, P_{aCO_2}, pH_a, P_{vO_2}, P_{vCO_2}, and pH_v were measured after the birds had been exposed to each mixture for 15 min. V_t, f, and $\dot{V}_{O_2}$ were measured in separate experiments following an experimental protocol identical except for the introduction of blood sampling catheters.

In separate experiments dissociation curves and Bohr effect factors ($\Delta\log P_{50}/\Delta pH$) were determined for each species using a mixing technique (Edwards and Martin, 1966), and $\dot{V}_E$, C_{aO_2}, C_{vO_2}, and $\dot{Q}$ were calculated.

The birds were then placed in the large hyperhypobaric chamber located at the Dartmouth Medical School (see Tenney and Ou, 1977 for detailed animal maintenance procedures) at an ambient pressure equivalent to 5640 m. After 4 weeks the experiments described above were repeated.

3. Results and Discussion

Although the two species have the same average weight (2.45 kg), mean normoxic ventilatory levels are lower in the Bar-headed Goose than in the Pekin duck (744 ml/min and 1035 ml/min respectively). In the Pekin Duck ventilation remains near the normoxic level until the F_{IO_2} becomes less than 0.14. Between F_{IO_2} values of 0.14 and 0.070, $\dot{V}_E$ increases by 50% and below 0.070, $\dot{V}_E$ increases rapidly until it reaches a maximum (4.1 × resting) at F_{IO_2} of about 0.05. Below this level ventilation fails rapidly. This general pattern is similar to that observed by Jones and Holton (1972) and by Colacino et al. (1977). After 4 weeks acclimatization at 5640 m, ducks hyperventilate at rest under normoxic conditions ($\dot{V}_E$ = 1393 ml/min), but the pattern of increase in $\dot{V}_E$ with increasing hypoxia is similar to that seen in the unacclimatized duck, except that maximum $\dot{V}_E$ is greater (7069 ml/min) and occurs at F_{IO_2} = 0.038 rather than 0.05.

In the unacclimatized Bar-headed Goose, $\dot{V}_E$ does not increase above resting levels until F_{IO_2} is less than 0.07, much lower than the 0.14 observed in the duck. Below this level $\dot{V}_E$ increases rapidly until maximum $\dot{V}_E$ (6077 ml/min) is reached at F_{IO_2} of 0.032 inspired O_2. Below this level of hypoxia, ventilation is depressed. Acclimatized Bar-headed Geese also hyperventilate under resting normoxic conditions ($\dot{V}_E$ = 1043 ml/min), and although the F_{IO_2} threshold at which ventilation begins to rise above the acclimatized resting value does not change, maximum $\dot{V}_E$ is greater (9756 ml/min) and occurs at a lower F_{IO_2} (0.028 rather than 0.032).

Clearly, the acclimatization process does not change the threshold at which the hypoxic ventilatory response begins in either species. Thus, there is no "blunting", as occurs in man (Severinghaus et al., 1966), or the cat (Tenney and Ou, 1977). However, the process of acclimatization does appear to decrease the depressant effects of extreme acute hypoxia, since the maximum $\dot{V}_E$ occurred in acclimatized individuals of both species at levels of hypoxia which cause ventilatory depression in the unacclimatized birds.

Along with the differences between the two species in their ventilatory response to hypoxia, there are also striking differences in hemoglobin-oxygen affinity. The P_{50} determined for the Pekin Duck (pH = 7.5, T = 41°C) was 42.6 torr, while that for the Bar-headed Goose was 27.2 torr. This value is somewhat lower than the 31.2 torr suggested by Petschow et al. (1977) for Bar-headed Goose hemoglobin under the same temperature and pH conditions. Bohr-effect factors were determined to be -0.44 and -0.53 for the duck and goose respectively, well within the range reported for other birds (cf. Scheipers et al., 1975).

Acclimatization caused the P_{50} of Pekin-duck blood to increase by 7.5 torr to 50.1 torr while the P_{50} for Bar-headed Goose hemoglobin increased by only 2.7 torr to 29.8 torr (pH = 7.5, T = 41°C). This small shift in P_{50} in the goose is probably due to the weak interaction between the hemoglobin in this species and IPP (inositol pentaphosphate) which was described by Petschow et al. (1977).

Under normoxic resting conditions, the Pekin Duck has arterial and venous P_{O_2} values of 93.5 and 44 torr respectively, while these values for the Bar-headed Goose are 92.5 and 35 torr. These values for the duck are similar to those reported by Jones and Holeton (1972). Arterial and venous P_{O_2} values for the two species show a similar pattern of decrease as the inspired gas is made more hypoxic, but due to the greater O_2 affinity of Bar-headed Goose hemoglobin, arterial and venous O_2 content values are higher. The slope and position of the goose dissociation curve are such that arterial O_2 content remains relatively high in the face of increasing hypoxia until F_{IO_2} drops below 0.053. This corresponds to an altitude of 9150 m, about the highest the birds would have to go to cross the highest Himalayan peaks. At this F_{IO_2} in the Bar-headed Goose, arterial P_{O_2} is 28.5 torr and saturation is 65%, while the corresponding values for the Pekin Duck are 30.0 torr and only 26% saturation. This in turn allows for a much greater arterial-venous content difference in the goose under hypoxic conditions.

At an F_{IO_2} of 0.070 (7620 m elevation), the a-v content difference is 55% greater in the goose (8.7 vol% vs. 5.6 vol%) and at an F_{IO_2} of 0.053 (9150 m) the a-v content difference is almost twice that of the duck (8.4 vol% vs. 4.3 vol%). This gives the goose a great advantage, since O_2 delivery to the tissues will be much greater for a given cardiac output level and arterial P_{O_2}.

Both species show increasing blood pH values as F_{IO_2} is decreased. Resting normoxic arterial pH values are 7.46 for the duck and 7.47 for the goose, but at an F_{IO_2} of 0.053, they have risen to 7.54 and 7.66 respectively. At F_{IO_2} values lower than this, however, both birds begin to become acidotic. (At F_{IO_2} = 0.038 arterial pH = 7.13 and 7.30 for the duck and goose respectively.)

Normoxic arterial P_{O_2} values in the acclimatized birds are only marginally higher at rest, and show no real difference from the unacclimatized birds as F_{IO_2} is decreased. Arterial blood in the Pekin Duck still becomes acidotic below F_{IO_2} levels of 0.053, but in contrast the acclimatized Bar-headed Goose is able to maintain a relatively high arterial pH (7.59) at an F_{IO_2} of 0.032. This in turn means higher arterial O_2 content, since the alkalotic pH will keep the dissociation curve to the left.

Mean hematocrit and hemoglobin values are 45.4% and 16.2 g% for the unacclimatized duck and 47.8% and 14.2 g% for the unacclimatized goose. Acclimatized ducks show a striking polycythemia (Hct = 55.9%, Hb = 22.5 g%), while both hematocrit and hemoglobin fall slightly in the acclimatized geese (43.1%, 12.9 g%). A similar lack of acclimatization polycythemia has also been observed in other high-altitude adapted species (llamas, Brooks and Tenney, 1968; alpacas, Sillau et al., 1976). Arterial and venous O_2 content data for the acclimatized duck show that the polycythemia is just enough to compensate for the loss of O_2-carrying capacity caused by the rightward shift in the dissociation curve.

The differences between these two species in hypoxic response and O_2 transport are not nearly as striking as one would expect from the differences in natural history. Clearly both species cope with extreme hypoxia remarkably well, at least under the resting conditions utilized in this study. The unacclimatized Bar-headed Goose has a slight but unmistakable advantage over the duck, mostly because of the position of its dissociation curve. The difference between the two species in the acclimatized state is more clear-cut: Although the duck is able to maintain O_2-carrying capacity in the face of a rightward shift in the dissociation curve, it does so at the cost of an increase in hematocrit and presumably blood viscosity and cardiac work. The acclimatized Bar-headed Goose is able to avoid this, yet still keep O_2-carrying capacity high by minimizing the rightward shift in the dissociation curve, and by keeping arterial pH high, particularly at levels of extreme hypoxia.

Acknowledgments. We gratefully acknowledge the technical assistance of Mrs. S. Knuth, Mrs. J. Areson, Miss H. Kim, and Miss A. Paulsen. This work was supported by grant number HL 02888-20 from the United States Public Health Service. The results from the study described here will be published elsewhere.

References

Ali, S.A., Ripley, S.D.: Handbook of the Birds of India and Pakistan Together with Those of Nepal, Sikkim, Bhutan, and Ceylon. Vol. I. Bombay: Oxford University Press, 1968

Brooks, J.G., Tenney, S.M.: Ventilatory response of the Llama to hypoxia at sea level and high altitude. Respir. Physiol. 5, 269-278 (1968)

Colacino, S.M., Hector, D.H., Schmidt-Nielsen, K.: Respiratory responses of ducks to simulated altitude. Respir. Physiol. 29, 265-281 (1977)

Delacour, J.T.: The Waterfowl of the World. London: Country Life, 1959

Drorbaugh, J.E., Fenn, W.O.: A barometric method for measuring ventilation in newborn infants. Pediatrics 16, 81-87 (1955)

Edwards, M.J., Martin, R.J.: Mixing technique for the oxygen-hemoglobin equilibrium and Bohr effect. J. Appl. Physiol. 21, 1898-1902 (1966)

Jones, D.R., Holeton, G.F.: Cardiovascular and respiratory responses of ducks to progressive hypocapnic hypoxia. J. Exp. Biol. 56, 657-666 (1972)

Petschow, D., Würdinger, I., Baumann, R., Duhm, J., Braunitzer, G., Bauer, C.: Causes of high blood O_2 affinity of animals living at high altitude. J. Appl. Physiol.: Respir. Environ. Exercise Physiol. 42, 139-143 (1977)

Scheipers, G., Kawashiro, T., Scheid, P.: Oxygen and carbon dioxide dissociation of duck blood. Respir. Physiol. 24, 1-13 (1975)

Severinghaus, J.W., Bainton, C.R., Carcellen, A.: Respiratory insensitivity to hypoxia in chronically hypoxic man. Respir. Physiol. 1, 308-334 (1966)

Sillau, A.H., Cueva, S., Volenzuela, A., Candela, E.: O_2 transport in the alpaca (*Lama pecas*) at sea level and 3300 meters. Respir. Physiol. 27, 147-155 (1976)

Swan, L.W.: The ecology of the high himalayas. Sci. Am. 205, 68-78 (1961)

Swan, L.W.: Goose of the himalayas. Natural History 79, 68-75 (1970)

Tenney, S.M., Ou, L.C.: Hypoxic ventilatory response of cats at high altitude: an interpretation of "blunting". Respir. Physiol. 30, 185-199 (1977)

Würdinger, I.: Breeding of bar-headed geese in captivity. Int. Zoo Yearbook 13 (1973)

Role of Arterial Chemoreceptors in Ventilatory Acclimation to High Altitude in Unanesthetized Pekin Ducks

P. Bouverot

Resting ventilation, pulmonary gas exchanges, arterial gas tensions and pH, and ventilatory responses to inhalation of three tidal volumes of pure N_2 (N_2-test) were studied in awake Pekin Ducks before (intact) and after chronic bilateral carotid body denervation (denervated) or after sham operation (control). The experiments were performed in an altitude chamber near sea level (barometric pressure P_b = 750 torr) and at various stages of a 3-week exposure to 3000 m (P_b = 530 torr).

1. Within a few seconds a N_2-test provoked a transient increase of ventilation in all intact and control ducks, a response which was abolished after bilateral section of the vagal branches from the nodose ganglion to the carotid body.

2. At low altitude, carotid body denervation was accompanied by a chronic respiratory acidosis with a 11-torr increase in arterial P_{CO_2} (from 33 to 44 torr) and a 0.05-unit decrease in arterial pH (from 7.48 to 7.43).

3. At 3000 m, all intact and control animals hyperventilated, with resulting arterial hypocapnia and increasing arterial P_{O_2}. The ventilatory response was rapid within the first 30 min of hypoxia, and was further augmented, but more slowly, during the first few days of chronic hypoxia. In contrast, denervated ducks did not hyperventilate: they remained markedly hypercapnic and they were more hypoxic and more polycythemic than control birds.

It is concluded that, in Pekin Ducks, the integrity of the carotid body innervation is essential in determining the eupneic level of ventilation at low altitude, as well as during acute or chronic exposure to high altitude.

Laboratoire de Physiologie Respiratoire, CNRS, 23 rue Becquerel, 67087 Strasbourg/ France.

Ventilation in the Humming Birds Colibri coruscans During Altitude Hovering

M. Berger

Summary

In hovering hummingbirds (*Colibri coruscans*) respiratory water loss was measured under sea-level and high-altitude conditions (about 4000 m). The calculation of ventilation showed an increase with height by 36% for body temperature and pressure (BTPS) volume and a decrease by 19% for standard temperature and pressure (STPD) volume. Oxygen extraction and convection requirement increased by 32% and 28% respectively.

1. Introduction

Studies of ventilation in flying birds have been performed only in a few species: pigeon (Hart and Roy, 1966), budgerigar (Tucker, 1968b), Evening Grosbeak, Ring-billed Gull, black duck (Berger et al., 1970), Glittering Emerald (Berger and Hart, 1972), Fish Crow (Bernstein, 1976). When combined on a body-mass basis the regression line $\dot{V} = 42 \cdot M_B^{0.73}$ ml/min (M_B, body mass in g) was calculated for the first six species (Berger and Hart, 1974). If this equation is brought into relation to the equation for O_2 consumption $\dot{V}_{O_2} = 1.01 \cdot M_B^{0.72}$ ml/min (Berger and Hart, 1974) it follows that the O_2 extraction $E = \dot{V}_{O_2}/0.209 \cdot \dot{V} = 0.115 \cdot M_B^{-0.01}$ is nearly independent of body mass. The ratio of extracted to ventilated O_2 appears to be very low when compared to resting data (cf. Lasiewski and Calder, 1971).

It should be mentioned that the individual data on which the regressions are based refer partly rather to maximal sustained effort or a special flight type than to "normal" flight. Therefore comparisons within a species are of greater informative value.

Flight under high-altitude conditions (reduced pressure) with its necessary physiological responses to changed power requirements and reduced O_2 pressure has scarcely been investigated. The present study provides first determinations of ventilatory adaptations.

2. Methods

Data on O_2 consumption were taken from previous studies (Berger, 1974). The experiments on ventilation were done in 3 specimens of the Sparkling Violetear, *Colibri coruscans* (body mass 7-9 g). With the assumption that

Westfälisches Landesmuseum für Naturkunde, D-4400 Münster.

(1) expired air is saturated with water vapor and (2) at high ambient temperatures (36°C) the temperature of expired air is similar to body temperature, the ventilation volume can be determined by measuring the respiratory water loss $\dot{M}_W$ (mg H_2O/g/min). With the humidity ratio HR = 622 · $P_{H_2O}/(P_B-P_{H_2O})$ (mg H_2O/g dry air) the ventilation is $\dot{M}_{vent}$ = $\dot{M}_W$/HR (g dry air/g/min). The O_2 extraction coefficient is given as E = $\dot{V}_{O_2}/0.209\ \dot{V}$, both volumes STPD, with ventilatory volume $\dot{V}$ calculated from $\dot{M}_{vent}$.

Determinations of respiratory water loss during hovering were done by sucking off all expired air through drying tubes filled with P_2O_5 and measuring the weight increase. The method of collecting expired air for hovering hummingbirds is described elsewhere (Berger and Hart, 1972). The experiments were performed at normal pressure and in a hypobaric chamber at simulated altitude of about 4000 m, most often at low humidity.

3. Results and Discussion

The results are presented in Table 1. The difference between sea-level and high-altitude water loss is significant. The ventilatory volume

Table 1. Expired water and respiration in hovering *Colibri coruscans* under sea-level and high-altitude conditions

		Sea-level	Altitude	Ratio
Barometric pressure (P_B)	mbar	1000	621	0.62
Respiratory water loss ($\dot{M}_W$)	mg/g/min	0.80	1.10	1.37
	n; SE	10;0.076	27;0.047	
Ventilation ($\dot{M}_{vent}$)	mg/g/min	16.0	13.0	0.80
($\dot{V}_{BTPS}$)	ml_{BTPS}/g/min	15.6	21.3	1.36
($\dot{V}_{STPD}$)	ml_{STPD}/g/min	12.4	10.0	0.81
Respiration frequency	min^{-1}	330	380	1.15
Tidal volume	ml_{BTPS}/g	0.047	0.056	1.18
Oxygen consumption ($\dot{V}_{O_2}$)	ml_{STPD}/g/min	0.63	0.67	1.06
Oxygen extraction (E)	$\dot{V}_{O_2STPD}/0.209\ \dot{V}_{STPD}$	0.24	0.32	1.32
Convection requirement	$\dot{V}_{BTPS}/\dot{V}_{O_2STPD}$	25	32	1.28

$\dot{V}_{BTPS}$ increases with altitude by 36% whereas $\dot{V}_{STPD}$ decreases by 19%. The high-altitude hyperventilation results from increases in respiration frequency (15%) and tidal volume (18%). However, no high-frequency respiration was observed, although in budgerigars frequency has been shown to increase from about 300/min to 960/min at ambient temperature > 34°C (body temperature > 42.0°C) (Aulie, 1975). In *C. coruscans* during flight no panting-like pattern was observed; in some cases panting started immediately after landing.

In fish crows the flight ventilation depends on ambient temperature (increase above 20°C, Bernstein, 1976), and this might also be true for hummingbirds. Moreover, the assumption that expired air is saturated with water vapor should be investigated. The first factor would reduce the ventilatory volume for normal temperatures, the latter - if it were not true - would lead to higher values. The data presented here therefore may be best approximations.

As found in resting birds under hypoxic conditions (e.g. sparrows, Tucker, 1968a; pigeons, Bouverot et al., 1976), the hummingbirds showed increased O_2 extraction values at high-altitude flight. It is not known so far whether increased effective ventilation through the air capillaries in the lung or/and changed circulatory performances are responsible. In any case, the low sea-level E-values can be considered as reserves on which the birds fall back under reduced O_2 pressure. It is clearly the O_2 supply that limits high-altitude flight capacities (Berger, 1974); aerodynamically these birds are able to hover at heights up to 8000 m (350 mbar) and possibly more.

The high amount of respiratory water loss of more than 1 mg/g/min is not likely to occur in nature. It is affected mainly by a high ambient temperature and it could lead to dehydration, if, in longer lasting flights, there were no water intake (either by respiration or drinking). Under natural altitude conditions temperatures are usually lower and it is possible that no water conserving problems arise. In a previous study (Berger and Hart, 1972), the respiratory water loss at 10°C ambient temperature was about 1/3 of that at 35°C, and including metabolic water production the net water losses were 0.7% and 8.2% of body weight/h respectively. Even, if by altitude-induced hyperventilation the net water loss in flight increased to slightly more than 1% of body weight/h, there should be no impairment of action, when compared to sea-level conditions. High-altitude flight is certainly affected by O_2 pressure, whereas flights up to 3000 or 4000 m appear to be little affected as far as heat balance and water budget is concerned. In *Oreotrochilus estella* long daily feeding flights from 3700-4800 m to 2400 m occur (Langner, 1973). The lowered ambient temperatures under natural high-altitude conditions appear to have the strongest influence on birds at rest.

Acknowledgment. This work was supported by the Deutsche Forschungsgemeinschaft.

References

Aulie, A.: Two respiratory patterns in the budgerigar during flight at different ambient temperatures. Comp. Biochem. Physiol. 52A, 81-84 (1975)

Berger, M.: Energiewechsel von Kolibris beim Schwirrflug unter Höhenbedingungen. J. Ornithol. 115, 273-288 (1974)

Berger, M., Hart, J.S.: Die Atmung beim Kolibri Amazilia fimbriata während des Schwirrfluges bei verschiedenen Umgebungstemperaturen. J. comp. Physiol. 81, 363-380 (1972)

Berger, M., Hart, J.S.: Physiology and energetics of flight. In: Avian Biology, Vol. IV. Farner, D.S., King, J.R. (eds.). New York-San Francisco-London: Academic Press, 1974, pp. 415-477

Berger, M., Hart, J.S., Roy, O.Z.: Respiration, oxygen consumption and heart rate in some birds during rest and flight. Z. vergl. Physiol. 66, 201-214 (1970)

Bernstein, M.H.: Ventilation and respiratory evaporation in the flying crow, *Corvus ossifragus*. Respir. Physiol. 26, 371-382 (1976)

Bouverot, P., Hildwein, G., Oulhen, Ph.: Ventilatory and circulatory O_2 convection at 4000 m in pigeon at neutral or cold temperature. Respir. Physiol. 28, 371-385 (1976)

Hart, J.S., Roy, O.Z.: Respiratory and cardiac responses to flight in pigeons. Physiol. Zool. 39, 291-306 (1966)
Langner, S.: Zur Biologie des Hochlandkolibris Oreotrochilus estella in den Anden Boliviens. Bonn. Zool. Beitr. 24, 24-47 (1973)
Lasiewski, R.C., Calder, W.A.: A preliminary allometric analysis of respiratory variables in resting birds. Respir. Physiol. 11, 152-166 (1971)
Tucker, V.A.: Respiratory physiology of house sparrows in relation to high-altitude flight. J. Exp. Biol. 48, 55-66 (1968a)
Tucker, V.A.: Respiratory exchange and evaporative water loss in the flying budgerigar. J. Exp. Biol. 48, 67-87 (1968b)

Respiration During Flight in Birds

J. R. Torre-Bueno

Summary

The gas composition of the anterior and posterior thoracic airsacs and the rate of CO_2 production were measured during flight in a starling. Metabolic rate is 8.9 times greater and minute volume is 8.2 times greater in flight than at rest. Oxygen extraction (E) is approximately 0.3 and does not change between rest and flight. The respiratory exchange ratio (R) after the first 15 min of flight is 0.7. By making certain simplifying assumptions it is possible to calculate E and R separately for the neo- and paleopulmo. The respiratory exchange ratio does not differ significantly between the neo- and paleopulmo. In the paleopulmo E is 0.32 at rest and during flight. In the neopulmo E is 0.08 at rest and declines to 0.06 during flight.

1. Introduction

Although it has been widely assumed that lung morphology plays an important role in bird flight, there have been very few studies of respiratory function during flight in birds. Lasiweski (1963), Tucker (1968, 1972), Berger (1974), Bernstein et al. (1973), and Torre-Bueno and Larochelle (1978) have measured respiratory exchange during flight, and Bernstein (1976) showed that the O_2 extraction does not change between rest and flight in crows at moderate ambient temperatures. Studies of the gas composition of the airsacs have been extremely valuable in promoting understanding of the function of the respiratory system in resting birds (Colacino et al., 1977; Piiper et al., 1970; Bouverot and Dejours, 1971; and others), but no measurements of the airsac composition of active birds have been reported. This paper describes the first study of airsac gas composition in a flying bird.

2. Methods

Experimental birds were wild starlings (*Sturnus vulgaris*, average weight 78 g), caught as adults; sexes were not distinguished. They were kept in a large outdoor aviary and fed mealworms, liver, and ground dog chow ad lib. Birds were trained to fly for periods of 45 to 90 min in a closed circuit wind tunnel with a working section 70 cm long: 40 cm high: 70 cm wide. The wind tunnel is described more fully in Linderoth (1975). Data were taken only from birds which would fly 45 min or longer. In a previous study (Torre-Bueno, 1975) it was found that

---

Department of Zoology, Duke University, Durham, North Carolina 27706, U.S.A.

starlings flew best at 14 m/s, and this speed was used in all experiments described here. Ambient temperature was 10°-14°C and P_{H_2O} was less than 5 mmHg.

Airsac samples were drawn through polyethylene cannulas with a perforated tip 7 mm long and 1.5 mm in diameter. The rest of the cannula was stretched out to a diameter of 1 mm to reduce air resistance. Cannulas were implanted in either the anterior thoracic - which in starlings is fused with the interclavicular sac, see Dunker (1971) and Hector (1974) - or the posterior thoracic sac by piercing the sac with a No. 14 gauge needle, threading the cannula through the needle, withdrawing the needle and securing the cannula with sutures. In order to avoid imposing excessive drag on a flying bird only one airsac was cannulated at a time. Each cannula was used daily until it clogged, usually after two or three days. Gas samples of 2 ml were collected in a syringe at the start of flight and every 5 min thereafter. In order to avoid increasing the ventilation of the airsac, samples were drawn at the rate of 0.2 ml/s; since the respiratory rate during flight is approximately 3/s (Torre-Bueno, 1975), the sampling rate is negligible compared to airsac ventilation. Calculations showed that the diffusion of O_2 and CO_2 through the walls of the cannula was negligible. Between flights cannulated birds were kept in a soundproof box and samples of airsac contents at rest were drawn through the cannulas during the night preceding or following an experimental flight.

Samples were analyzed either on a Scholander apparatus or a Radiometer BMS3MK2 gas analyzer. Samples analyzed on the Radiometer gas analyzer were equilibrated with water at 42°C, the presumed body temperature during flight (Torre-Bueno, 1976). The rate of CO_2 production ($\dot{V}_{CO_2}$) during flight was determined by measuring the rate of change of CO_2 pressure (P_{CO_2}) in the sealed wind tunnel. Details of this method are given elsewhere (Torre-Bueno and Larochelle, 1978). $\dot{V}_{CO_2}$ was measured both with and without the cannula; the rate was 14% higher when the bird was cannulated, indicating that the drag of the cannula slightly increased the work of flight.

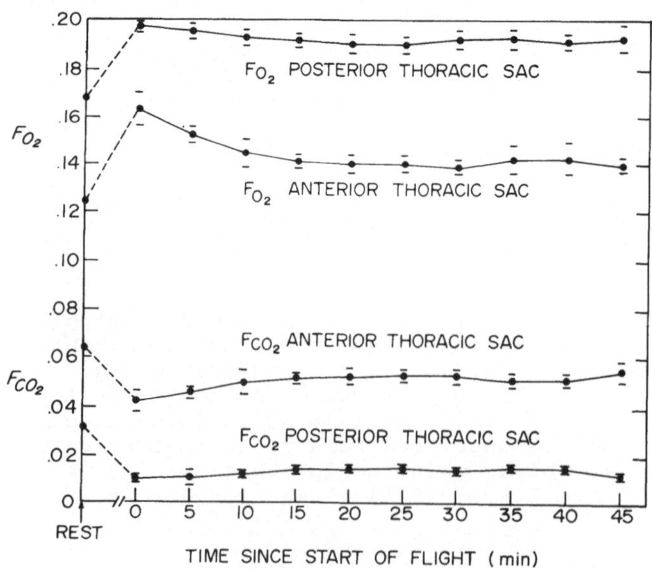

Fig. 1. Fractional gas composition of the airsacs in a starling during rest and flight

3. Results

The rate of CO_2 production ($\dot{V}_{CO_2}$) between the 30th and 90th min of flight was 19 ml/min ± 1 (± S.D.) in a cannulated starling. Figure 1 shows the mean fractional concentration of O_2 and CO_2 (F_{O_2}, F_{CO_2}) in airsacs over all flights as a function of time since start of flight. The first samples taken during a flight always had more O_2 and less CO_2 than a resting sample. During the first 15 min of flight F_{O_2} decreased and F_{CO_2} increased, reaching values between the resting and initial flight values which were stable for the rest of the flight. In Table 1 the F_{CO_2} and F_{O_2} values given are the mean over all flights of the samples collected between the 30th and 45th min of flight.

Table 1. Respiratory parameters during rest and flight in a starling

	Rest mean	± S.D.	(N)	Flight mean	± S.D.	(N)	Flight/Rest
$F_{i\ CO_2}$	0.0011	0.0004	(4)	0.0011	0.0004	(4)	
$F_{i\ O_2}$	0.208	0.001	(4)	0.208	0.001	(4)	
$F_{ant\ CO_2}$	0.064	0.004	(3)	0.053	0.002	(5)	
$F_{ant\ O_2}$	0.125	0.004	(3)	0.140	0.005	(5)	
$F_{post\ CO_2}$	0.033	0.005	(2)	0.014	0.002	(3)	
$F_{post\ O_2}$	0.168	0.007		0.192	0.004		
$\dot{V}_{CO_2}$ ml STPD/min	2.1[a]			19	1		8.9
$\dot{V}_r$ ml BTPS/min	41			472	30		11.4
f min^{-1}	92[b]			180[c]			2
V_r ml BTPS	0.45			2.6	0.16		5.8
V_{tr} ml BTPS	0.22[d]			22[d]			1
V_T	0.67			2.8	0.16		4.2
$\dot{V}$ ml BTPS/min	61.3			504	30		8.2
$F_{i'CO_2}$	0.021	0.001		0.0049	0.0004		
$F_{i'O_2}$	0.182	0.001		0.203	0.001		
$F_{e\ CO_2}$	0.044	0.003		0.049	0.002		
$F_{e\ O_2}$	0.152	0.003		0.146	0.004		
R	0.72	0.01		0.71	0.02		
R_{pi}	0.72	0.01		0.71	0.02		
R_{ni}	0.81	+0.2 −0.4		0.77	+0.1 −0.3		
R_{le}	0.69	0.1		0.70	0.03		
E	0.28	0.02		0.31	0.02		
E_{pi}	0.32	0.02		0.32	0.02		
E_{ni}	0.08	0.04		0.06	0.02		
E_{le}	0.27	0.03		0.28	0.03		

[a] Tel-Zur (1977); [b] Calder (1968); [c] Torre-Bueno (1975); [d] Hinds and Calder (1971).

4. Discussion

In the absence of data on the ventilation of individual airsacs, it is not possible to determine oxygen extraction (E) or other respiratory parameters from these data. However, they can be used to estimate several respiratory parameters if we make certain simplifying assumptions similar to those made by Colacino et al. (1977).

The most basic assumption is that expired gas consists of a mixture of fresh air from the tracheal dead space and gas with the same composition as the gas in the anterior thoracic sac. This assumption implies that all anterior sacs have the same gas composition, that gas leaving the parabronchi has the same composition during inspiration and expiration, and that there is no flow in the primary bronchus during expiration. Bouverot and Dejours (1971) observed that P_{CO_2} in the trachea rises rapidly to a plateau at the beginning of expiration and remains constant until the onset of inspiration. This indicates that either the fraction of the expired gas from different sources is maintained constant during expiration or the composition of the various sources of expired gas is equal. If the expired gas is effectively a mixture of inspired air (i) and anterior thoracic sac (ant) gas it is possible to calculate the ventilation of the respiratory system ($\dot{V}_r$) by the following formula:

$$\dot{V}_r = \frac{\dot{V}_{CO_2}\ [760 \cdot T_b/(273 \cdot (P_b - P_{H_2O}))]}{F_{ant\ CO_2} - F_{i\ CO_2} \cdot (F_{ant\ N_2}/F_{i\ N_2})} \qquad (1)$$

where $F_{ant\ X}$ is the fraction of gas X in the anterior thoracic sac [T_b = body temperature (absolute); P_b, barometric pressure].

Although this is the same as the formula used to calculate alveolar ventilation, it does not imply that this much gas has passed over the exchange surfaces, since fresh air which enters the anterior thoracic sac directly will be included in $\dot{V}_r$. Respiratory system ventilation ($\dot{V}_r$) should rather be considered as the difference between minute volume ($\dot{V}$) and ventilation of the trachea.

$$\dot{V} = \dot{V}_r + V_{tr} \cdot f \qquad (2)$$

where V_{tr} is tracheal volume and f is respiratory frequency.

Using Equation (2) and estimates of tracheal volume and respiratory frequency available from other studies (see footnote, Table 1), minute volume and tidal volume can be computed. The respiratory exchange ratio (R) can be computed as:

$$R = \frac{F_{ant\ CO_2} - F_{i\ CO_2} \cdot (F_{ant\ N_2}/F_{i\ N_2})}{F_{i\ O_2} \cdot (F_{ant\ N_2}/F_{i\ N_2}) - F_{ant\ O_2}} \qquad (3)$$

Oxygen extraction (E) can be computed as:

$$E = 1 - \frac{F_{e\ O_2} \cdot F_{i\ N_2}}{F_{i\ O_2} \cdot F_{e\ N_2}} \qquad (4)$$

where expired fractional concentration for gas X is:

$$F_{e\,X} = \frac{V_{tr} \cdot F_{i\,X} + (\dot{V}_r/f) \cdot F_{ant\,X}}{V_{tr} + (\dot{V}_r/f)} \qquad (5)$$

Given the further assumption that inspired air and tracheal gas are not differentially partitioned to different parts of the lung, the fractional composition of gas entering both the neopulmo and the paleopulmo is:

$$F_{i'X} = \frac{V_{tr} \cdot F_{ant\,X} + [(\dot{V}_r/f) \cdot F_{i\,X} \cdot (F_{ant\,N_2}/F_{i\,N_2})]}{V_{tr} + (\dot{V}_r/f) \cdot (F_{ant\,N_2}/F_{i\,N_2})} \qquad (6)$$

where $F_{i'X}$ is the fraction of gas X in the air entering the neo- or paleopulmo.

Oxygen extraction and respiratory exchange ratio in the paleopulmo during inspiration (E_{pi} and R_{pi}) can then be calculated with the following formula:

$$E_{pi} = 1 - \frac{F_{ant\,O_2} \cdot F_{i'N_2}}{F_{i'O_2} \cdot F_{ant\,N_2}} \qquad (7)$$

$$R_{pi} = \frac{F_{ant\,CO_2} - [F_{i'CO_2} \cdot (F_{ant\,N_2}/F_{i'N_2})]}{[F_{i'O_2} \cdot (F_{ant\,N_2}/F_{i'N_2})] - F_{ant\,O_2}} \qquad (8)$$

The same formulas can be used to calculate the neopulmal O_2 extraction (E_{ni}) and exchange ratio (R_{ni}) if posterior thoracic sac gas fractions are substituted for anterior thoracic sac gas fractions (assuming that the posterior thoracic sac is representative of all posterior sacs).

During expiration, air from the posterior sacs must flow either through the neopulmal parabronchi or the large first laterobronchus. If, as was assumed above, the unidirectional flow model is correct and there is no flow in the primary bronchus during expiration the gas must then pass through the paleopulmo. Respiratory exchange ratio and O_2 extraction for the whole lung during expiration (R_{le} and E_{le}) can therefore be calculated using Equations (7) and (8) with posterior thoracic gas fractions substituted for inspired (i') gas fractions.

Observed and calculated respiratory parameters during flight and rest are summarized in Table 1. The gas composition of the airsacs during rest in the starling is within the range found in other species (Bouverot and Dejours, 1971). No significant differences in R between different parts of the lung were observed. However, because the composition of the gas in the posterior thoracic sac was close to that of inspired air, there are large uncertainties in R for the neopulmo (R_{ni}) and differences in R between the neo- and paleopulmo could not be detected using these data.

Table 1 shows that the flight:rest ratio for respiratory system ventilation ($\dot{V}_r$) is greater than for $\dot{V}_{CO_2}$, indicating that the relative ventilation of the respiratory system is increased during flight. However, $\dot{V}$ does increase in proportion to $\dot{V}_{CO_2}$ because the increased tidal volume during flight leads to a relative decrease in the ventilation of the dead space in the trachea. The overall O_2 extraction does not decrease during flight because of the reduction in the ratio of tracheal volume to tidal volume mentioned above. This constancy of overall O_2 extraction between flight and rest was also found by Bernstein (1976) in crows. Bouverot et al. (1976) found that O_2 extraction did not change when the metabolic rate of pigeons was doubled by exposure to low temperatures.

Acknowledgments. Supported by NIH Fellowship F32 HL-03151 and NIH Research Grant HL-02228. The author wishes to acknowledge the expert technical assistance of Shlomit Lipton.

References

Berger, M.: Oxygen consumption and power of hovering hummingbirds at varying barometric and oxygen pressures. Naturwissenschaften 61, 407 (1974)

Bernstein, M.H.: Ventilation and respiratory evaporation in the flying crow. Respir. Physiol. 26, 371-382 (1976)

Bernstein, M.H., Thomas, S.P., Schmidt-Nielsen, K.: Power input during flight of the fish crow, *Corvus ossifragus*. J. Exp. Biol. 58, 401-410 (1973)

Bouverot, P., Dejours, P.: Pathway of respired gas in the air sacs-lung apparatus of fowl and ducks. Resp. Physiol. 13, 330-342 (1971)

Bouverot, P., Hildwein, G., Oulhen, P.H.: Ventilatory and circulatory O_2 convection at 4000 m in pigeon at neutral or cold temperature. Resp. Physiol. 28, 371-385 (1976)

Calder, W.A.: Respiratory and heart rates of birds at rest. Condor 70, 358-365 (1968)

Colacino, J.M., Hector, D.H., Schmidt-Nielsen, K.: Respiratory responses of ducks to simulated altitude. Respir. Physiol. 29, 265-281 (1977)

Duncker, H.R.: The lung air sac system of birds. Adv. in Anatomy, Embryology and Cell Biology 45, 1-171 (1971)

Hector, D.H.: Functional anatomy of the respiratory apparatus of the starling. Thesis, Ohio State Univ., 1974

Hinds, D.S., Calder, W.A.: Tracheal dead space in the respiration of birds. Evolution 25, 429-440 (1971)

Lasiewski, R.C.: Oxygen consumption of torpid, resting, active and flying humming birds. Physiol. Zool. 36, 122-140 (1963)

Linderoth, L.S.: A wind tunnel for bird flight studies at high altitudes. J. Appl. Physiol. 39, 501-502 (1975)

Piiper, J., Drees, F., Scheid, P.: Gas exchange in the domestic fowl during spontaneous breathing and artificial ventilation. Resp. Physiol. 9, 234-245 (1970)

Tel-Zur, D.: A comparative study on the heat balance of two starlings from different habitats: Starling (Sturnus vulgaris) and Tristram's starling (Onycognathus tristrami). M. Sc. Thesis, Tel-Aviv Univ., 1977

Torre-Bueno, J.R.: Thermoregulatory adjustments to flight in birds. Thesis, The Rockefeller Univ., N.Y. University Microfilms (Diss. Abstr. 37(4), 76-23, 275, 1975)

Torre-Bueno, J.R.: Temperature regulation and heat dissipation during flight in birds. J. Exp. Biol. 65, 471-482 (1976)

Torre-Bueno, J.R., Larochelle, J.: The metabolic cost of flight in unrestrained birds. J. Exp. Biol. in press (1978)

Tucker, V.A.: Respiratory exchange and evaporative water loss in the flying budgerigar. J. Exp. Biol. 48, 67-87 (1968)

Tucker, V.A.: Metabolism during flight in the laughing gull. Am. J. Physiol. 222, 237-245 (1972)

The Contribution of Arterial Chemoreceptors and Baroreceptors to Diving Reflexes in Birds

D. R. Jones and N. H. West

Summary

The nature of the cardiovascular adjustments to forced and voluntary submergence in diving birds is described. In voluntary diving suprabulbar nervous structures clearly play a role in modifying the responses observed during forced submergence. In forced submergence, bradycardia and increased peripheral resistance (PR) are reflexly maintained by stimulation of cardiovascular receptors. Whether carotid body chemoreceptors cause an increase in peripheral resistance has not been demonstrated, but they play a predominant role in maintaining bradycardia. Arterial baroreceptors have effects on both heart rate (HR) and PR in a dive, balancing cardiac output and resistance so that arterial blood pressure (BP) is maintained.

1. Introduction

During submergence diving birds exhibit apnoea and a group of cardiovascular adjustments which serve to conserve O_2 for tissues, such as the myocardium and brain, which are most vulnerable to O_2 lack. Other vascular beds exhibit a pronounced vasoconstriction although some flow occurs in major arteries supplying, for example, muscular vascular beds but it is shunted through non-nutritive channels thus facilitating improved utilisation of the venous blood O_2 stores by preventing venous stagnation (Djojosugito et al., 1969). That these adjustments are effective is inferred from the prolonged underwater survival times of diving birds compared with forced submergence of terrestrial species (Bond et al., 1961). The efficacy of the diving adjustments may be assessed more directly by measuring NADH accumulation in an O_2-dependent tissue such as the cerebral cortex. A two-minute apnoea in a duck (*Anas platyrhynchos*) causes a rise in cortical NADH, measured fluorometrically, to about one-fifth the maximum fluorescence level observed after death, whereas in a chicken (*Gallus domesticus*) a one-minute apnoea causes NADH fluorescence to increase to about one-half the death induced maximum (Fig. 1, Bryan and Jones, 1978).

2. Cardiovascular Responses to Forced and Voluntary Submergence

In forced laboratory dives the cardiovascular adjustments follow a characteristic pattern in which both heart rate (HR) and arterial O_2 tension ($P_{a\,O_2}$) fall biphasically. On submergence, ducks (*A. platyrhyn-*

Department of Zoology, University of British Columbia, Vancouver, B.C., Canada V6T 1W5.

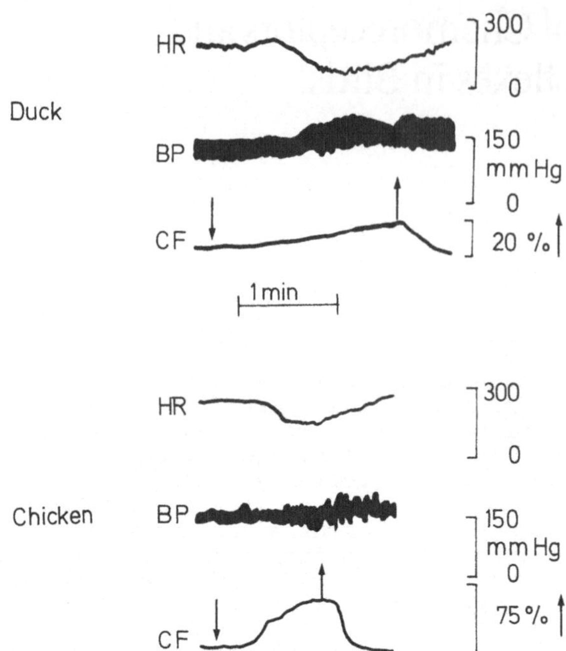

Fig. 1. Comparison between increase in NADH fluorescence in the cerebral cortex of a duck and chicken (*lower three traces*) in response to asphyxia caused by stopping air flow in paralysed unidirectionally ventilated preparations. *HR* heart rate, beats/min; *BP*, arterial blood pressure, mmHg; *CF*, corrected NADH fluorescence, expressed as a % of the maximum value obtained after death. The *downward pointing arrow* on the CF trace indicates the start of the asphyxic period and the *upward pointing arrow* the termination of asphyxia. A *time bar* divides the sets of traces and applies to both (Bryan and Jones, 1978)

chos var) show an initial decrease in HR, which occurs rapidly with a halftime of 8 s, followed by a slow HR adjustment with a halftime of 94 s (Jones and Purves, 1970). A similar pattern in HR adjustment has been described for other birds in forced dives (Bond et al., 1961; Murrish, 1970; Catlett and Johnston, 1974; Butler and Woakes, 1976). Both the initial rapid bradycardial phase as well as the secondary gradual fall to the more or less stable HR ultimately attained during diving are accompanied by adjustments in peripheral resistance (Djojosugito et al., 1969; Butler and Jones, 1971; Jones, 1973; Blix et al., 1975) and mean arterial blood pressure (BP) is seldom more than 20% changed from the predive level (Hollenberg and Uvnäs, 1963; Butler and Jones, 1971; Jones, 1973). Arterial oxygen tension also falls biphasically (Jones and Purves, 1970; Butler and Jones, 1971) but, since large changes in P_{aO_2} can occur with only small changes in blood O_2 content, it is not certain that the relatively slow onset of the cardiovascular adjustments is as significant in depleting the blood O_2 store as might first seem to be the case.

There is little doubt that in a single species both the rate of onset and intensity of diving bradycardia varies during forced dives depending on the condition of the animal and the experimental protocol. Since most experimental protocols call for diving periods which are usually only 0.2 to 0.5 times the birds' maximum capability the effect of these variations in development of bradycardia on survival is unknown. Even studying HR during voluntary dives has added little to our

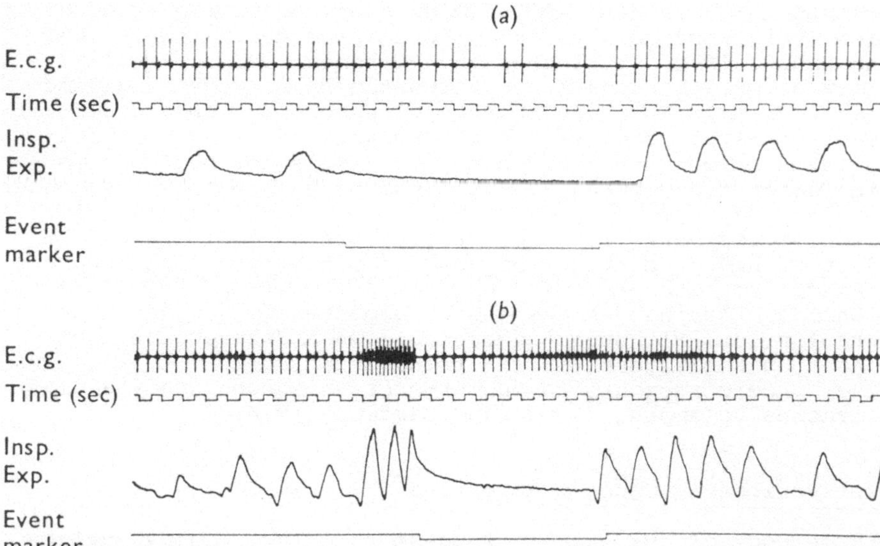

Fig. 2. Changes in HR and RF of a pochard accompanying (a) forced submersion in the laboratory, and (b) natural submersion while swimming gently on an outside pond. Each trace shows reading *down*, electrocardiogram (E.c.g), time marker (s), changes in air temperature in trachea (inspiration upwards), event marker (*down* indicates submersion, *up* indicates surfacing). From Butler and Woakes (1975), by permission

knowledge of the relationship between the rate of onset of and degree of cardiovascular adjustments and underwater survival time, although these studies have emphasised the variability in the diving responses. For instance, bradycardia develops more rapidly in the dipper (*Cinclus mexicanus*) during voluntary than enforced submergence (Murrish, 1970); while in the penguin (*Pygoscelis papua*) HR is more labile in restrained dives, but can attain lower levels than those observed in voluntary dives (Millard et al., 1973). Butler and Woakes (1975) found a marked difference in the responses of pochards (*Aythya ferina*) to restrained and voluntary diving (Fig. 2). In forced dives HR fell to about 40% the predive rate after 10 s submergence (Fig. 2a). When the same bird was swimming on a pond HR and respiratory frequency (RF) were somewhat higher than laboratory resting values. In the immediate pre-submersion period RF and HR more than doubled; the latter fell rapidly on submergence to a value only slightly below the free-range swimming value and then increased progressively throughout the dive (Fig. 2b). Furthermore in voluntary dives by the Tufted Duck (*Aythya fuligula*) not only was there tachycardia and tachypnoea before the first spontaneous dive of a series but also an anticipatory bradycardia which preceded the diving apnoea (Butler and Woakes, 1976).

3. Nervous Control of the Cardiovascular Responses to Diving

a) Higher Nervous Centres

The striking divergence between the cardiovascular responses to voluntary and to forced diving implies that they are controlled or modified by the higher nervous centres. However, brain stimulation at suprabulbar levels has failed to elicit a diving-type series of cardiovascular adjustments in the absence of apnoea in either diving or non-diving species (Feigl and Folkow, 1963; Folkow and Rubinstein, 1965; Macdonald and Cohen, 1973; Kotilainen and Putkonen, 1974). Equally, ponto-medullary stimulations have not given the full range of diving cardiovascular adjustments in the absence of apnoea (Cohen and Schnall, 1970). Since diving-type cardiovascular responses are only seen following brain stimulations which cause apnoea, it is not possible to separate stimulation from apnoea as the cause of the observed responses.

The cardiovascular responses to forced diving are undoubtedly reflex and independent of higher nervous centres since they are fully expressed in ponto-medullary decerebrate ducks (*A. platyrhynchos* var) (Andersen, 1963). On the other hand the apnoeic response may depend on suprabulbar integration for, although Andersen's (1963) decerebrate ducks displayed apnoea on submergence, brain transection rostral to the pons eliminated the apnoeic response to submergence, although bradycardia accompanied tracheal clamping (Jones and Bamford, 1978).

b) The Role of Peripheral Sensory Receptors

There is a large body of evidence to suggest that the cardiovascular responses to forced submergence are of reflex origin resulting from stimulation of carotid body chemoreceptors or cardiovascular mechanoreceptors. There is little agreement, however, about the respective roles of these receptors in causing or maintaining the bradycardia or increase in peripheral resistance. In many cases the evidence concerning the role of these receptors is conflicting and the delimitation of a definitive sensory limb of the reflex pathway (if such exists) must await further research.

It is known that the integrity of arterial chemoreceptors is vital for the full development of the cardiovascular responses to submergence. Diving bradycardia is promoted by experimental interventions leading to increased chemoreceptor stimulation during submergence and retarded by interventions which achieve the opposite end (Feigl and Folkow, 1963; Cohn et al., 1968; Blix and Berg, 1974; Blix, 1975). However, diving bradycardia is initiated before there is any fall in P_{aO_2} in a dive, which indicates that other receptors are making a contribution to the response, at least in the early stages (Jones and Purves, 1970).

From measurement of HR's in chronically chemoreceptor denervated ducks both Jones and Purves (1970) and Holm and Sørensen (1972) concluded that most of the diving bradycardia resulted from arterial chemoreceptor stimulation. This observation has recently been confirmed by selective perfusion of the carotid-body chemoreceptors before, during, and after a dive with hypoxic or hyperoxic blood in acute, baroreceptor denervated ducks. Perfusion with hyperoxic blood (Fig. 3a) prevented much of the bradycardia normally observed after 60-s diving (Fig. 3b) while perfusion with blood withdrawn after one minute of submergence in a previous dive (before surfacing) caused an extremely rapid and pronounced bradycardia when the ducks were submerged (Fig. 3c; Jones and West, 1978). However, the interplay of other receptors in causing diving reflexes is indicated, particularly in acute experiments, by the fact that chemoreceptor stimulation alone is never able to mimic the total response seen in intact ducks during forced submergence (Fig. 3).

Acute or chronic denervation of arterial baroreceptors has an effect on the cardiovascular responses to forced submergence, but the precise

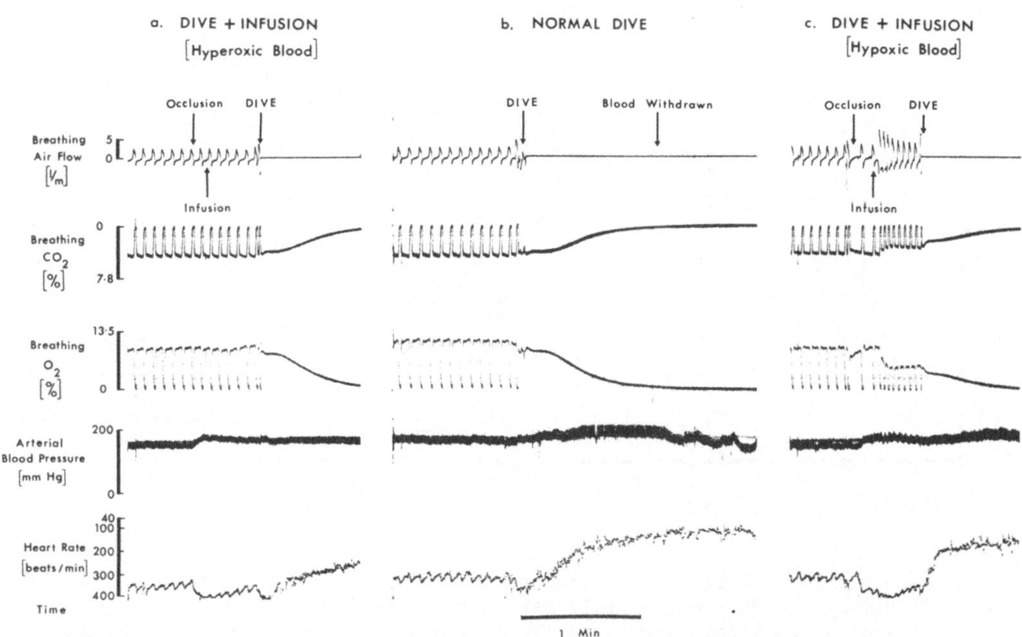

Fig. 3. The effect of selective perfusion of the carotid body chemoreceptors on HR and arterial BP in a baroreceptor denervated duck. (a) perfusion with hyperoxic blood (P_{aO_2} > 350 mmHg). At the *downward pointing arrow* the common carotid artery was occluded and blood infusion started at the *upward pointing arrow* and was continued throughout the rest of the trace. The *second downward pointing arrow* indicates the start of the dive. (b) normal dive in which the carotid bodies were auto-perfused. The *first downward pointing arrow* indicates the start of the dive while at the *second arrow* blood was withdrawn to use in perfusing the carotid body in a subsequent dive (c). (c) perfusion with hypoxic blood, (P_{aO_2} < 40 mmHg), withdrawn after 60 s submergence in a previous dive. The start of infusion was marked by a prominent stimulation of breathing. *Arrows* indicate occlusion, infusion, and dive as described for (a). In (a), (b), and (c) the traces *from top to bottom* are as indicated in *left hand column* and the *time bar* (1 min) under (b) also applies to (a) and (c) (Jones and West, 1978)

role of the barostatic reflex in diving is not clear. In a recent series of papers the view has been expressed that chemoreceptor stimulation during the dive causes an increase in peripheral resistance so that bradycardia follows as an expression of the barostatic reflex (Andersen and Blix, 1974; Blix et al., 1974; Blix, 1975; Blix et al., 1975). Unfortunately, most of the evidence for this hypothesis has been obtained indirectly from experiments using various autonomic depleting or blocking drugs (Andersen and Blix, 1974; Blix et al., 1974). These experiments gave similar results to those obtained earlier by Kobinger and Oda (1969) and Butler and Jones (1971), who had interpreted them as showing that baroreceptors play only a minor role in controlling diving bradycardia. In chronically baroreceptor denervated ducks, Jones (1973) found that the bradycardia attained after 60-120 s submergence was identical to that in normal ducks although BP fell greatly in denervates because of a relatively smaller increase in PR. Even so, leg blood flow was markedly reduced in both normal and denervated ducks showing that more than half of the normally observed increase in flow

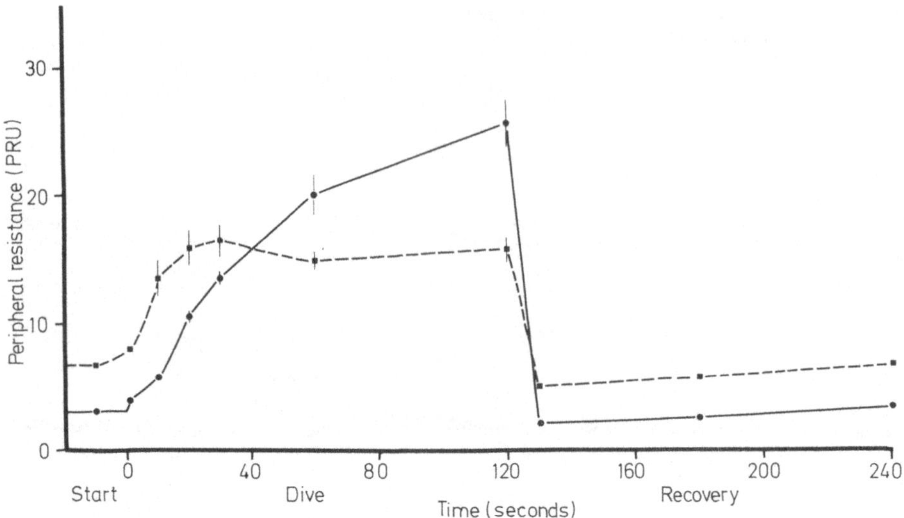

Fig. 4. Mean changes in sciatic vascular resistance (*PRU*) during 2-min dives in four intact (21 dives) and four chronically baroreceptor denervated (23 dives) ducks. Submergence and emersion at time 0 and 120 s respectively. ●–●, intact ducks; ■--■, denervated ducks. From Jones (1973), by permission

resistance was achieved independently of baroreceptor stimulation (Fig. 4). On the other hand, in acute denervates bradycardia was not as pronounced as in normal ducks while BP usually rose during a dive (Fig. 3b; Jones and West, 1978).

In an attempt to gain more information about the role of the baroreceptors in diving, we have implanted stimulating electrodes on the central end of a cut depressor nerve in chronic preparations. The baroreceptors innervated by the other depressor nerve were denervated by vagotomy. Nerve stimulation caused a fall in arterial BP both before and during diving (Fig. 5). Stimulation provoked a similar fall in leg vascular resistance both before and during diving although the absolute fall in HR on stimulation was, of necessity, far smaller in the later stages of a dive (Fig. 5) since the prestimulation rate was lower. These results show that baroreceptors exert their effects on both HR and PR both before and during diving, although their action on HR, in absolute terms, is variable (Jones and West, 1978).

4. Conclusions

The roles of the chemoreceptors and baroreceptors in the cardiovascular response to forced submergence are summarised in Figure 6. The chemoreceptors are of great importance in initiating and maintaining bradycardia, while the baroreceptors operate to balance the fall in cardiac output to the change in PR. Whether these reflexes are similarly activated in voluntary dives is an intriguing problem, although present evidence suggests that their contribution in short dives is slight (Butler and Woakes, 1975, 1976).

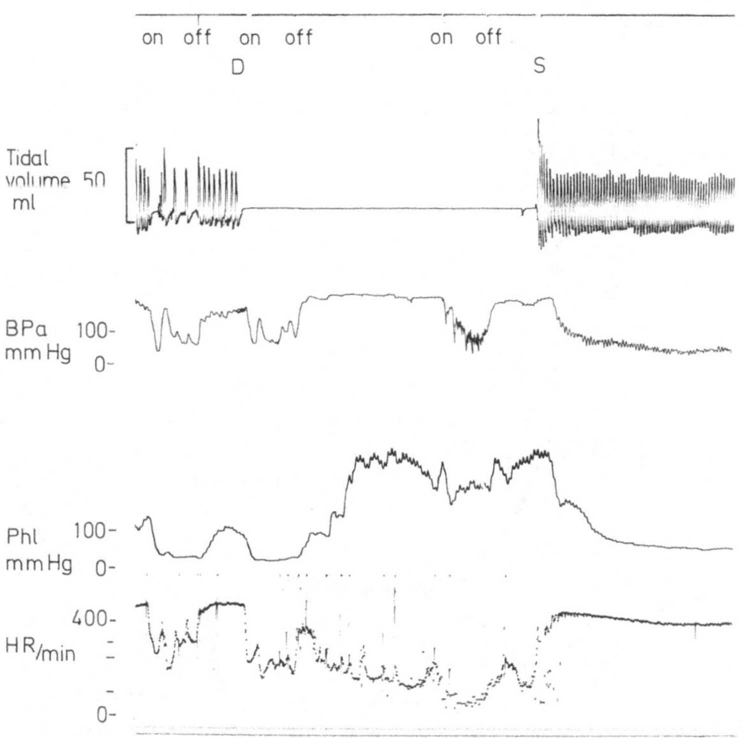

Fig. 5. The effect of electrical stimulation (20 Hz, 4 msec duration, 2 V) of the central end of a cut depressor nerve in an awake duck on arterial BP, HR, and leg vascular resistance before and during a dive of 3.5-min duration. The baroreceptors supplied by the remaining intact depressor nerve were denervated by vagotomy. The electrodes were implanted on the nerve 3 days before the experiment. Traces, from *top to bottom,* (1) event marker, on and off refer to periods of stimulation and *D,* start of dive, *S,* end of dive; (2) tidal volume (ml) - note the slowing of RF during the stimulation period before the dive; (3) BP_a (mmHg), arterial BP; (4) P_{hl} (mmHg), pressure generated by constant blood flow perfusion of a leg, therefore indicating leg vascular resistance; (5) HR (beats/min); (6) time (s) (Jones and West, 1978)

Also included in Figure 6 are our ideas concerning the role of two other peripheral sensory receptor groups: the nasal receptors and pulmonary receptors. The former, which we take to include not only receptors on the beak but also those in the nasopharyngeal region and the wall of the glottis, are instrumental in causing apnoea in forced dives (see Jones, 1976 for review) and also appear to have specific or nonspecific cardiovascular effects (Andersen, 1963; Feigl and Folkow, 1963; Rey, 1971; Bamford and Jones, 1974; Blix et al., 1975).

The cessation of rhythmic input from pulmonary receptors and the activity of central respiratory neurones in diving is important for initiation of diving responses in a negative rather than positive sense (Fig. 6; Bamford and Jones, 1976). If breathing ducks are subjected to hypercapnic-hypoxia then, due to the increase in breathing, diving-type cardiovascular responses are overruled (Feigl and Folkow, 1963; Cohn et al., 1968; Butler and Taylor, 1973). However, artificially preventing an increase in ventilation causes bradycardia when an hypoxic-hypercapnic gas mixture is breathed (Butler and Taylor, 1973).

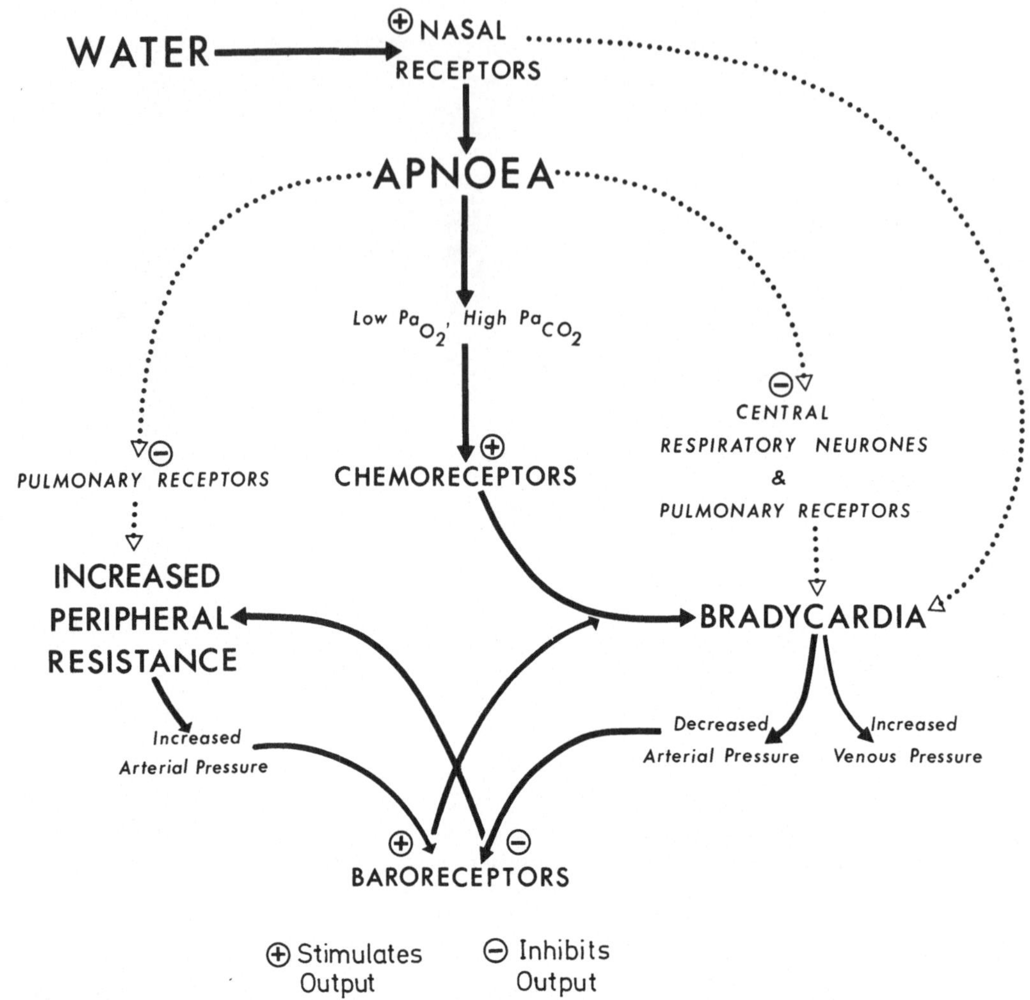

Fig. 6. Diagrammatic representation of the role of arterial chemoreceptors and baroreceptors in initiating and maintaining the cardiovascular responses to forced submergence in ducks. *Solid lines* and *arrows* refer to pathways which have been discussed in the text and whose involvement appears to be firmly established. *Dotted lines* and *open arrows* indicate the possible roles nasal receptors, pulmonary receptors, and the central respiratory neurones may play in the cardiovascular responses

Total cessation of ventilation, combined with hypoxaemia and hypercapaemia, gives the diving-type response even without water immersion (Butler and Jones, 1968; Butler and Taylor, 1973; Bamford and Jones, 1974). It is the cessation of breathing alone which is instrumental in any "permissive" (Feigl and Folkow, 1963) or "facilitatory" role (Holm and Sørensen, 1972) claimed for water immersion per se.

Acknowledgments. This work was supported by grants from the B.C. Heart Fund and National Research Council of Canada. We are extremely grateful to Graham and Margaret Shelton for scientific and literary criticism of the penultimate draft of this manuscript.

References

Andersen, H.T.: The elicitation of physiological responses to diving. Norwegian Research Council of Science and Humanities, Universitetsforlaget (1963)

Andersen, H.T., Blix, A.S.: Pharmacological exposure of components in the autonomic control of the diving reflex. Acta Physiol. Scand. 90, 381 (1974)

Bamford, O.S., Jones, D.R.: On the initiation of apnoea and some cardiovascular responses to submergence in ducks. Respir. Physiol. 22, 199 (1974)

Bamford, O.S., Jones, D.R.: Respiratory and cardiovascular interactions in ducks: The effect of lung denervation on the initiation of and recovery from some cardiovascular responses to submergence. J. Physiol. (London) 259, 575 (1976)

Blix, A.S.: The importance of asphyxia for the development of diving bradycardia in ducks. Acta Physiol. Scand. 95, 41 (1975)

Blix, A.S., Berg, T.: Arterial hypoxia and the diving responses of ducks. Acta Physiol. Scand 92, 566 (1974)

Blix, A.S., Gautvik, E.L., Refsum, H.: Aspects of the relative roles of peripheral vasoconstriction and vagal bradycardia in the establishment of the "diving reflex" in ducks. Acta Physiol. Scand. 90, 289 (1974)

Blix, A.S., Lundgren, O., Folkow, B.: The initial cardiovascular responses in the diving duck. Acta Physiol. Scand. 94, 539 (1975)

Bond, C.F., Douglas, S.D., Gilbert, P.W.: Effects of submergence on cardiac cycle and rate in aquatic and terrestrial birds. Am. J. Physiol. 200, 723 (1961)

Bryan, R.M., Jr., Jones, D.R.: NADH fluorescence in the brain of ducks and chickens during asphyxia (in preparation) (1978)

Butler, P.J., Jones, D.R.: Onset of and recovery from diving bradycardia in ducks. J. Physiol. (London) 196, 255 (1968)

Butler, P.J., Jones, D.R.: The effect of variations in heart rate and regional distribution of blood flow on the normal pressor response to diving in ducks. J. Physiol. (London) 214, 457 (1971)

Butler, P.J., Taylor, E.W.: The effect of hypercapnic hypoxia, accompanied by different levels of lung ventilation, on heart rate in the duck. Respir. Physiol. 19, 176 (1973)

Butler, P.J., Woakes, A.J.: Changes in heart rate and respiratory frequency associated with natural submersion of ducks. J. Physiol. (London) 256, 73P (1975)

Butler, P.J., Woakes, A.J.: Changes in heart rate and respiratory frequency associated with spontaneous submersion of ducks. Biotelemetry III. Fryer, T.B., Miller, H.A., Sandler, H. (eds.). New York, San Francisco, London: Academic Press, 1976, p. 215

Catlett, R.H., Johnston, B.L.: Cardiac response to diving in wild ducks. Comp. Biochem. Physiol. 47A, 925 (1974)

Cohen, D.H., Schnall, A.M.: Medullary cells of origin of vagal cardioinhibitory fibers in the pigeon. II. Electrical stimulation of the dorsal motor nucleus. J. Comp. Neurol. 140, 321 (1970)

Cohn, J.E., Krog, J., Shannon, R.: Cardiopulmonary responses to head immersion in domestic geese. J. Appl. Physiol. 25, 36 (1968)

Djojosugito, A.M., Folkow, B., Yonce, L.R.: Neurogenic adjustments of muscle blood flow, cutaneous A-V shunt flow and of venous tone during "diving" in ducks. Acta Physiol. Scand. 75, 377 (1969)

Feigl, E., Folkow, B.: Cardiovascular responses in "diving" and during brain stimulation in ducks. Acta Physiol. Scand. 57, 99 (1963)

Folkow, B., Rubenstein, E.H.: Effect of brain stimulation on "diving" in ducks. Hvalradets. Skr. 48, 30 (1965)

Hollenberg, N.K., Uvnäs, B.: The role of the cardiovascular response in resistance to asphyxia of avian divers. Acta Physiol. Scand. 58, 150 (1963)

Holm, B., Sørensen, S.C.: The role of the carotid body in the diving reflex in the duck. Respir. Physiol. 14, 302 (1972)

Jones, D.R.: Systemic arterial baroreceptors in ducks and the consequences of their denervation on some cardiovascular responses to diving. J. Physiol. (London) 499 (1973)

Jones, D.R.: The control of breathing in birds with particular reference to the initiation of diving apnea. Federation Proc. 35 (1975)

Jones, D.R., Bamford, O.S.: The effect of brain transection, at various levels, on the diving adjustments in birds (in preparation) (1978)

Jones, D.R., Purves, M.J.: The carotid body in the duck and the consequences of its denervation upon the cardiac responses to immersion. J. Physiol. (London) 211, 279 (1970)

Jones, D.R., West, N.H.: The role of chemoreceptors and baroreceptors in initiation and maintenance of cardiovascular adjustments to diving in the duck (*Anas platyrhynchos* var). In preparation (1978)

Kobinger, W., Oda, M.: Effects of sympathetic blocking substances on the diving reflex of ducks. Eur. J. Pharmac. 7, 289 (1969)

Kotilainen, P.V., Putkonen, P.T.S.: Respiratory and cardiovascular responses to electrical stimulation of the avian brain with emphasis on inhibitory mechanisms. Acta Physiol. Scand. 90, 358 (1974)

Macdonald, R.L., Cohen, D.H.: Heart rate and blood pressure responses to electrical stimulation of the central nervous system in the pigeon (*Columbia livia*). J. Comp. Neurol. 150, 109 (1973)

Millard, R.W., Johansen, K., Milsom, W.K.: Radiotelemetry of cardiovascular responses to exercise and diving in Penguins. Comp. Biochem. Physiol. 46A, 227 (1973)

Murrish, D.E.: Responses to diving in the dipper *Cinclus mexicanus*. Comp. Biochem. Physiol. 34, 853 (1970)

Rey, .N.: Afferent structures involved in heart response to diving reflex. Acta Physiol. Latinoam. 21, 235 (1971)

Adjustment of the Regional Pulmonary Circulation to the Profile of Oxygen Pressure Along the Parabronchus in the Duck

P. Scheid and J. P. Holle

Summary

Radioactively labeled microspheres were used in the duck in order to study the distribution of pulmonary blood flow. In the awake, spontaneously breathing animal, blood flow to the parabronchial gas exchange tissue decreased in the same direction as gas flows through, and thus as P_{O_2} decreased along, the parabronchi. Hypoxia could be shown to exert a vasoconstriction in the lung of the duck and thus provides a potential mechanism for the blood flow pattern in the awake duck. However, when the direction of parabronchial gas flow was artificially reversed, the pattern of perfusion did change only slightly, suggesting that the P_{O_2} profile is not the main determinant of the perfusion profile within the parabronchi of the spontaneously breathing animals. It can be shown in theory that any inhomogeneity of lung perfusion in birds results in a reduction of the gas exchange efficiency. However, at the level of inhomogeneity observed in this study the impairment is expected to be insignificant.

1. Introduction

In contrast to the mammalian lung, the lung of birds shows a substantial structural heterogeneity. The neopulmo and the paleopulmo have been recognized as morphologically distinct subunits of the lung (Duncker, 1972).

Moreover, due to the peculiar arrangement of the blood capillaries within the mantle of parabronchial exchange tissue (Piiper and Scheid, 1973) gas exchange is not evenly allotted to the length of the parabronchus, and as a result there exist complicated profiles of both O_2 and CO_2 concentrations in the gas along the parabronchial length (Meyer et al., 1976). We have studied the regional distribution of pulmonary blood flow in the duck to investigate if the structural and functional heterogeneity of the lung is connected with regional inhomogeneity of perfusion and to determine if an observed variability in blood flow could be due to the variability of O_2 and CO_2 concentrations within the lung.

Max-Planck-Institut für experimentelle Medizin, Abteilung Physiologie, 3400 Göttingen, Federal Republic of Germany.

2. Methods

We have used plastic microspheres (average diameter, 25 µm) labeled with either of the four radioactive isotopes: ^{46}Sc, ^{85}Sr, ^{141}Ce, ^{51}Cr. The activity of the isotopes and the number of spheres injected were chosen to yield sufficiently high count rates at minimal blockade of pulmonary vessels. In each experiment, the four labels were injected successively into the right heart and thus allowed to image the blood flow distribution in each animal during four independent experimental situations.

After temination of the experiment, the lungs were subdivided according to a fixed scheme which allowed determination of the γ activity of the four labels in four subunits of the lungs: (1) the neopulmo (NP), and, within the paleopulmo, parts adjoining (2) the medioventral (MV), (3) the mediodorsal secondary bronchi (MD), and (4) the middle region of the paleopulmonic parabronchi (Pb). The γ activity of a given tissue sample was converted to the perfusion of this sample, expressed as the fraction of total lung perfusion per fresh weight of tissue.

This technique of measuring regional lung perfusion was applied in three series of experiments in order to evaluate the blood flow distribution in awake ducks at rest (Series I, 5 animals); to determine the effects of O_2 and CO_2 on regional lung blood flow (Series II, 12 animals); and to test if reversing the gas flow direction, and thereby reversing the P_{O_2} profiles in lung gas, can effectively change the blood flow distribution (Series III, 3 animals).

3. Results and Discussion

a) Series I. Blood Flow Distribution in Awake, Spontaneously Breathing Ducks

Except for the right heart catheter, introduced under local anesthesia, the animals were not operated upon and were kept in a frame which restrained them only slightly and did not impair breathing. The four microsphere labels were injected at intervals of 20 min when the animals appeared to be at steady resting conditions.

The results obtained with the four labels were statistically not different. Similarly, corresponding areas of both lungs appeared to receive the same blood flow per tissue weight. The average results, plotted as deviation of the fractional perfusion per tissue weight ($\dot{q}$) in the four subunits of the lung, reveal no major inhomogeneity in the regional distribution of pulmonary blood flow. Since at rest O_2 saturation of arterial blood has been measured at about 90% (Kawashiro and Scheid, 1975), the nearly homogeneous blood flow is presumably matched by a similarly homogeneous ventilation of all parabronchi, which contrasts with the hypothesis of Duncker (1974) that mainly only neopulmo is ventilated for gas exchange at rest.

It is evident from Figure 1 that the decrease in perfusion along the paleopulmonic parabronchi coincides with the direction in which respired gas flows through the parabronchial tubes (Piiper and Scheid, 1973). As the air passes through the parabronchi from MD towards MV, P_{O_2}

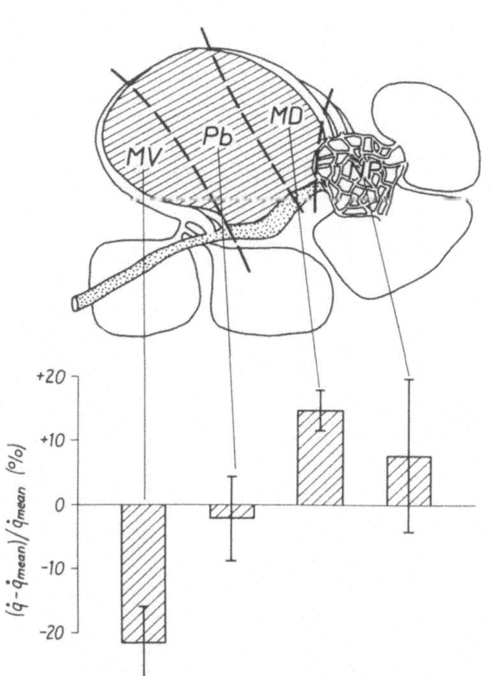

Fig. 1. Mean blood flow pattern (± SD) obtained in unanesthetized, spontaneously breathing ducks. The schema represents the avian lung with neopulmo, *NP*, and with three regions discerned within the paleopulmo: parts adjoining mediodorsal (*MD*) or medioventral secondary bronchi (*MV*) and parts in middle region of the parabronchi (*Pb*)

decreases and P_{CO_2} increases, and these profiles of both respired gases are characteristic of the cross-current arrangement of gas exchange in avian parabronchial lungs (Meyer et al., 1976). If the hypoxia exerts a vasoconstriction in avian lungs, as in the mammalian pulmonary vessels (cf. Fishman, 1976), the reduction in perfusion at the MV end of the parabronchi could be a result of the low O_2 (and possibly the high CO_2) tension in the gas at this site. The effects of respired gas on regional lung perfusion were tested in the next series of experiments.

b) Series II. Effects of Respired Gases on Regional Lung Perfusion

These experiments were performed under general anesthesia and with artificial ventilation. Wide lumen catheters were introduced into the caudal thoracic airsacs of both body sides and blocking balloons were advanced through the trachea to block the main bronchi on both sides between the origins of the medioventral and mediodorsal secondary bronchi. In this preparation, continuous streams of gas could be insufflated into the airsacs which passed over the lung exchange tissue and exited via a tracheostomy. Thus O_2 or CO_2 concentrations could be varied in one lung (test lung) while the other was held at control levels (control lung). Microspheres were injected after attainment of steady state conditions with the gas ventilating the test lung at high or low O_2 or CO_2 concentrations.

Figure 2 shows mean results expressed as differences in the perfusion of the test lung between control condition (both lungs ventilated with identical gas mixtures which were chosen to result in normal arterial blood gases) and test conditions. It is apparent that hypoxia results in a significant reduction in blood flow whereas no significant changes were obtained from hyperoxia or from CO_2.

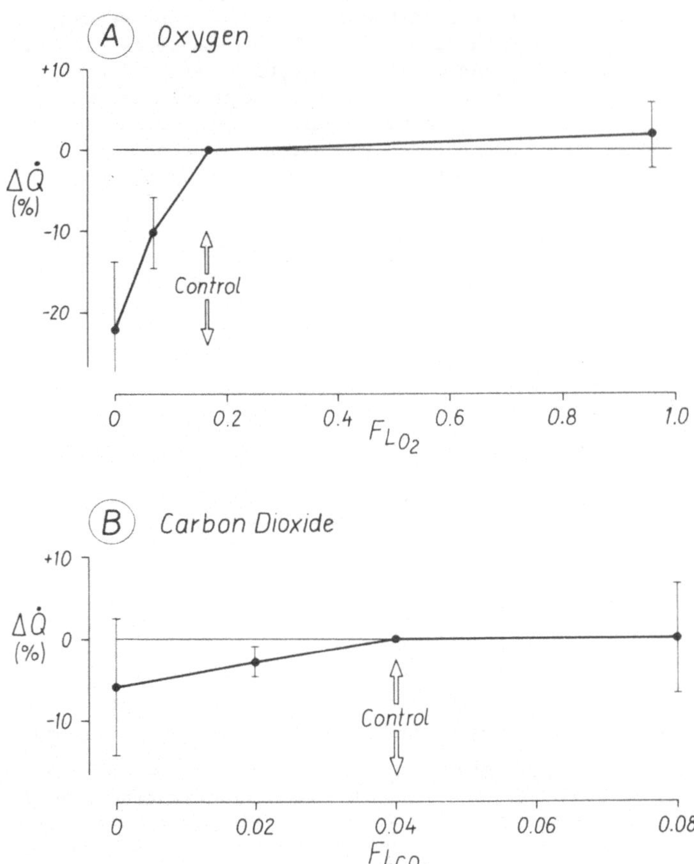

Fig. 2. Effects of respiratory gases on pulmonary perfusion. *A*, O_2; *B*, CO_2. $\Delta\dot{Q}$, percentage change in perfusion of test lung. F_L, fractional concentration of O_2 and CO_2 in test lung gas (values for control lung marked by double arrows)

These results provide thus a potential mechanism to explain the blood flow pattern observed in the resting animal (Fig. 1). The next series of experiments was performed to test if this blood flow pattern is sensitive to changes in the pattern of lung gas P_{O_2}.

c) Series III. Reversal of Parabronchial Gas Flow

The same preparation was used as in Series II. However, identical gas mixtures were administered to both lungs and the blood flow pattern was studied when ventilation was from the caudal thoracic airsacs towards the trachea or in the reversed direction, from the trachea to the caudal thoracic airsacs. Thereby the P_{O_2} profile along the parabronchi was reversed, and either the MD region or the MV region could be made hypoxic.

The results of this series showed substantial variability. On the average the blood flow patterns observed along the parabronchi were only slightly affected by the reversal of gas flow. However, the direction of this change was as expected, viz. when gas flow was from MD towards MV, the MD regions received somewhat more, and the MV region somewhat less perfusion than with the opposite parabronchial gas flow. However, most of the inhomogeneity in the blood flow pattern could not be affected by changes in parabronchial gas flow and therefore appeared to be unaffected by acute changes in lung gas composition. This inhomogeneity must therefore be produced by other mechanisms, for example, by the inhomogeneity in vascular supply to the various lung regions, for which there is, however, no obvious anatomical basis (Abdalla and King, 1976).

In summary to Series I-III, the blood supply to the exchange tissue is not homogeneous along the parabronchus and this pattern can essentially not be affected by changing parabronchial P_{O_2}. It is of interest to investigate in which sense and to what degree an unequal distribution of perfusion along the parabronchial length may affect the gas exchange efficiency of the parabronchus.

d) Model Analysis. Effects of Inhomogeneous Parabronchial Perfusion on Gas Exchange

Figure 3A depicts a model parabronchus with its blood supply. For the calculations, this parabronchus was subdivided along its length into

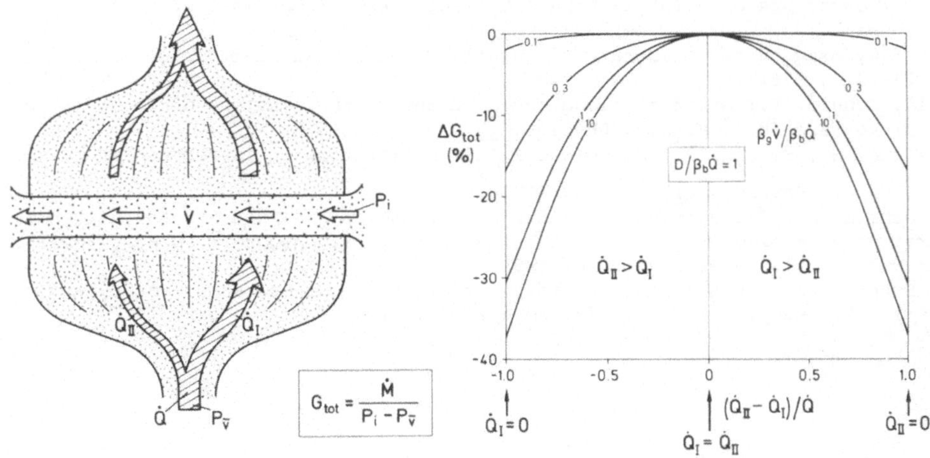

Fig. 3. Model calculations of the effects of inhomogeneous blood flow distribution on total gas exchange conductance (G_{tot}). For details, see text

two parts (of equal diffusing capacity, D) and the relative blood flow to the first ($\dot{Q}_I$) and the second half ($\dot{Q}_{II}$) was varied. The gas exchange efficiency was expressed as the total gas conductance, G_{tot}, which can be visualized as a measure of gas exchange efficiency (Piiper and Scheid, 1975).

Figure 3B is a plot of the change in G_{tot} with varying relative perfusion of both parabronchial halves. Calculations were made for various values of the total ventilation/perfusion ratio (expressed as $\beta g\dot{V}/\beta b\dot{Q}$; cf. Piiper and Scheid, 1975) and for a nearly normal ratio of diffusing capacity to blood flow (expressed as $D/\beta b\dot{Q}$; cf. Piiper and Scheid, 1975).

It is apparent that G_{tot} is decreased for any degree of blood flow inhomogeneity and for any value of the ventilation/perfusion ratio. Thus shunting blood from the gas outflow end, where P_{O_2} is low and P_{CO_2} high, to initial parts of the parabronchus is detrimental to gas exchange. The symmetry in the curves of Figure 3B shows moreover that the impairment of gas exchange resulting from a given blood flow pattern is independent of the direction of parabronchial gas flow, that is, it is independent of whether the gas inflow end or the gas outflow end receives relatively more blood flow.

It can be assessed from Figure 3B that the blood flow inhomogeneity observed in the spontaneously breathing ducks (Fig. 1) corresponds to an only slight impairment of the total gas exchange conductance, G_{tot}, of less than 2%. It may therefore be concluded that the observed pattern of blood flow is functionally not significantly different from a homogeneous blood flow distribution in the unanesthetized spontaneously breathing duck.

References

Abdalla, M.A., King, A.S.: Pulmonary arteriovenous anastomoses in the avian lung: Do they exist? Respir. Physiol. 27, 187-191 (1976)

Duncker, H.-R.: Structure of avian lungs. Respir. Physiol. 14, 44-63 (1972)

Duncker, H.-R.: Structure of avian respiratory tract. Respir. Physiol. 22, 1-19 (1974)

Fishman, A.P.: Hypoxia on the pulmonary circulation. How and where it acts. Circ. Res. 38, 221-231 (1976)

Kawashiro, T., Scheid, P.: Arterial blood gases in undisturbed resting birds: measurements in chicken and duck. Respir. Physiol. 23, 337-342 (1975)

Meyer, M., Worth, H., Scheid, P.: Gas-blood P_{CO_2} equilibration in parabronchial lungs of birds. J. Appl. Physiol. 41, 302-309 (1976)

Piiper, J., Scheid, P.: Gas exchange in avian lungs: models and experimental evidence. In: Comparative Physiology. Bolis, L., Schmidt-Nielsen, K., Maddrell, S.H.P. (eds.). Amsterdam: North Holland, 1973, pp. 161-185

Piiper, J., Scheid, P.: Gas transport efficacy of gills, lungs and skin: theory and experimental data. Respir. Physiol. 23, 209-221 (1975)

Session III

The Avian Respiratory System: Structure and Function of Lungs

Chairman: G. M. Hughes

The Structure of the Intrapulmonary Vasculature of the Domestic Fowl

A. S. King, D. Zoe King, and M. A. Abdalla

Summary

The walls of the small veins and arteries in the exchange tissue of the paleopulmo of the domestic fowl (*G. domesticus*) have been investigated by light and electron microscopy. The intra- and interparabronchial veins had one or two layers of thin smooth muscle cells devoid of nerve fibres. These vessels would have little capacity for regulating blood flow. The terminal arterioles possessed a single layer of non-innervated stout smooth muscle cells. The small interparabronchial arteries had two to four stout smooth muscle cells, innervated regularly but sparsely by noradrenergic, and perhaps also by cholinergic axons, in the adventitia. These arteries appeared capable of the local regulation of the microcirculation. One muscular valve was found in the opening of a small interparabronchial vein. The valve comprised two strong muscular components guarding a slit-like opening. The smooth muscle fibres in each component were arranged radially (dilator), with circular fibres (constrictor) at the base. This valve could have regulated the venous drainage from the vascular territory of about 5% to 10% of the length of two or three adjacent parabronchi of the longest type.

1. Introduction

The arrangement of the main branches of the pulmonary artery and vein in several species of domestic bird was first established by Radu and Radu (1971). The general organization of the smaller branches within the exchange tissue was the subject of conflicting analyses by Akester (1971, Fig. 26) and Duncker (1971, 1972, 1974), which were largely resolved for the domestic fowl by Abdalla and King (1975) and for the duck and pigeon by West et al. (1977).

Various attempts have been made to find an anatomical basis for the regulation of the pulmonary microcirculation. It was suggested by Duncker (1971, p. 146, 1972) that, during low ventilation of the parabronchus in a resting bird, a large, variable arteriovenous shunt should occur in the last part of each parabronchus where venous blood has come into equilibrium with end-expiratory gas tensions so that no further gaseous exchanges are possible. Anatomical and experimental studies by Abdalla and King (1975, 1976) have indicated that arteriovenous anastomoses are absent from the avian lung. Later, Duncker (1974) suggested that the sheath of circular smooth muscle surrounding the wall of the atrial veins, which he regarded as "very well-developed", might

University of Liverpool, Department of Veterinary Anatomy, P.O. Box 147, Liverpool L69 3BX, Great Britain, and University of Khartoum, Faculty of Veterinary Science, Department of Anatomy, P.O. Box 32, Khartoum North, Sudan.

react directly to the changing gas tensions in the adjacent parabronchial lumen; contraction of these muscle cells in response to end-expiratory gas in the lumen could increase the resistance to flow within the atrial veins, thereby allowing autoregulation of perfusion along the length of the parabronchus. Arteriolar sphincters, such as those tentatively identified histologically by Abdalla and King (1975) in *Gallus domesticus* and with the scanning electron microscope in *Anas platyrhynchos* and *Columba livia* by West et al. (1977), may offer an alternative mechanism for regulating the pulmonary microcirculation. The existence of some such mechanism for regulating parabronchial blood flow seems consistent with experiments with radioactive microspheres, which showed a decrease in blood flow per unit tissue volume along the parabronchus in the direction of gas flow in the conscious resting duck especially during hypoxia (Holle et al., this vol., Chap. 14). The innervation of the avian pulmonary vaculature has been studied by fluorescence techniques and histochemistry. Adrenergic fibres were observed by Bennett (1971), who described the pulmonary artery as sparsely innervated over "its whole length"; the larger branches of the pulmonary vein within the lung were very sparsely innervated, but the smaller branches appeared to be altogether devoid of an adrenergic innervation. Noradrenergic and acetylcholinesterase-positive fibres were reported by Akester (1971, p. 808) as occurring in both the pulmonary artery and vein in more or less equal numbers, but not profusely. The nerve fibres in the pulmonary artery were distributed "throughout its entire length", mainly in the adventitia; the number of fibres in the pulmonary vein diminished as the vessels became smaller. Akester (1971, p. 830) could find no information in the literature about the ultrastructure of the pulmonary vasculature of birds.

This paper reports an investigation, with the light and electron microscopes, of the walls of the small veins and arteries in the exchange tissue of the lung of the domestic fowl, to assess the structural evidence for the regulation of the microcirculation in the paleopulmo.

2. Materials and Methods

Two adult female domestic fowls (*Gallus domesticus*, Hubbards Golden Comet hybrid) were killed by an intravenous injection of barbiturate. Glutaraldehyde in sodium-cacodylate buffer at room temperature was injected into the trachea by gravity at a height of 30 cm. Small pieces were dissected from the right lung between the first and second costal sulci and between the third and fourth costal sulci, routinely processed, and embedded in resin. Thick and thin ultratome sections were examined with the light and electron microscope. The terminology for the vasculature is that of the *Nomina Anatomica Avium* (Baumel, 1978).

3. Results

a) The Venous Wall

The walls of the atrial veins, which adjoin the lumen of the atria or infundibula (Fig. 6), usually possessed only one or two layers of smooth muscle cells (Fig. 1); in places the cells of the single layer were intermittent, leaving gaps with no muscle cells at all (Fig. 2). Postcapillary venules, about 10-20 μm in diameter, had no smooth muscle cells (Fig. 3). The intraparabronchial veins (Fig. 4) possessed one or

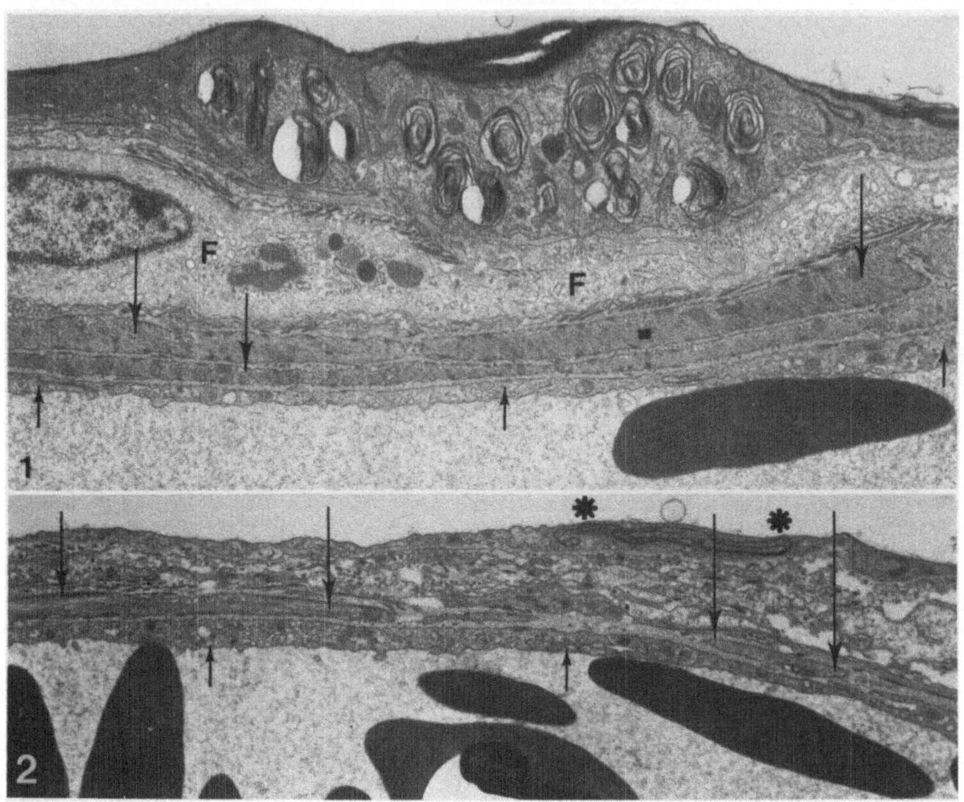

Fig. 1. The venous wall. The wall of an atrial vein, showing two layers of smooth muscle cells (*long arrows*) which are similar in thickness to the endothelial cells (*short arrows*). The *upper surface*, facing the atrial airway, is lined by an atrial epithelial cell containing osmiophilic bodies. Between the smooth muscle cells and the atrial epithelial cell is a fibrocyte *F*, ×7900

Fig. 2. The venous wall. The wall of an atrial vein, showing a single layer of intermittent smooth muscle cells (*long arrows*) which are thinner than the endothelial lining cells (*short arrows*). The upper surface facing the atrial airway is lined by an atrial or infundibular epithelial cell, recognisable by the *dark folded line* (*) which is probably trilaminar substance. Other constituents of the wall, lying between the smooth muscle cells and the atrial epithelial cell, are slender fibrocyte processes and collagen, ×6600

or two layers of smooth muscle cells (Fig. 5). One to two layers of smooth muscle cells (Figs. 11 and 12) also occurred in the walls of interparabronchial veins such as the one labelled *V4* in Figure 6. In all of these venous structures the smooth muscle cells appeared much thinner than those in the terminal arterioles (Fig. 13) and small muscular arteries (Fig. 14). No nerve fibres were seen in the wall of any vein. Ordinary membranous semilunar pocket valves were absent.

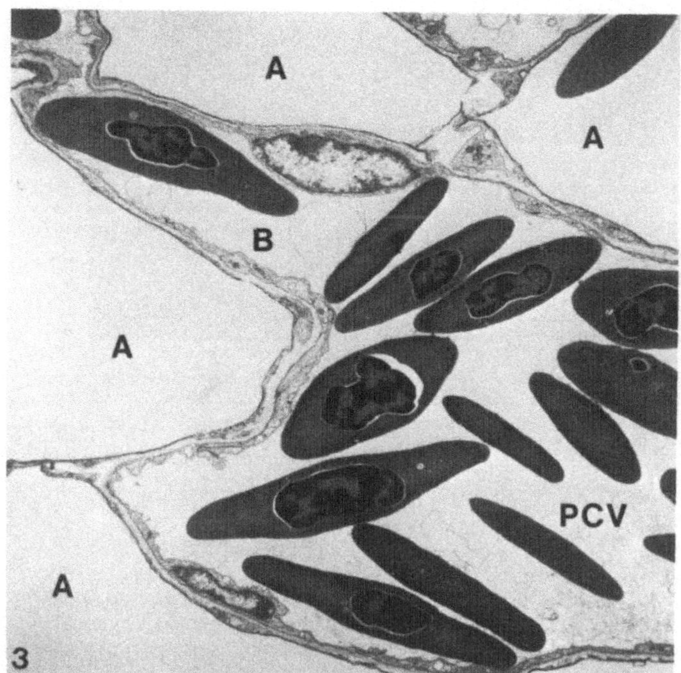

Fig. 3. The venous wall. A blood capillary (*B*) is draining into an intraparabronchial postcapillary venule (*PCV*) with a diameter of about 17 µm. The wall of the venule consists only of endothelial cells and basal lamina. *A* air capillaries. The air capillary in the *top right corner* contains an extraneous red blood cell, ×3600

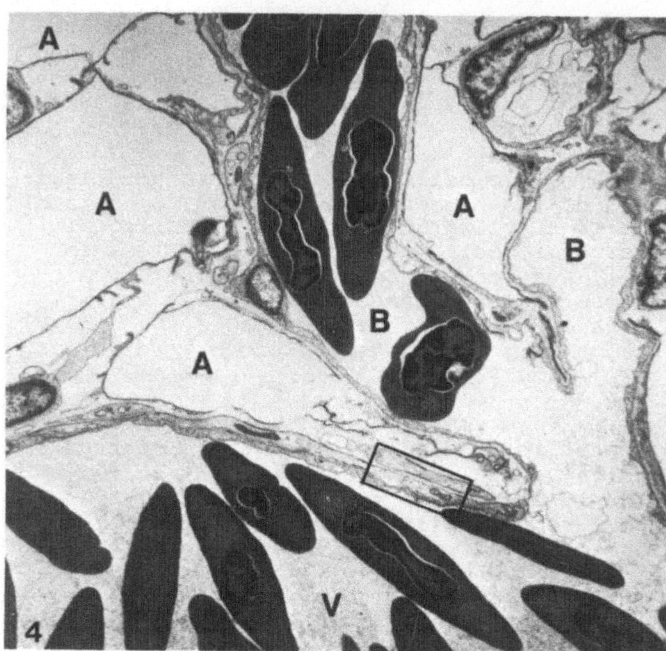

Fig. 4. The venous wall. Blood capillaries (*B*) are emptying into an intraparabronchial vein (*V*). The wall of the vein is reinforced by one, and in places two, smooth muscle cells. The rectangular area is enlarged in Figure 5 to show these muscle cells. *A* air capillaries. Electron micrograph, ×3600

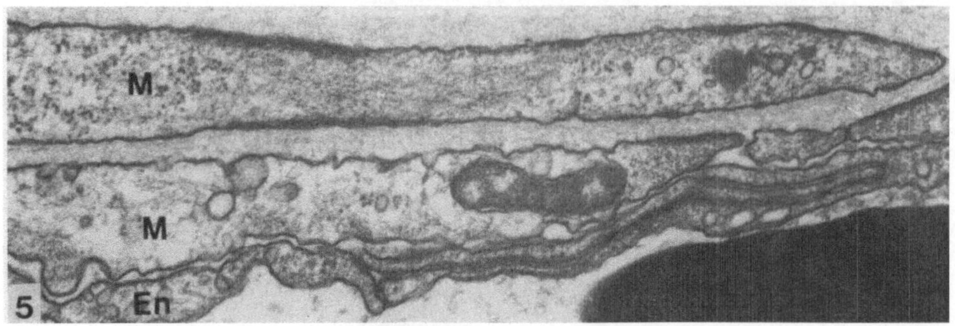

Fig. 5. The venous wall. An enlargement of the rectangular area of the wall of the vein in Figure 4, to show two layers of slender smooth muscle cells (*M*) in this part of the wall. *En* endothelial cell, ×35,625

b) Muscular Venous Valve

A muscular venous valve was discovered at the opening of a small interparabronchial vein (*V3*, Figs. 6 and 7) into a larger interparabronchial vein (*V4*, Figs. 6 and 7). The valve was observed with the light microscope in four adjacent resin sections, representing a total depth of about 4-6 µm. The tissue for this block was obtained near the costal surface of the lung, between the third and fourth costal sulci.

In all four sections the valve consisted of two tapered muscular processes (Figs. 8, 9, and 10). The longer process was about 26 µm wide at the base, and about 65 µm in length; the smaller one measured about 26 µm and 45 µm. The two processes were separated at their tips by a gap about 4 µm wide. The diameter of the whole valve, from base to base of each process was about 120 µm. The bulk of each process was formed by five or six layers of smooth muscle cells, orientated longitudinally parallel with the long axis of the process itself (Figs. 8, 9, and 10). At the base of each process there were two to three additional layers of smooth muscle cells; these cells were cut more or less in cross-section (Figs. 9 and 11), and were therefore approximately circular in relation to the process. Thin nucleated endothelial cells were visible on the outer surface of each process (Figs. 8, 9, and 10). The adjoining venous wall was thin, containing only one or two layers of smooth muscle cells (Figs. 11 and 12).

If this valve was a diaphragm, with a small central circular aperture like that of the human eye, the first or last of the series of four sagittal sections through the valve could have shown an intact membrane passing right across the lumen of the vein. Alternatively if the valve had a slit-like aperture, resembling the pupil of a cat's eye, a sequence of sagittal sections at right angles to the long diameter of the slit would have resembled each other quite closely. A slit, between two muscular components, seems the more likely arrangement. Since the orientation of the longitudinal smooth muscle cells in each component was radial, their contraction could have opened the aperture: the circular muscle cells at the base could have constricted it.

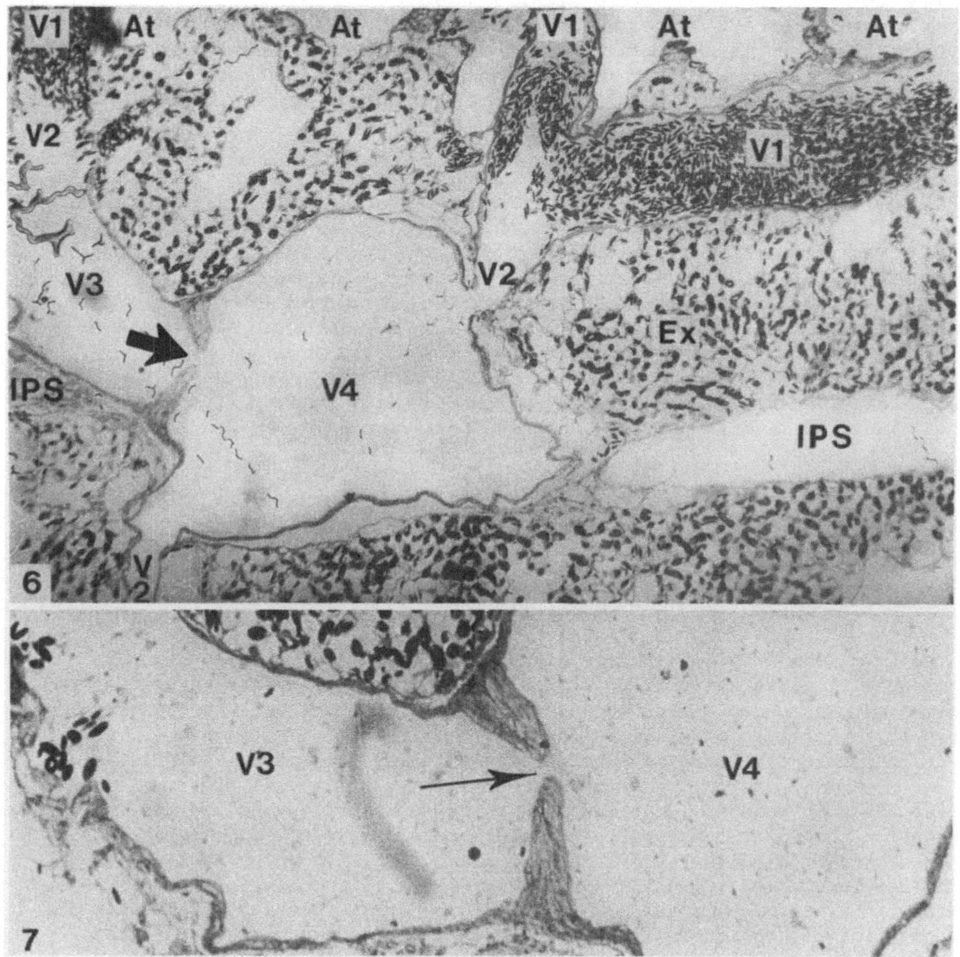

Fig. 6. Muscular venous valve. *Arrow* points to a muscular valve situated at the opening of a small interparabronchial vein (*V3*) into a larger interparabronchial vein (*V4*). *V1* Atrial veins; *V2* intraparabronchial veins; *At* atria; *Ex* exchange tissue; *IPS* interparabronchial septum. Resin section, methylene blue stain, ×152

Fig. 7. Muscular venous valve. A section different from, but close to, that of Figure 6, showing the two tapered components of the valve and its aperture (*arrow*). The width of each component of the valve is much greater than the thickness of the wall of the smaller interparabronchial vein (*V3*) which opens into the larger interparabronchial vein (*V4*). Resin section, methylene blue stain, ×305

c) The Arterial Wall

Blood capillaries arose in the exchange tissue from small terminal intraparabronchial arterioles about 10 μm in diameter (Fig. 13). Such arterioles possessed a single layer of stout smooth muscle cells, without axons.

Small interparabronchial arteries had two to three (Fig. 14) and occasionally four layers of stout smooth muscle cells. There were fairly numerous zones of close apposition between the muscle cells, apparently with direct contact at some points (Fig. 15). Nerve fibres occurred regularly but not commonly in the adventitia, just outside the outer elastic lamina, being loosely or only partly invested by Schwann cells (Figs. 14, 15, and 16). Some axons showed beaded varicosities (Fig. 15). Two examples were found of an axonal enlargement associated with the projection of a smooth muscle cell through the outer elastic lamina, as in Figure 16, the axon here being bare of Schwann cell with a gap between cell and axon of about 100 nm. Most of the axonal enlargements contained predominantly small dense-cored granular vesicles about 50 nm in diameter, and a number of larger dense-cored vesicles up to 100 nm, as in Figure 16. These were interpreted as noradrenergic fibres. A few axonal enlargements appeared to contain numerous small clear vesicles and were regarded as possibly cholinergic.

No arteriolar sphincters of the type tentatively identified by Abdalla and King (1975) and West et al. (1977) were seen.

Fig. 8. The lower component of the muscular valve in Figure 7, and the tip of the upper component. A different section from Figures 6 and 7. The structure consists of elongated smooth muscle cells, the nuclei of some of these cells being visible (*long arrow*). There is an outer lining of endothelial cells (*short arrows*), which is thin except where the cells are nucleated. Resin section, methylene blue, ×1000

Fig. 9. The upper component of the muscular valve in Figure 7, and the tip (*bottom left corner*) of the lower component. The structure consists of smooth muscle cells most of which are longitudinally orientated, the typical elongated nuclei of some of them being visible (*medium-sized arrows, lower side*). The base of the structure consists of circular smooth muscle cells cut in transverse section, the nuclei of some being visible (*large arrow, upper side*). The same section as Figure 8. *Small arrows*, endothelial cell cytoplasm. Resin section, methylene blue, ×1780

Fig. 10. The lower component of the muscular valve in Figure 7. Elongated smooth muscle cells (*large arrows*) form the bulk of the structure. *Small arrows*, nuclei of endothelial cells. This came from a fourth section, different from Figures 6, 7, and 8. Resin section, methylene blue, ×1510

Fig. 11. The base of the component in Figure 9, but a different section from Figure 9. *Lower large arrow*, the middle cell of three smooth muscle cells which are cut in transverse section, and therefore were approximately circular in relation to the projection. Similar circular cells are present across the whole of the base. *Upper large arrow*, a smooth muscle cell contributing to a single muscle layer in the wall of the smaller interparabronchial vein (*V3* in Figs. 6 and 7). *Small arrow*, nucleus of an endothelial cell. Resin section, methylene blue, ×1750

Fig. 12. The wall of the larger interparabronchial vein (*V4*, Figs. 6 and 7), near the valve. This was not the same section as Figure 11, but it continues the wall of the same vein; *small arrow*, the same nucleated endothelial cell as the small arrow in Figure 11. The wall of this vein probably had two smooth muscle layers; *arrow head*, a nucleated muscle cell in transverse section. Resin section, methylene blue, ×1520

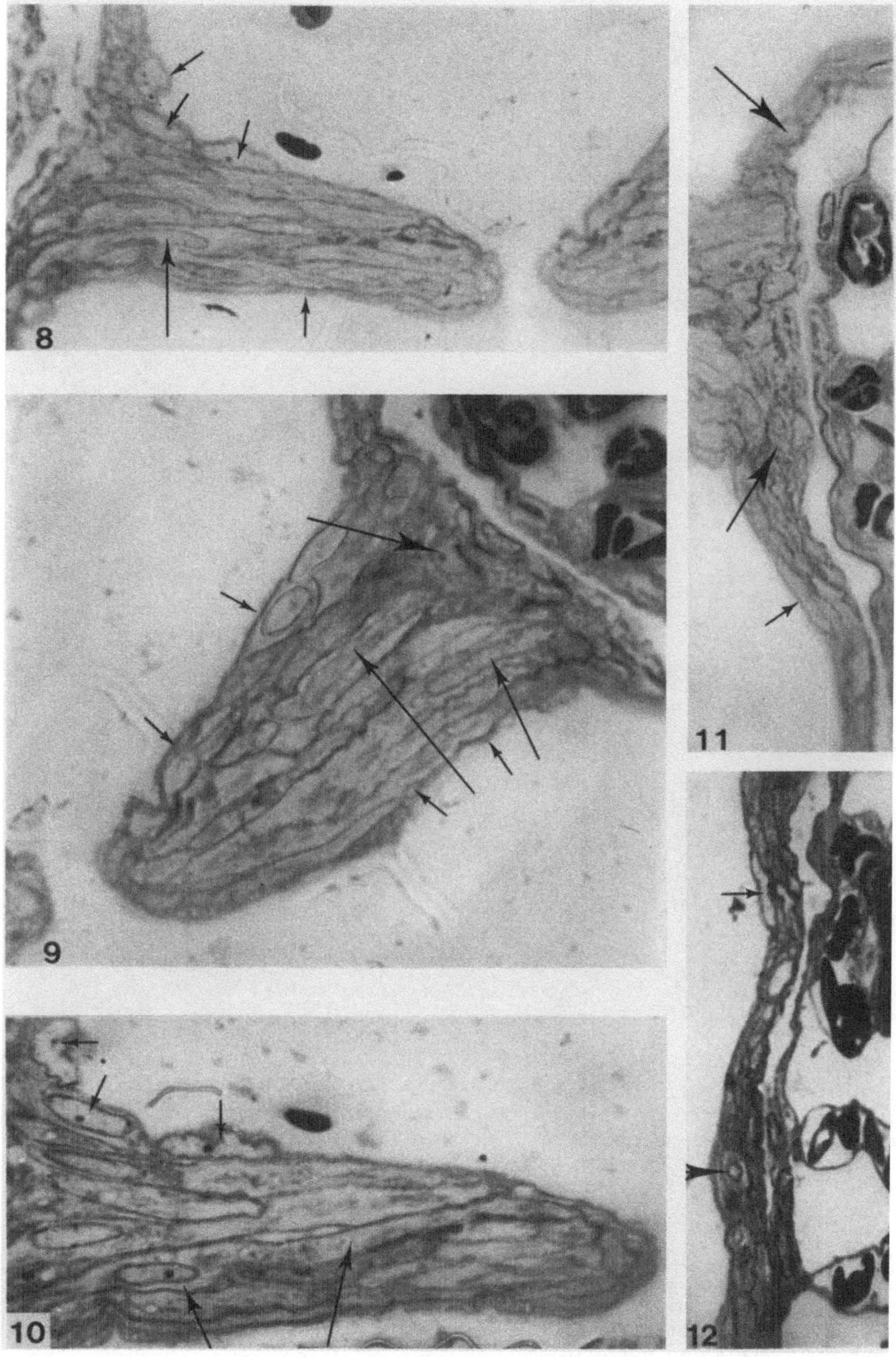
8
9
10
11
12

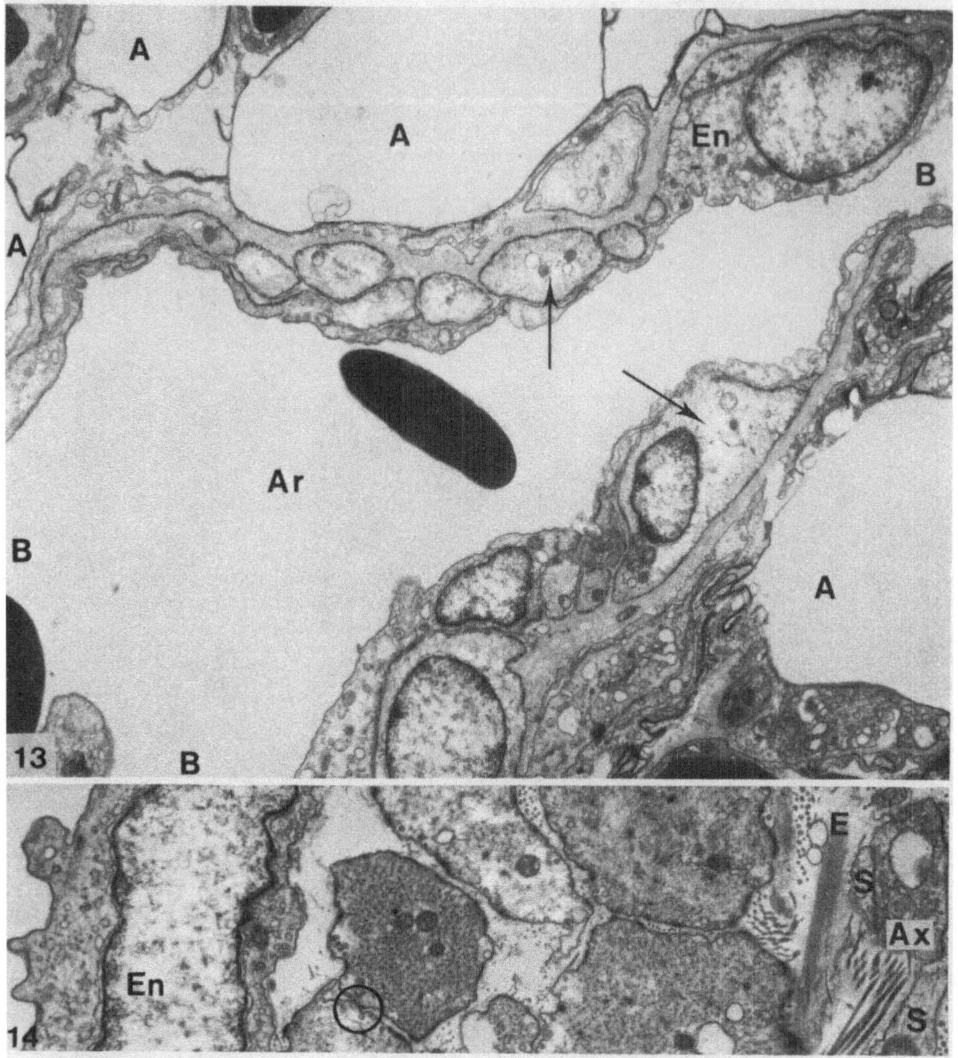

Fig. 13. The arterial wall. An intraparabronchial terminal arteriole (*Ar*), about 10 µm in diameter, is forming three blood capillaries (*B*). The wall of the arteriole is formed by endothelial cells (*En*) and a single layer of smooth muscle cells (*arrows*) most of which are much wider than the adjacent endothelial cells. Elastic elements have almost completely disappeared. No axons are present. *A*, air capillaries, ×4920

Fig. 14. The arterial wall. A segment of the wall of a small muscular artery, of the interparabronchial type. Beneath the endothelial lining cell (*En*) are two to three layers of large smooth muscle cells, with (*circled*) a zone of close apposition. Between the media and adventitia, part of the outer elastic lamina (*E*) is visible. In the adventitia there is an unmyelinated axon (*Ax*) partly enclosed by a Schwann cell (*S*), but bare at one point where it faces the media, ×10,700

4. Discussion

Our histological and ultrastructural observations on two birds (*Gallus domesticus*) do not support the suggestion that the atrial veins regulate parabronchial blood flow. The walls of these capacious veins possessed smooth muscle cells, but there was sometimes only one incomplete layer and two layers were the most that were seen; moreover the muscle cells were thin. Intra- and interparabronchial veins (up to the size shown by *V4* in Fig. 6) had only one or two layers of attenuated smooth muscle cells. No axons were seen, confirming Bennett's conclusion from fluorescence studies that the smaller branches of the pulmonary vein have no adrenergic innervation. None of these venous walls seemed to be adequately constructed for regulating the venous drainage of the exchange tissue by increasing the resistance to blood flow.

No evidence for arteriovenous anastomoses has been found in the domestic fowl by histological methods (Abdalla and King, 1975), or by the intravenous injection of spores and non-radioactive microspheres about 15-35 μm in diameter (Abdalla and King, 1976). This negative result has been confirmed in our laboratory by Parry and Yates (1978) using radioactive microspheres about 13 μm in diameter.

Although the presence of arteriolar sphincters as tentatively identified by Abdalla and King (1975) and West et al. (1977) could not be confirmed, the walls of the intraparabronchial arterioles and small interparabronchial arteries contained stout smooth muscle cells, arranged in a single layer in the arterioles and in two or three layers in the small arteries. Axons were regularly though not commonly found in the adventitia of the small arteries; no axons were seen in the walls of terminal arterioles.

The features of the axonal enlargement in Figure 16 were characteristic of the noradrenergic fibres which commonly innervate vascular smooth muscle in mammals (Burnstock, 1970, pp. 56-64). Some of the axons in the adventitia may have been cholinergic, but convincing examples were scarce. The absence of axons in the terminal arterioles contrasts with the profuse innervation of mammalian terminal arterioles (Rhodin, 1967). Zones of close apposition suitable for the electrotonic spread of current between the muscle cells (Burnstock, 1970, pp. 16, 54) were apparently present in the small interparabronchial arteries (Fig. 15). In general, therefore, these arteries appeared capable of the local regulation of blood flow across the parabronchial wall.

The muscular valve that was discovered in the opening of a small interparabronchial vein appeared to consist of a slit-like aperture guarded by two strong muscular components. The radial smooth muscle cells of each component could have widened the aperture of the slit: the circular smooth muscle cells could have constricted it. We are not aware of any similar muscular valve in any other blood vessel. It was found in the paleopulmo, between parabronchi which would have connected approximately the third or fourth mediodorsal secondary bronchus to the second or third medioventral secondary bronchus (see King, 1975, Fig. 64-23); it would have been about one third of the way along these parabronchi, in the direction of normal air flow.

Careful consideration is necessary of any histological artefact which could have created this structure. Possibilities are a tangential section through a bend in the wall of the vein, or an invagination of the wall of the vein into the lumen. The first of these possibilities can be eliminated by the following observations: 1. the wall of the vein on either side of the structure was so thin that four successive tangential sections would be unlikely to remain virtually unchanged; 2. endothelial cells should have been visible passing beneath the smooth muscle cells; 3. a tangential section would not show two distinct groups of smooth muscle cells, one longitudinal and the other circular.

An invagination of the venous wall can also be ruled out. Each projection consisted of a homogeneous muscle mass, five or six times as thick as the adjacent venous wall: an invagination of the thin muscular wall of the vein must have carried with it an axial core of connective tissue in order to produce such a wide structure.

We are confident that this was a muscular valve. Since it lay at the end of a small interparabronchial vein, it could have regulated the territory drained by several intraparabronchial veins, each draining several atrial veins; moreover two or three parabronchi could have been involved (Abdalla and King, 1975). We estimate from our venous casts that the length of the areas drained from these parabronchi could have been 2 to 3 mm, i.e., about 5-10% of the longest (about 30 mm) parabronchi in this species (King, 1975, p. 1905). But since only one valve has so far been found it remains uncertain whether or not such structures play a significant role in the physiology of the avian pulmonary circulation as a whole: this question can only be answered by finding out whether they are a regular feature of the avian lung.

References

Abdalla, M.A., King, A.S.: The functional anatomy of the pulmonary circulation of the domestic fowl. Respir. Physiol. 23, 267-290 (1975)

Abdalla, M.A., King, A.S.: Pulmonary arteriovenous anastomoses in the avian lung: do they exist? Respir. Physiol. 27, 187-191 (1976)

Akester, A.R.: The blood vascular system. In: Physiology and Biochemistry of the Domestic Fowl, Vol. II. Bell, D.J., Freeman, B.M. (eds.). London: Academic Press, 1971

Fig. 15. Arterial innervation. A further section of the interparabronchial artery in Figures 14 and 16, at the junction of the adventitia and media. An unmyelinated axon (*Ax*) is running in the adventitia, parallel with the outer elastic lamina (*E*). The axon, which is loosely invested by Schwann cells (*S*) but bare at the *arrow*, shows beaded varicosities the largest being about 0.6 µm in its narrow diameter. The varicosities contain a few small dense=cored granular vesicles about 50 nm in diameter, and some larger ones (about 80 nm). The axon is probably noradrenergic. Each muscle cell is surrounded by a basal lamina, but adjacent cells break through the basal lamina and establish zones of close apposition (*circles*); in some zones, e.g., the one in the centre of the field, the cell membranes appear at some points to be touching, ×24,000 ▶

Fig. 16. Arterial innervation. Another section of the interparabronchial artery in Figures 14 and 15, at the junction of the adventitia with the media. A terminal axonal enlargement, 1.3 µm in its narrow diameter, lies in the adventitia. The axon is largely bare of Schwann cell processes (*S*) on the side which faces the smooth muscle cells. A process of a smooth muscle cell has passed through the outer elastic lamina (*E*) and lies close to the axon, the narrowest gap between them being about 80-100 nm and the basal lamina of axon and muscle cell being partly fused. The axon contains numerous small dense-cored granular vesicles about 50 nm in diameter and some larger ones up to about 100 nm in diameter. The axon is interpreted as being noradrenergic, ×37,500

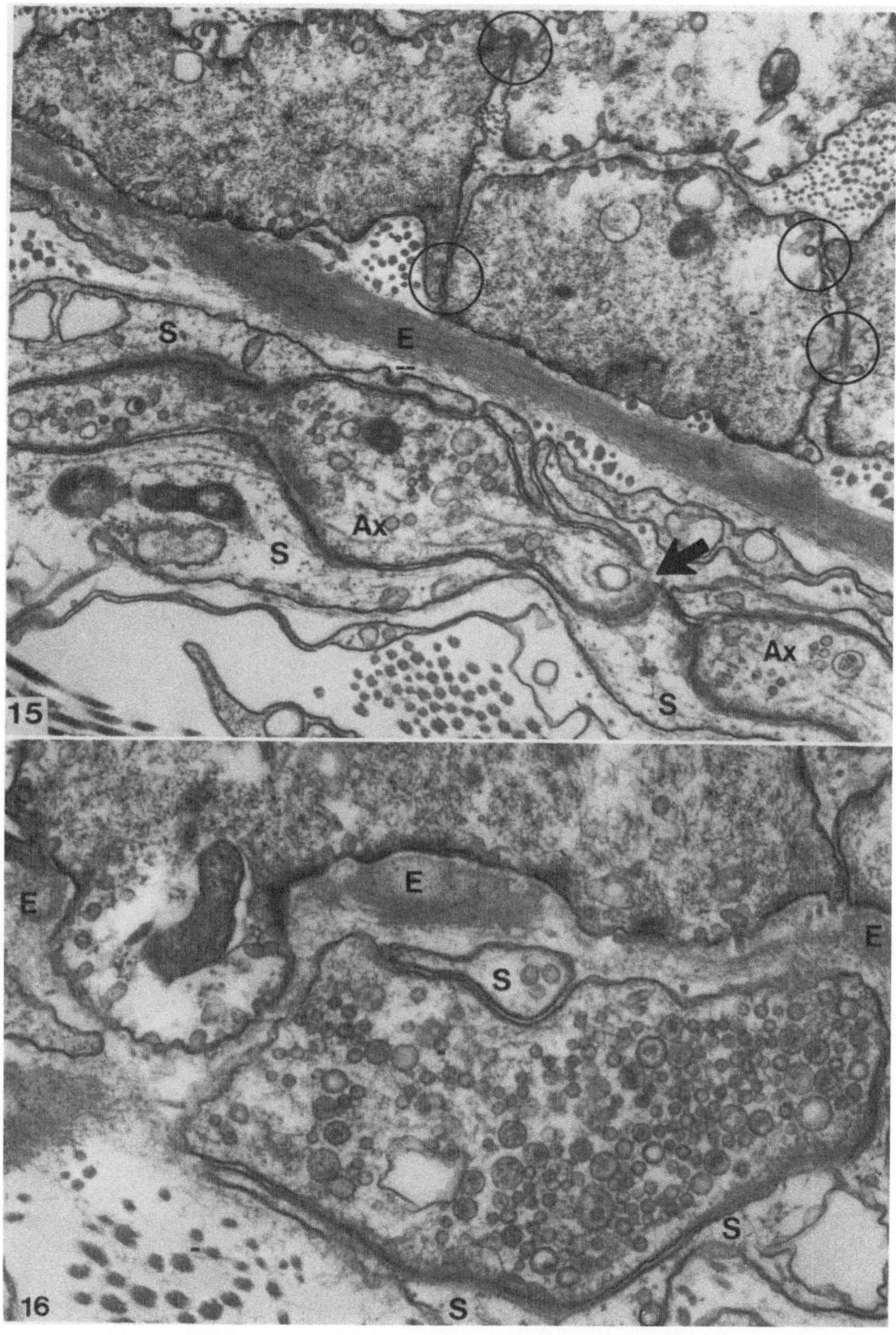
S
E
Ax
S
Ax
S
15
E
E
E
S
S
S
16

Baumel, J.J.: Systema cardiovasculare. In: Nomina Anatomica Avium. Baumel, J.J., King, A.S., Lucas, A.L. (eds.). London: Academic Press, 1978

Bennett, T.: The adrenergic innervation of the pulmonary vasculature, the lung and the thoracic aorta, and on the presence of aortic bodies in the domestic fowl (*Gallus gallus domesticus* L.). Z. Zellforsch. 114, 117-134 (1971)

Burnstock, G.: Structure of smooth muscle and its innervation. In: Smooth Muscle. Bulbring. E., Brading, A.F., Jones, A.W., Tomita, T. (eds.). London: Arnold, 1970

Duncker, H.-R.: The lung air sac system of birds. A contribution to the functional anatomy of the respiratory apparatus. Ergeb. Anat. Entwickl.-Gesch. 45, Heft 6, 1-171 (1971)

Duncker, H.-R.: Structure of avian lungs. Respir. Physiol. 14, 44-63 (1972)

Duncker, H.-R.: Structure of the avian respiratory tract. Respir. Physiol. 22, 1-19 (1974)

Holle, J.P., Heisler, N., Scheid, P.: Effects of O_2 and CO_2 on regional pulmonary blood flow in ducks. Pflüg. Arch. 365, R20 (1976)

King, A.S.: Aves respiratory system. In: Anatomy of Domestic Animals, Vol. 2. Getty, R., Sisson, S., Grossman, J.D. (eds.). Philadelphia: Saunders, 1975

Parry, K., Yates, M.S.: Observations with radioactive microspheres on vascular shunting across the avian pulmonary bed. J. Anat. Lond., in press (1978)

Radu, C., Radu, L.: Le dispositif vasculaire du poumon chez les oiseaux domestiques (coq, dindon, oie, canard). Rev. Méd. Vét. 122, 1219-1226 (1971)

Rhodin, J.A.G.: The ultrastructure of mammalian arterioles and precapillary sphincters. J. Ultrastr. Res. 18, 181-223 (1967)

West, N.H., Bamford, O.S., Jones, D.R.: A scanning electron microscope study of the microvasculature of the avian lung. Cell Tiss. Res. 176, 553-564 (1977)

A Model of the Capillary Zone of the Avian Tertiary Bronchus

J. Brackenbury and A. R. Akester

Summary

The capillary zone of the tertiary bronchus is interpreted spructurally as two identical cubic lattices, mutually interlocking, one representing the air capillaries, the other the blood capillaries. Using this model the specific area of capillary walls in the chicken is calculated to be 11 cm^2/g body weight. This figure is 33% less than that predicted by a parallel tube model. The lattice model also predicts a cross-current relationship between blood and air flow at every point within the network.

Transmission electron micrographs (E.M.s) (Fig. 1) and scanning E.M.s of corrosion casts (Fig. 2) suggest that the two sets of blood and air capillaries in the exchange zone of the tertiary bronchus are approximately equal in volume and that each forms a highly branched three-dimensional network or lattice interdigitating with the other. Blood capillaries originate from the terminal branches of the pulmonary artery, air capillaries from the atria of the tertiary bronchi. Each blood capillary is about 6 μ in diameter and anastomoses with adjacent blood capillaries to form a homogeneous three-dimensional lattice in which the distance between neighbouring anastomoses is no greater than the diameter of the capillary (Fig. 2). The air capillaries occupying the interstices between the blood capillaries thus form an identical lattice which is the three-dimensional mirror-image of the blood capillary network.

It is suggested that the smallest structural unit of such a double lattice approximates to a six-way junction corresponding to a point of anastomosis between six neighbouring capillary (air or blood) branches (Fig. 3). Assume for the moment that the outer surfaces of the walls of the unit in Figure 3 are coloured white and the inner surfaces are coloured black. By joining many such units together by their open ends (each unit being joined to six other units) a structure results consisting of two identical, independent through-continuous systems of channels one lined by white walls, representing air capillaries, the other lined by black walls, representing blood capillaries.

In order to calculate the area of the (double) walls separating the air and blood spaces, and thus the area available for gaseous exchange, it was assumed that the capillary zone of each tertiary bronchus was completely filled with capillaries. Each tertiary bronchus was regarded as a straight tube 2 cm in length, 1 mm in diameter, and with

Sub-Department of Veterinary Anatomy, University of Cambridge, Tennis Court Road, Cambridge, Great Britain.

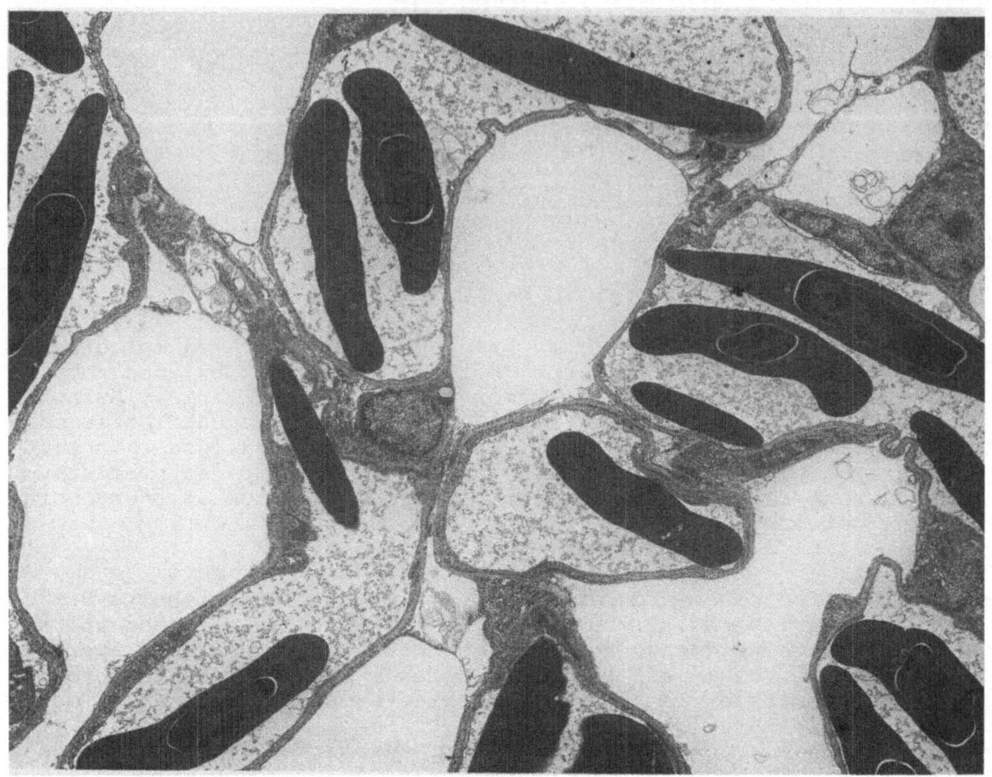

Fig. 1. Transmission electron micrograph of tertiary bronchial capillary zone of chicken showing air spaces and blood spaces occupied by blood cells

a 200 μ wide capillary zone containing 30-35 layers of capillaries. A chicken's lung contains about 500 such tertiary bronchi.

For the purposes of calculation, the model can be interpreted geometrically as consisting of three sets of parallel sheets intersecting mutually at right angles. One such sheet in the horizontal plane is shown in Figure 4. Each sheet consists of a chequer-board of double capillary wall (coloured black) and blood or air capillary lumen (coloured white). Within each of the three sets the sheets are parallel and separated by a distance of 6 μ. Half the total area of each sheet thus represents wall, and half the total area of all the sheets making up the zone of a single tertiary bronchus represents the exchange area of such a bronchus. This area is 25 cm^2. With a total of

---

Fig. 2. Scanning electron micrograph of latex cast of blood capillaries showing extensive three-dimensional branching. The spaces between the blood capillaries would be occupied in nature by air capillaries

Fig. 3. Lattice unit of model of capillary zone consisting of a junction of six capillary (blood or air) branches. In the model each such unit is joined to six similar units, at its six open ends

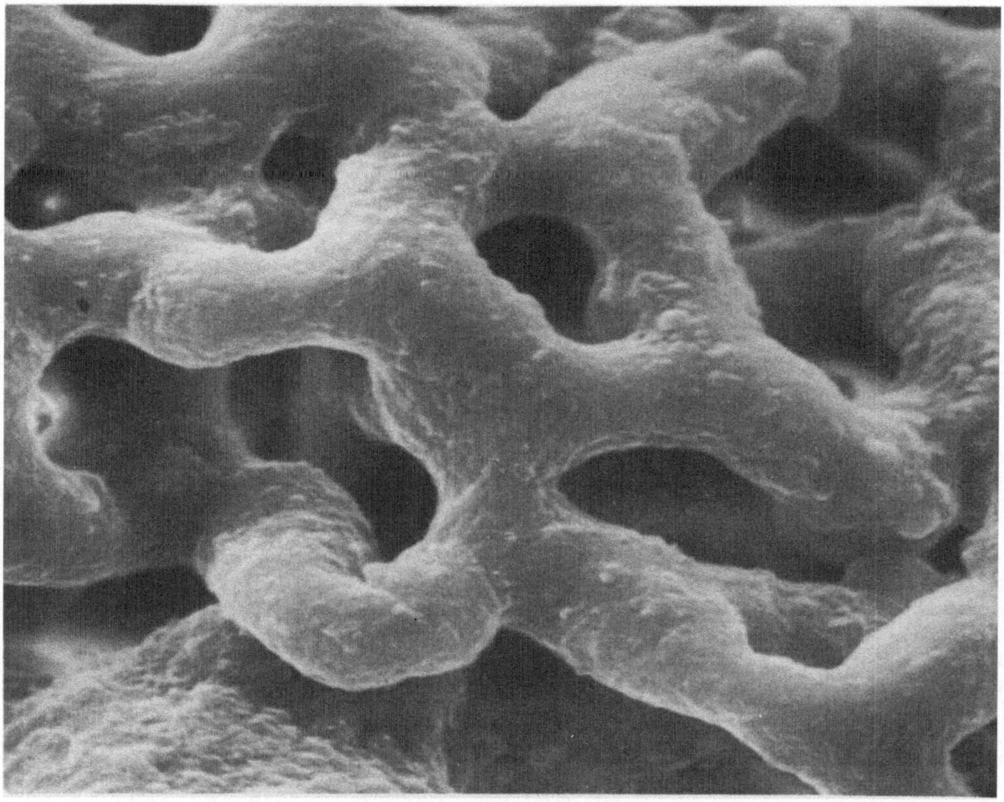

Fig. 2

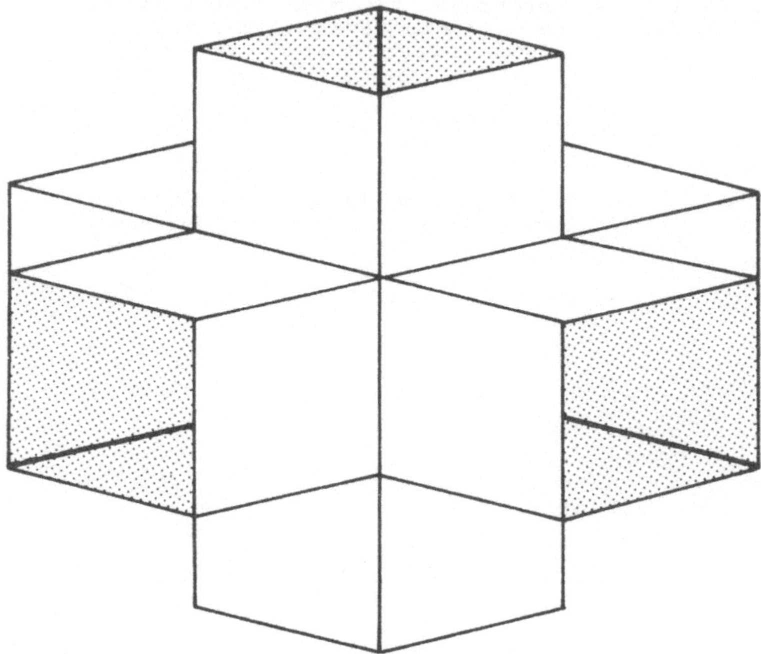

Fig. 3

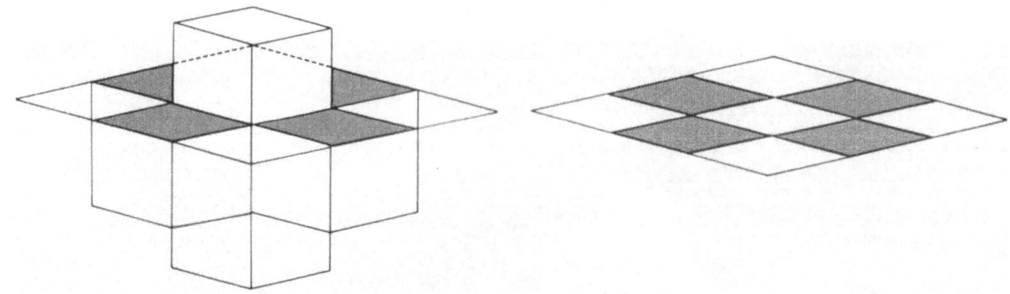

Fig. 4. In calculating the total area of walls in the capillary zone, the model is interpreted as a series of intersecting sheets divided into equal areas of wall (*black*) and capillary lumen (*white*). One part of one such sheet is shown here

1000 tertiary bronchi in both lungs, this makes a total area of 2.5 m^2. The average weight of the chickens on which we based our measurements was 2.3 kg. The specific exchange area of the lung is therefore 11 cm^2/g body weight. This figure is considerably less that the 17.9 cm^2/g suggested by Duncker (1972), although, since in this case details of calculation are not given, a proper comparison cannot be made. It is noted, however, that the exchange surface area of both lungs based on a model of the capillary zone consisting of parallel tubes is 33% greater than in the case of the present model. There is however, little anatomical evidence for such a parallel tube model.

Close examination of the precise way in which adjacent sections of capillary are related in our model shows that at every point the flow of air on one side of a double wall and the flow of blood on the other side, are orthogonal or cross-current. Scheid and Piiper (1972) demonstrated experimentally the probability of a cross-current relationship between blood and air at the level of the pulmonary artery terminal branches and the tertiary bronchus; the model which has been proposed extends this idea to the ultimate level of the capillary.

References

Duncker, H.R.: Structure of avian lungs. Respir. Physiology 14, 44-63 (1972)

Scheid, P., Piiper, J.: Cross-current gas exchange in avian lungs: effects of reversed parabronchial air flow in ducks. Respir. Physiology 16, 304-312 (1972)

Ventilatory ^{133}Xenon Distribution Studies in the Duck (Anas platyrhynchos)

B. Burns, A. Everette James, G. Hutchins, G. Novak, and R. R. Price

Summary

The distribution of an inspired bolus of the ^{133}Xe in air was measured in lightly anesthetized, spontaneously breathing ducks using a scintillation camera. Inspired gas enters the avian lungs and airsacs nearly simultaneously. The respiratory dead space appeared initially preferentially distributed to the posterior (abdominal) sacs. The pathway of gas from the posterior thoracic airsac during expiration was observed by injecting radioactive gas into this sac at end-inspiration. The posterior thoracic sac contents entered the lung during expiration which supports the unidirectional airflow hypothesis of avian respiration in so far as the lungs fill during inspiration and expiration. The relative compartmental ventilation rates were measured by determining at 0.1-s intervals the disappearance of ^{133}Xe when breathing room air. The lungs and posterior thoracic sac cleared more rapidly than the other compartments.

1. Introduction

The anatomic complexities of the avian respiratory system have made it difficult to determine in vivo the pathways taken by inspired or expired gases (Fig. 1). The purpose of this study was to evaluate the

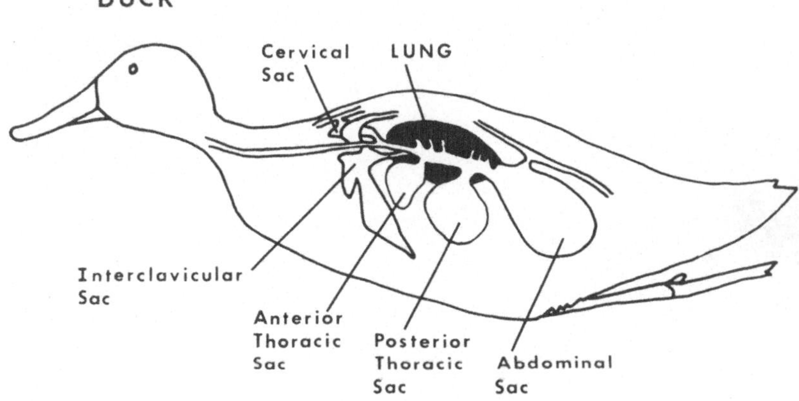

Fig. 1. Diagram of avian respiration system

Johns Hopkins University, Medical Institutions, Baltimore, Maryland 21205, and Department of Radiology and Radiological Sciences, Vanderbilt University Hospital, Nashville, Tennessee 37232/USA.

flow patterns of respired air using non-invasive imaging techniques. We wished to compare the results of a completely non-invasive method with those previous studies in which airflow probes, with their potential for disrupting flow patterns, etc., were inserted into various portions of the respiratory system. The ability to measure semiquantitatively the regional distribution of the radioactive gas while simultaneously acquiring anatomical images in an intact bird was felt to offer certain advantages over other methods of analysis.

2. Methods

The respiratory system was visualized by the addition of radioactively labelled gas (^{133}Xenon 15-30 mCi) to the inspired air. The image was recorded on a scintillation camera* and successive 0.1-s frames stored on computer tape for subsequent retrieval and analysis. The actual area in the image field of views occupied by the lungs was determined by intravenous injection of 1-2 mCi ^{99m}Tc labelled albumin microspheres (15-30 microns diameter)** and recording this image with the same system. The ducks were lightly anesthetized (Ketamine, 20 mg/kg I.M.), incubated, and positioned upright on a V-shaped foam block. Respiration was monitored by a Fleisch pneumotach in line with the trachea. Body temperature averaged 41°C.

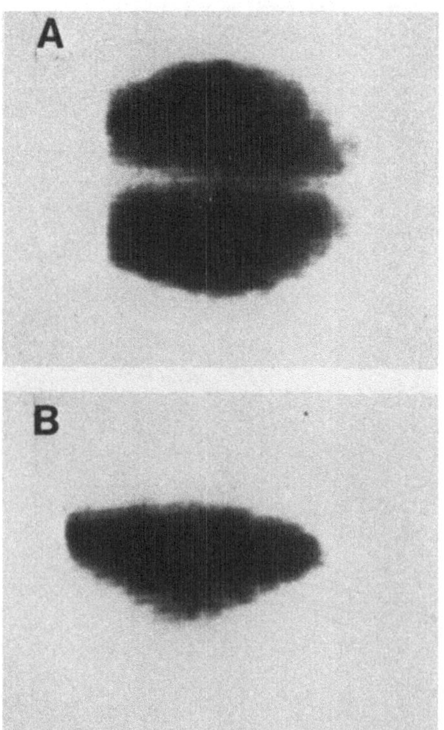

Fig. 2. Perfusion lung scan (duck). (A) Posterior view showing outline of both lungs. (B) Lateral view of scan with ventral *below* and cranial *right*. Note the homogenous perfusion depicted by uniform distribution of radioactivity

* Anger Type. Searle LFOV. Searle Coporation, Des Plaines, Illinois.
** Minnesota Mining and Manufacturing Assoc., St. Paul, Minn.

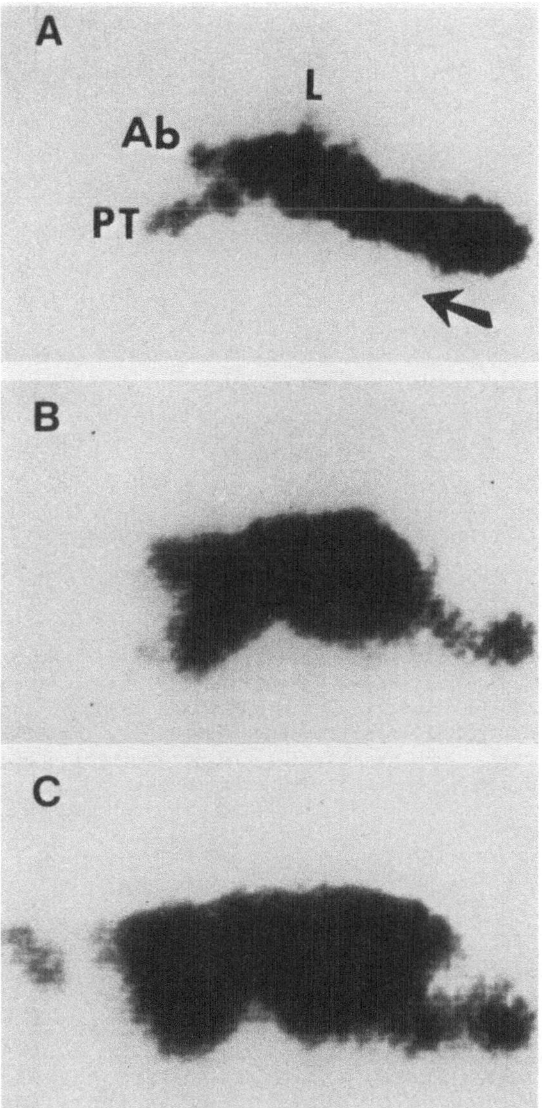

Fig. 3. Inhalation lung scan (^{133}Xe in air). Duck A. Refer to methods section for procedure. *Top* of figure is dorsal; *bottom* is ventral and *right* is cranial. (A) Direction of airflow given by *arrow*. This image frame from the display monitor is very early during inspiration and represents complete filling of respiratory dead space with inspiration. Note the preferential filling of the posterior thoracic sac (*PT*). Entry of gas into the abdominal sac (*Ab*) and lungs (*L*) via the dorsomedial secondary bronchi can also be seen. (B) This frame was recorded 0.25 s following (A) and shows filling of the lungs in a radial manner (center outward) about midway through inspiration. A small amount of residual gas can be seen trapped in the syrinx (*circular area to extreme right*). (C) Late inspiration [0.25 s after (B) above]. The lungs and posterior sac system are completely filled, with some gas reaching the posterior recess of the abdominal sac (*extreme left*). The lungs appear very similar in shape, when filled, with the perfusion lung scan (Fig. 2B) from a different duck in the same lateral view

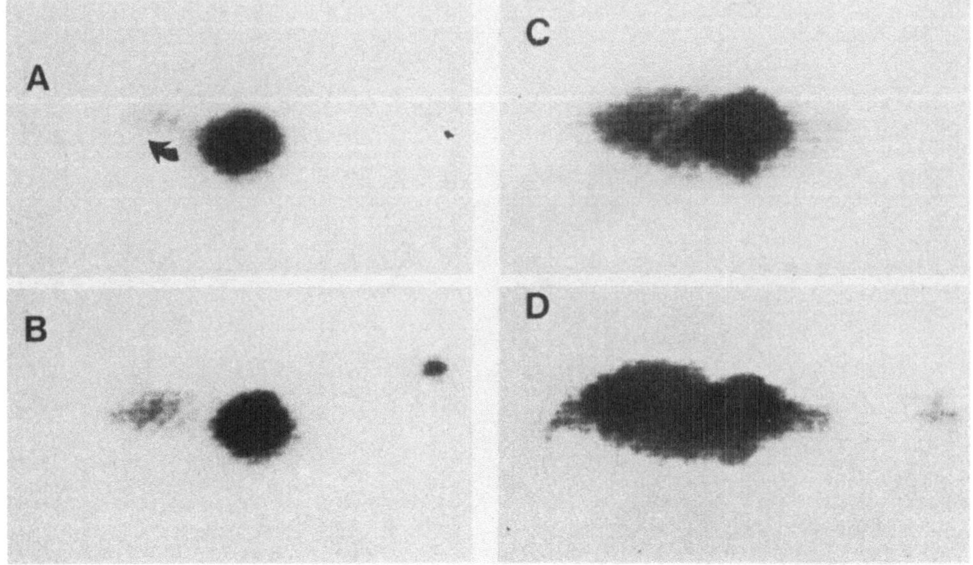

Fig. 4. Exhalation series of xenon gas dynamic lung scan. Duck B. Radioactive tracer (15 mCi) was added to the posterior thoracic sac at end-inspiration. (A) Initial image of the posterior thoracic sac (*center*) with the direction of airflow indicated by the arrow. *Top* of figure is dorsal, the head is to the *left* (reverse of Fig. 3). (B)(C) Successive stages of filling in the lung during expiration. (D) End of expiration and beginning of inspiratory effort. Figures 3 and 4 represent different ducks

3. Results

The perfusion lung scans (Fig. 2A, B) show the relative distribution of labeled particles in the lung after intravenous injection. The distribution of radioactivity is homogenous; vertical gradients would not be expected in a lung this small and no radioactivity was present outside the confines of the lungs. Essentially all of the microspheres were trapped in the precapillary arterioles or capillaries in the first pass through the lungs.

Inhalation scan. The inspiratory dead space distal to the trachea was filled with air containing ^{133}Xe. The path taken by this bolus during inspiration is depicted in Figure 3. The lungs, posterior thoracic and the anterior portion of the abdominal sac simultaneously filled during early inspiration. The major portion of the dead space appeared initially directed to the posterior thoracic and abdominal airsacs.

Exhalation scan. Movement of gas from the posterior thoracic sac during expiration was followed by injecting a small amount (0.5 cc) of ^{133}Xe into this sac through a percutaneous catheter (Intracath, 16 gauge) at end-inspiration. The gas mixture was directed entirely to the lungs with the first subsequent expiration (Fig. 4). Figures 3 and 4 represent different ducks.

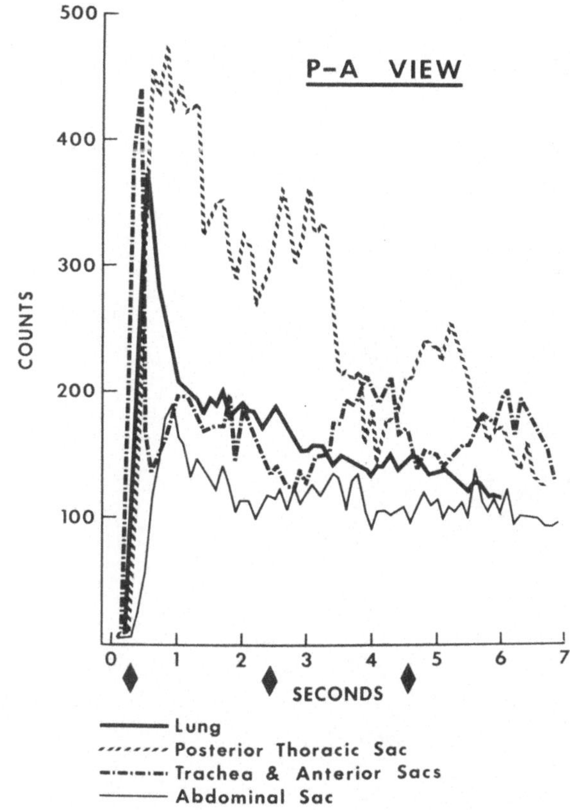

Fig. 5. Dorsoventral or posterior view showing the initial distribution and clearance of a bolus of ^{133}Xe in air in the duck A. Data were recorded at 0.1-s intervals. The tracer first appears in the tracheal area (see legend) and then simultaneously enters the lungs and posterior thoracic sac. The initial peak in lung concentration falls sharply (perhaps due to uptake of tracer by blood or dilution by continued entry of room air). The approximate times for the beginning of inspiration are marked by solid diamonds on the x-axis. There were large swings in tracer content in the posterior thoracic sac with continued ventilation - possibly reflecting the re-inspiration of dead space gas since the tracheal record is a mirror image of the posterior thoracic sac. The changes in signal strength in the lung field show smaller fluctuations, indicating a more even flow of gas with minimal entry of dead space during inspiration. The abdominal sac is apparently not as well ventilated as the lungs or posterior thoracic airsac

Washout. The major identifiable compartments were electronically flagged on the computer image display screen, and the changes in radioactivity recorded at 0.1-s intervals for several successive breaths of room air following the inhalation of a bolus of ^{133}Xe (*vide supra*). The initial distribution of the bolus and the washout during continued ventilation are depicted in Figures 5 and 6. The fastest washout rates were seen in the region corresponding to the lungs. The anterior sacs were not clearly separable from the trachea and the recordings of regional distribution have been combined in this preliminary computer analysis.

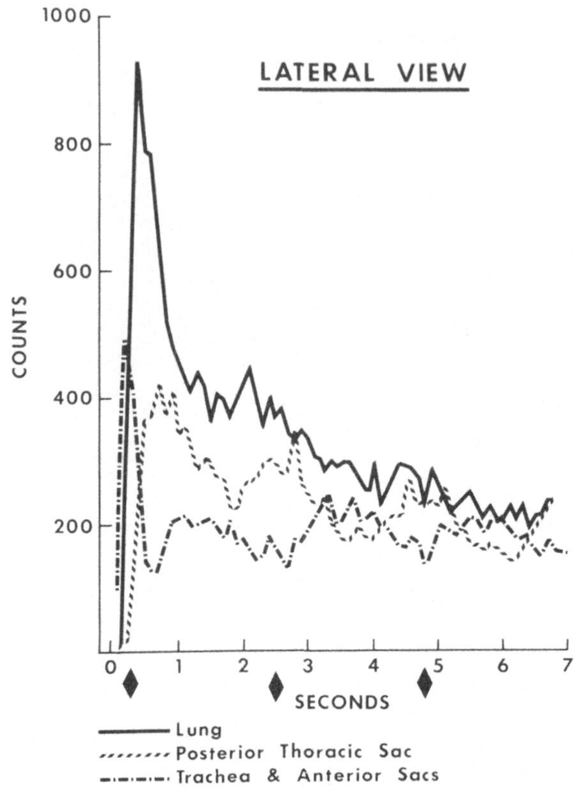

Fig. 6. This figure represents a second study of the same duck (A) used in Figures 3 and 5, as seen from the lateral view (the duck is upright and the detector is repositioned). The results are essentially the same as given in Figure 5. It should be emphasized that, although the anterior sacs have been electronically flagged in the record of tracheal activity changes, their filling cycles *cannot* be distinguished in these images. A reciprocal relationship is seen between the lungs and posterior thoracic sac on breaths after the first, representing movement of tracer from this sac into the lung field. (Beginning of inspiration marked with solid diamonds)

4. Discussion

Our findings may help to explain the relatively high (3%) CO_2 present in the posterior airsac system (Piiper et al., 1970). We interpret Figure 3A as showing preferential shunting of the dead space to the posterior sacs prior to filling of the lungs during inspiration. During later inspiration (Fig. 3B), the lungs appear to fill "radially" via the dorsomedial secondary bronchi. We did not observe filling of the anterior sacs via the craniomedial secondary bronchi during inspiration - afinding consistent with those of Bretz and Schmidt-Nielsen (1971).

We observed that much of the gas from the posterior thoracic sac is directed to the lungs during expiration. These data confirm reports that the lungs fill both during inspiration and expiration (Scheid and Piiper, 1971; Bethe, 1925; Hazelhoff, 1973; Bretz and Schmidt-Nielsen, 1971, and Shepard et al., 1959). In the absence of meso-bronchial filling by gas originating from the posterior thoracic sac during expiration (Fig. 4), our preliminary results do not support the loop hypothesis (Hazelhoff, 1943).

The primary advantage of the image analysis method is that many compartments may be observed simultaneously by a non-invasive technique in the bird during normal respiration. The findings in this report extend our earlier observations on several species of birds during normal respiration and simulated flying in which true steady states were reached during the "wash-in" and "wash-out" phases of tracer equilibration (James et al., 1976). In the earlier study we were unable to show the initial distributional pattern of tracer depicted in Figure 3. Studies employing these techniques are currently being performed to assess the pathways of gas movement from the remaining airsacs during the respiratory cycle.

Acknowledgments. This research was supported in part by funds from the USPHS. We also gratefully acknowledge additional support from the Deutsche Forschungsgemeinschaft for attendance to this symposium. The resources of the Department of Environmental Medicine (T.K. Natarajan) and the Department of Radiology (R.M. Bush) were utilized for this study.

References

Bethe, A.: Atmung: Allgemeines und Vergleichendes. In: Handbuch der normalen und pathologischen Physiologie. Bethe, A., Bergmann, G.v., Embden, G., Ellinger, A. (eds.). Berlin: Springer, 1925, pp. 1-36

Bretz, W.L., Schmidt-Nielsen, K.: Bird respiration: flow patterns in the duck lung. J. Exp. Biol. 54, 103-118 (1971)

Hazelhoff, Ė.H.: Bouw en functie van de vogellong. Versl. gewone Vergad. Afd. Natuurk. Kon. Ned. Akad. Wet. 52, 391-400 (1943)

James, A.E., Jr., Hutchins, G., Bush, M., Natarajan, T.K., Burns, B.: How birds breathe: correlation of radiographic with anatomical and pathological studies. J. Am. Vet. Radiol. Soc. 17(2), 77-86 (1976)

Piiper, J., Drees, F., Scheid, P.: Gas exchange in the domestic fowl during spontaneous breathing and artificial ventilation. Resp. Physiol. 9, 234-245 (1970)

Scheid, P., Piiper, J.: Direct measurement of the pathway of respired gas in duck lungs. Resp. Physiol. 11, 308-314 (1971)

Shepard, R.H., Sladen, B.K., Peterson, N., Enns, T.: Path taken by gases through the respiratory system of the chicken. J. Appl. Physiol. 14, 733-735 (1959)

Diffusion in Avian Pulmonary Gas Exchange: Role of the Diffusion Resistance of the Blood-Gas Barrier and the Air Capillaries

P. Scheid, R. E. Burger, M. Meyer, and W. Graf

Summary

Gas exchange was studied in anesthetized, artificially ventilated ducks. Blood and gas flow shunts were measured and accounted for in estimating lung diffusing capacity for O_2 (D_{O_2}), as was the functional lung inhomogeneity. The average ratio of D_{O_2}/O_2 uptake ($\dot{M}_{O_2}$) was 0.13 torr^{-1} which is comparable with the value expected for a mammal of similar size at rest.

At rest, the diffusion of O_2 in the gas phase of the air capillaries contributes an insignificant resistance to the overall O_2 uptake resistance of the avian lung. During exercise this factor may become limiting.

1. Introduction

As a result of the combined efforts of anatomists and physiologists many basic features of the avian lung as a gas exchange organ have been clarified (cf. Piiper and Scheid, 1973). Blood capillaries contact air capillaries in the tissue surrounding the parabronchial lumen, and both anatomical and physiological investigation suggest that the serial-multicapillary or cross-current system is an appropriate model for the quantitative analysis of parabronchial gas transfer (Scheid and Piiper, 1972; Piiper and Scheid, 1975). Earlier attempts to measure the limitation of gas exchange imposed by diffusion resistances in the exchange tissue (Scheid and Piiper, 1970) are burdened with a number of assumptions regarding parameters, which could not be measured directly, like parabronchial ventilation and concentrations of respired gases in the lung. The parabronchi can be ventilated at a constant and known flow rate and with gas of known composition by the method of unidirectional ventilation (Scheid and Piiper, 1972). We have applied this method to measure the pulmonary diffusing capacity (D) in the duck.

Only in the parabronchial lumen (and possibly in the atria) is there convective gas movement, whereas transport of O_2 and CO_2 between the parabronchial lumen and the blood-gas membrane is by diffusion in the gas phase of the air capillaries. The resistance offered by this diffusion in air capillaries is usually assumed to be negligible as compared with the diffusion resistance across the membrane, although the contribution of each separately has not yet been elucidated. In this

Max-Planck-Institut für experimentelle Medizin, Abteilung Physiologie, 3400 Göttingen, Federal Republic of Germany.

paper, attempt will be made to estimate the importance of the limitation to gas transfer offered by the air capillaries, by mathematically analyzing a suitable model of the arrangement between air capillaries and blood capillaries. A more detailed account of this approach will be given elsewhere (Scheid, 1978).

2. Measurement of D_{O_2}: Experiments in the Duck

These experiments were performed on ten anesthetized muscovy ducks. Normocapnic hypoxia was produced by a continuous flow of gas (inspired P_{O_2} < 55 torr), insufflated into the cannulated caudal thoracic airsacs on both body sides. Balloon catheters blocked the main bronchus between the origins of the medioventral and mediodorsal secondary bronchi (Graf et al., 1976), insuring that ventilatory gas passed through the parabronchi of the lung. Partial pressures of O_2 and CO_2 were measured, in ventilatory gas at its entrance to and exit from the bird by mass spectrometry, and in blood samples withdrawn from a peripheral artery and the right ventricle by electrodes. Additionally, O_2 content was measured in blood samples for calculation of the cardiac output by the direct Fick principle.

The model used for calculating D_{O_2} from measurement of P_{O_2} values in blood and lung gas and $\dot{M}_{O_2}$ is schematically shown in Figure 1. This model has been used earlier by Piiper and Scheid (1975) who have also discussed the assumptions and limitations in calculating D_{O_2} from this model.

Before D_{O_2} was calculated, the measured values of expired gas (P_E) and arterial blood (P_a) partial pressures were corrected for the effects of blood (venous admixture) and gas shunts. The blood shunt was estimated by measurements at high ventilation of a hyperoxic gas mixture. The deviation of arterial blood from lung gas P_{O_2} allows the

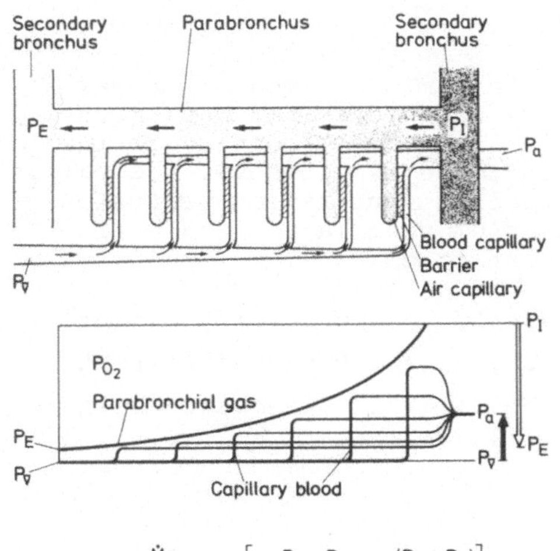

Fig. 1. Model used for quantitative analysis of gas exchange. Below, profiles of P_{O_2} in gas passing through the parabronchi and in the capillary blood at various sites along the parabronchial length. The formula relates the pulmonary O_2 diffusing capacity to metabolic rate ($\dot{M}_{O_2}$), and to partial pressures in lung gas and blood. ln, natural logarithm

assessment of blood shunt much as in the alveolar lung of mammals. Similarly the ventilation of non-exchanging parallel pathways, the gas shunt, can be calculated when a gas mixture containing a highly soluble inert gas (chloroform, $CHCl_3$, in our study) is flown through the lung at low rate. The assumption is that the inert gas is cleared from the gas phase by the blood and that the concentration of this species in exhaled gas derives from gas shunts.

If D is determined primarily by diffusion across the blood-gas membrane, the ratio of D_{CO_2}/D_{O_2} should equal the ratio between Krogh's diffusion constant for both gases, about 15 to 20. In these experiments, the ratio D_{CO_2}/D_{O_2} averaged below 5. The simplest and most likely explanation for this low ratio is regional inequality in the distribution of blood flow and ventilation to parallel lung units. This inhomogeneity was estimated in a simplified model consisting of only two parallel units, the ventilation to the first exceeding that to the second by a factor of λ, but with an inverse proportion of blood flow to both units. The true value of D_{CO_2} was assumed to be infinity (as in the Riley model used for alveolar gas exchange) and λ was estimated from the measured, apparent value of D_{CO_2}. λ then was utilized to correct the measured, apparent value of the diffusing capacity for O_2 to yield a better estimate of D_{O_2} in which shunts and inhomogeneity had been accounted for.

Table 1. Results of measuring gas exchange in ten anesthetized ducks. Mean values ±S.D.

Weight	1.8 kg
O_2 uptake, $\dot{M}_{O_2}$	0.83 ± 0.11 mmol/min
Cardiac output, $\dot{Q}$	370 ± 100 ml/min
Diffusing capacity, D_{O_2}	105 ± 10 µmol/min/torr
Perfusive shunt	2.7 ± 0.2%
Ventilatory shunt	10 ± 5%
Inhomogeneity, λ	1.6 ± 0.4

The average results are shown in Table 1. Only minor blood shunt was found, on the average 2.7%. The ventilatory shunt does not have much physiological significance, since its value is closely coupled to the artificial type of ventilation employed in these experiments. The value of λ indicates that on the average the parallel ventilation-perfusion inhomogeneity is equivalent to a dispersion of ventilation and perfusion by a factor of 1.6 in a two-compartment model.

The O_2 diffusing capacity, D_{O_2}, may be related to the resting metabolism, $\dot{M}_{O_2}$. The ratio of $D_{O_2}/\dot{M}_{O_2} = 0.13\ torr^{-1}$ is not much different from the ratio in a mammal (Piiper and Scheid, 1973), allowance being made for differences in body size (Weibel, 1972).

3. Diffusion Resistance in the Gas Phase of Air Capillaries: Theory and Application

It is commonly held that the air capillaries do not change their volume in the respiratory cycle. In fact, the extending pressure of the air capillaries would be relatively large due to their small radial dimensions unless surface tension was drastically reduced. There is no direct evidence for the presence of surface active material in the air capillaries (Pattle, this symposium). The air capillaries may be visualized as small but rigid tubules along which respiratory gases diffuse in transport between the blood-gas membrane and the convected gas in the parabronchial lumen.

Zeuthen (1942) and Hazelhoff (1943) estimated the limitation imposed by diffusion in air capillaries on gas exchange and have concluded its contribution to the gas transfer resistance to be negligible. However, their model may be questioned from an anatomical point of view and so may the structural dimensions which they used to derive their conclusion (cf. Scheid, 1978).

Piiper and Scheid (1973) have used the arrangement of air capillaries and blood capillaries suggested by Duncker (1972, 1974) to propose a model for gas exchange between air capillary gas and capillary blood which is schematically shown in Figure 2. In this model, O_2 diffuses from the bulk stream in the parabronchial lumen into the air capillaries, whereas blood is convectively transported along the radially extending air capillary, from the periphery of the parabronchus towards the lumen. Oxygen may be taken up through the blood-gas membrane

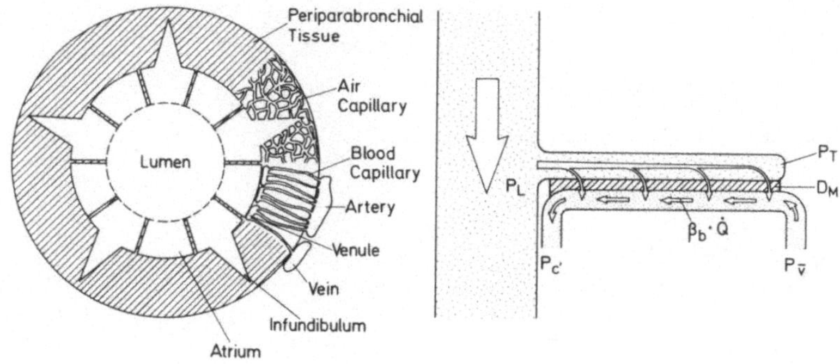

Air Capillaries (Anatomical)		Minimum	Maximum
Total	Number	$20 \cdot 10^6$	$200 \cdot 10^6$
	Cross-Sectional Area	30 cm²	300 cm²
Single	Length	200 µm	500 µm
	Diameter	10-20 µm	

Conductances (Physiological)	Rest	Exercise
D_M $\left(\frac{mmol}{min \cdot torr}\right)$	0.067	0.67
$\beta_b \cdot \dot{Q}$ $\left(\frac{mmol}{min \cdot torr}\right)$	0.034	0.34
$\dot{M}_{O_2}$ $\left(\frac{mmol}{min}\right)$	1	20

Fig. 2. Model used to estimate the diffusion resistance offered by the air capillaries. P_L, P_T, $P_{\bar{v}}$, $P_{c'}$ are partial pressures of the gas under study in the parabronchial lumen; in gas at the end of the air capillary, in mixed venous, and in end-capillary blood. D_M, membrane diffusing capacity

all along the air capillary. Piiper and Scheid (1973) have used the term *counter-current-like mechanism* to indicate functional similarities with the wellknown counter-current system, realized, for example, in fish gills, in which transport on both sides of a membrane is by convection. Scheid (1978) has analyzed this counter-current-like arrangement together with similar models that could be propounded on the basis of the anatomical work of Akester (1971) and Abdalla and King (1976). However, no major functional differences appear to derive from the slightly different anatomical descriptions.

The diffusion limitation offered by the air capillary gas phase both during rest and exercise can be estimated if morphometrical and physiological data are used. In the left-hand table of Figure 2, morphometrical data (Duncker, personal communication) are given. The range in data can be used to estimate maximal or minimal diffusional resistances. In the right-hand table values of the membrane component of the O_2 diffusing capacity, D_M, the blood convective conductance, $\beta b\dot{Q}$ (cf. Piiper et al., 1971), and the O_2 uptake rate, $\dot{M}_{O_2}$, were those used by Scheid (1978).

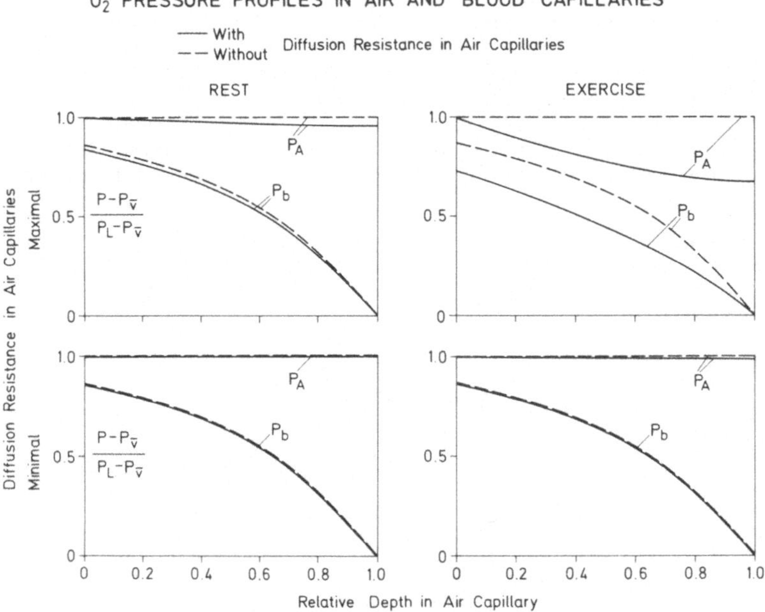

Fig. 3. Effects of diffusion resistance in air capillaries on O_2 uptake. P_L, O_2 pressure in parabronchial lumen; $P_{\bar{v}}$, O_2 pressure in mixed venous blood

Figure 3A shows the calculated P_{O_2} profiles (continuous lines) in air capillary gas (PA), and in capillary blood (Pb) during rest (left), and exercise (right) for the maximal (upper), and the minimal diffusion resistances (lower) that were calculated from morphometrical data. The dashed lines are calculated assuming no diffusion resistance. The disparity between the continuous and the dashed lines in profiles of blood P_{O_2} constitute a measure of the limitation imposed by diffusion in the gas phase (Scheid, 1978).

As can be seen, the limitation is minimal for resting conditions, in agreement with the results of Zeuthen (1942) and Hazelhoff (1943). Thus, diffusion in the gas phase of the air capillaries could not contribute much to the overall D_{O_2} value (Table 1). This assumption was directly tested by comparing measurements of D_{O_2}, by the techniques in the second section, when the background gas in the ventilatory gas was N_2 and helium (He). If diffusion in the gas phase contributes to O_2 uptake resistance, D_{O_2} would be greater when He than when N_2 was the background gas, since diffusivity of O_2 is about 2.5 times larger, and thus the diffusion resistance in the air capillaries 2.5 times smaller, in He than in N_2 (Worth et al., 1978). There was no significant difference between the two values of D_{O_2} suggesting a minor impairment by air capillary diffusion.

The profiles at exercise suggest that a major impairment to gas exchange from air capillaries could occur during activity. However, more anatomical and physiological data are needed to evaluate this suggestion.

References

Abdalla, M.A., King, A.S.: The functional anatomy of the pulmonary circulation of the domestic fowl. Respir. Physiol. 23, 267-290 (1976)

Akester, A.R.: The blood vascular system. In: Physiology and Biochemistry of the Domestic Fowl. Bell, D.J., Freeman, B.M. (eds.). London, New York: Academic Press, 1971, pp. 783-839

Duncker, H.-R.: Structure of avian lungs. Respir. Physiol. 14, 44-63 (1972)

Duncker, H.-R.: Structure of the avian respiratory tract. Respir. Physiol. 22, 1-19 (1974)

Graf, W., Molony, V., Scheid, P.: A technique for study of lung function in birds by blocking the primary bronchus. Respir. Physiol. 26, 325-331 (1976)

Hazelhoff, E.H.: Bouw en functie van de vogellong. Versl. gewone Vergad. Afd. Natuurk. kon. Ned. Akad. Wet. 52, 391-400 (1943). English translation: Structure and function of the lung of birds. Poultry Sci. 30, 3-10 (1951)

Piiper, J., Dejours, P., Haab, P., Rahn, H.: Concepts and basic quantities in gas exchange physiology. Respir. Physiol. 13, 292-304 (1971)

Piiper, J., Scheid, P.: Gas exchange in avian lungs: models and experimental evidence. In: Comparative Physiology. Bolis, L., Schmidt-Nielsen, K., Maddrell, S.H.P. (eds.). Amsterdam: North Holland, 1973, pp. 161-185

Piiper, J., Scheid, P.: Gas transport efficacy of gills, lungs and skin: theory and experimental data. Respir. Physiol. 23, 209-221 (1975)

Scheid, P.: Analysis of gas exchange between air capillaries and blood capillaries in avian lungs. Respir. Physiol., in press (1978)

Scheid, P., Piiper, J.: Analysis of gas exchange in the avian lung: theory and experiments in the domestic fowl. Respir. Physiol. 9, 246-262 (1970)

Scheid, P., Piiper, J.: Cross-current gas exchange in avian lungs: effects of reversed parabronchial air flow in ducks. Respir. Physiol. 16, 304-312 (1972)

Weibel, E.R.: Morphometric estimation of pulmonary diffusing capacity. V. Comparative morphometry of alveolar lungs. Respir. Physiol. 14, 26-43 (1972)

Worth, H., Nüsse, W., Piiper, J.: Determination of binary diffusion coefficients of various gas species used in respiratory physiology. Respir. Physiol., in press (1978)

Zeuthen, E.: The ventilation of the respiratory tract in birds. Kgl. Danske Videnskab. Selskab Biol. Medd. 17, 1-50 (1942)

Airway Resistance

V. Molony

Summary

A brief review of the two main methods used for measurement of airway resistance and the available results is followed by a summary of the evidence available for its control.

1. Introduction

Recent developments in understanding the physiology of breathing in birds have emphasised the importance of the resistance to flow of air through the various airways of the respiratory tract. Resistance is of particular importance: first, in the part it plays in the work done in breathing; second, for its part in dictating the pattern of airflow within the system; and finally, for the effect that changes in resistance may have on the balance between ventilation for respiratory exchange and ventilation for thermoregulation or vocalisation.

Bretz and Schmidt-Nielsen (1971), Scheid and Piiper (1971) and Brackenbury (1971, 1972a, b) have shown that the pattern of airflow in inspiration and expiration depends on the resistance presented by alternative pathways within the system. Brackenbury (1972), and Baldwin (1973) have shown that airway resistance depends on the phase of breathing and that, unlike mammals, resistance is higher in inspiration than in expiration. Zeuthen (1942) and King and Farner (1961) recognised that modification of airway resistance could act as a mechanism for controlling ventilation by redistribution of airflow within the system, and King and Cowie (1969) showed that changes in the tone of the bronchial smooth muscle are capable of achieving such changes in resistance. Ray and Fedde (1969) and Molony et al. (1976) found that changes in respiratory gases, particularly CO_2, can modify airway resistances both directly and indirectly by acting upon bronchial smooth muscle. According to Molony et al. (1976) these changes in resistance can occur quickly enough to influence the airflow pattern within a breath. Kampe and Crawford (1973) and Bouverot (1977) have commented upon the importance of the time constant of the respiratory system in determining the optimal rate and depth of breathing.

Department of Physiology, Royal (Dick) School of Veterinary Studies. Edinburgh E.H.9 1QH, Scotland.

2. Methods Used for the Measurement of Airway Resistance

Two main methods have been used. The first depends upon establishing a constant flow of gas through the system or through a part of the system and then measuring the pressure gradient across the required resistance. This pressure gradient is obtained using a differential manometer and cannulae located upstream and downstream of the resistance, the main difficulty being to position cannulae of appropriate sizes in the chosen positions (see Brackenbury, 1971, 1972a, b; Molony et al., 1976; Graf et al., 1976 for details).

In the second method, simultaneous measurements of flow and pressure gradients are made in spontaneously breathing birds. Since it is difficult to measure flow accurately in isolated parts of the system, this method has only been applied to the investigation of the whole system. The pressure gradient is measured between the airsacs and the trachea or atmosphere. Flow is measured by means of a pneumotachometer attached to the trachea, to a hood sealed over the head of the bird, or to a whole body plethysmograph (see Baldwin, 1973 for details).

$$\text{Resistance (cm } H_2O/l/s) = \frac{\text{Pressure gradient (cm } H_2O)}{\text{Flow (l/s)}}$$

By measuring the resistance at different flow rates, an estimate can be made of the conditions of flow within the section. Flow can be laminar, turbulent, or a mixture of both. Turbulence in the system is produced mainly by irregularities in the contours and abrupt changes in diameter of the airways.

3. Results Available for Airway Resistances

The results obtained using one or other of the above methods are included in Table 1.

Cohn and Shannon found that their measurements were independent of the position of the bird.

Brackenbury (1972a) showed that resistance increased with increasing flow in spontaneously breathing and unidirectionally ventilated birds, and thus indicated the contributions of aerodynamic and viscous resistances within the system. He proposed that in general the resistance pathway leading to the posterior groups of airsacs is greater than that to the anterior group, and that the sequence of decreasing resistance is cervical, posterior thoracic, abdominal, interclavicular, and anterior thoracic.

Kampe and Crawford (1973) concluded that the resistance of the pigeon's respiratory system is low compared with that of mammals of similar size. Their measurements include tissue resistance.

Baldwin (1973) found that inspiratory resistance was higher than expiratory resistance in both euthermia and hyperthermia.

Molony et al. (1976) measured the dependence of resistance upon flow in the right lung. Their results supported Brackenbury's (1972a) proposal that resistance increases more than linearly with increasing flow in any part of the system. The resistance measured was subdivided,

Table 1. Airway resistance in birds

Species (No.)	Sites used in measurement	Part of system measured	IE[g]	Resistance (R) ($cm/H_2O/l/s$)	Flow ($\dot{V}$) (l/s)	Pressure gradient ($cm\ H_2O$)	Ventilation
Goose[a]	Trachea and airsac	Lower resp. tract and syrinx	I	7.7	0.08 max	+1.0 max	unassisted
			E	6.6	0.08 max	-1.0 max	unassisted
Goose[b]	Trachea and airsac	Lower resp. tract and syrinx	IE	7.0 (viscous)	0.00	+0.8 max -0.6 min	unassisted
Goose[b]	Trachea and airsac	Lower resp. tract and syrinx	IE	20.0 (aerodynamic)	0.05	+0.8 max -0.6 min	unassisted
Chicken[b]	Trachea and airsac	Lower resp. tract and syrinx	IE	16.0 (viscous)	0.00	+1.0 max -1.0 min	unassisted
Goose[b]	Trachea and atmosphere	One side and syrinx	-	2.0 (viscous)	0.00	-	unidirectional[i]
Goose[b]	Trachea and atmosphere	One side and syrinx	-	100-400 (aerodyamic)	-	-	unidirectional[i]
Goose[b]	Mid-trachea and post.syrinx	Syrinx	-	1.0 (viscous)	0.00	-	unidirectional
		Syrinx	-	3.6 (aerodynamic)	0.083	+0.3 max	unidirectional
Goose[b]	Post syrinx and caud. thoracic a.s.	Lower resp.tract and syrinx	-	0.8 (viscous)	0.00	-	unidirectional
			-	15.0 (aerodynamic)	0.083	+1.3 max	unidirectional
Chicken[c] (9)	Trachea and airsac	Lower resp. tract and syrinx	IE	26.8 ± 0.8	0.038 max	+1.1 max	unassisted
			IE	23.9 ± 1.9	0.038 max	+1.1 max	unassisted
			IE	23.0 ± 1.8	0.065 max	+1.7 max	unassisted
			IE	20.8 ± 0.9	0.065 max	+1.7 max	unassisted
Pigeon[d] (9)	Driving pressure and plethysmograph	Whole bird-syrinx and upper airways	IE	40.86 ± 3.75	0.059 max	+6.0 max	assisted
Goose[e]	Trachea and airsac	Lower resp. tract and syrinx	IE	8.7 ± 1.1	-	-	unassisted
				11.9 ± 0.93[h]	-	-	unassisted

[a] Cohn and Shannon (1968) unanaesthetised birds; [b] Brackenbury (1972); [c] Baldwin (1973) body temperatures respectively 41.6°, 41,6°, 43.5°, and 43.5°C; [d] Kampe and Crawford (1973); [e] Callanan et al. (1974); [f] Molony et al. (1976), body temperatures 41.5°C; [b-f] Experiments on an aesthetised birds; [g] I, inspiration, E expiration, IE, average of inspiration and expiration; [h] post Vagotomy; [i] One side, all sacs open.

Table 1. Continued

Species (No.)	Sites used in measurement	Part of system measured	IE[g]	Resistance (R) ($cm/H_2O/l/s$)	Flow ($\dot{V}$) (l/s)	Pressure gradient ($cm\ H_2O$)	Ventilation
Duck[f] (10)	Primary bronchus distal to block and proximal to block	Paleopulmo	-	4.46	0.008	0.04	unidirectional[j]
			-	4.46	0.017	0.07	unidirectional
			-	6.28	0.023	0.21	unidirectional
			-	8.56	0.050	0.47	unidirectional
Duck[f] (6)	Primary bronchus distal to block and clavicular airsac	Parabronchi of paleopulmo	-	1.51	0.008	0.02	unidirectional
			-	1.76	0.017	0.05	unidirectional
			-	2.37	0.023	0.16	unidirectional
			-	2.98	0.050	0.34	unidirectional
Duck[f] (6)	Clavicular airsac and primary bronchus proximal to block	Medioventral bronchi	-	2.46	0.008	0.01	unidirectional
			-	2.85	0.017	0.03	unidirectional
			-	4.88	0.023	0.07	unidirectional
			-	6.70	0.050	0.15	unidirectional

[f] Molony et al. (1976), body temperatures 41.5°C; [b-f] Experiments on aesthetised birds.
[g] I, inspiration, E, expiration, IE, average of inspiration and expiration.
[j] Caudal thoracic airsac to trachea with blocking catheter.

and it was shown that the orifices of the medioventral secondary bronchi contributed about twice as much to this resistance as did the parabronchi of the paleopulmo.

It is evident from this brief summary that, to establish the contributions of resistance to respiratory mechanics in birds, it will be necessary for basic measurements to be confirmed and for the methods used to be standardised.

4. Evidence for Control of Airway Resistance

To demonstrate such control it is necessary to show that the resistances of particular airways change in a predictable way in response to physiological disturbances imposed upon the bird.

Direct evidence for parasympathetic efferent control of airway resistance was obtained by King and Cowie (1969) who changed the resistance to gas flow by electrically stimulating the peripheral stump of the vagus nerve. Molony et al. (1976) showed that vagal afferent and/or efferent reflex pathways are involved in the control of early changes in airway resistance in response to changes in airway CO_2. Callanan et al. (1974) have shown that vagotomy increases the resistance of the lower respiratory tract in the goose. Fedde et al. (1961), King and Cowie (1969) and Molony et al. (1976) did not find any evidence for the existence of vagal tone.

No studies have been found of the direct effects of sympathetic stimulation or blockade upon airway resistance, but sympathomimetic drugs and blocking agents have been shown to affect bronchial smooth muscle tone (King and Cowie, 1969).

Some effects of respiratory gases upon airway resistance have been demonstrated. Ray and Fedde (1969) have shown that the overall resistance of the lower respiratory tract of unidirectionally ventilated anaesthetised chickens decreases linearly with increasing fractional concentration of CO_2 (F_{CO_2}) between 2% and 12%. Leitner and Roumy (1974) have reported constriction of secondary bronchi in anaesthetised ducks in response to increased inspired F_{CO_2}. Molony (1972) showed that in spontaneously breathing unanaesthetised (decerebrated) chickens, increases in airsac pressure were the most sensitive indication of rapid changes in breathing occuring in response to small increases in F_{ICO_2}. Molony et al. (1976), however, found that decreasing F_{ICO_2} increased the resistance of the right lung and that most of this increase appeared to be a direct effect on smooth muscle, since it occurred after bilateral vagotomy and after killing the bird. They found no evidence that changing F_{IO_2} affects airway resistance.

The only physiological or pharmacological study found of the effects of irritant, anaesthetic, or toxic gases upon airway resistance or upon bronchial smooth muscle is that of Callanan et al. (1974). They showed that SO_2 decreased the resistance of the lower respiratory tract and that this effect did not depend upon the vagal innervation.

It is concluded that, while there is some evidence available for selective control of airway resistance, the amount of confirmed quantitative data is too limited, the consequence being that speculation continues to be the main feature of this field of study.

References

Baldwin, J.K.: Ph.D.thesis, Univ. of California, Davis (1973)

Bouverot, P.: Control of breathing in birds as compared with mammals. Physiol. Rev. in press (1977)

Brackenbury, J.H.: Airflow dynamics in the avian lung as determined by direct and indirect methods. Respir. Physiol. 13, 319-329 (1971)

Brackenbury, J.H.: Physical determinants of air flow pattern within the avian lung. Respir. Physiol. 15, 384-397 (1972a)

Brackenbury, J.H.: Lung-air-sac anatomy and respiratory pressures in the bird. J. Exp. Biol. 57, 543-550 (1972b)

Bretz, W.L., Schmidt-Nielsen, K.: Bird respiration: flow patterns in the duck lung. J. Exp. Biol. 54, 103-118 (1971)

Callanan, D., Dixon, M., Widdicombe, J.G., Wise, J.C.M.: Responses of geese to inhalation of irritant gases and injections of phenyl diguanide. Respir. Physiol. 22, 157-166 (1974)

Fedde, M.R., Burger, R.E., Kitchell, R.L.: The influence of the vagus nerve on respiration. Poultry Sci. 40, 1401 (1961)

Graf, W., Molony, V., Scheid, P.: A technique for study of lung function in birds by blocking the primary bronchus. Respir. Physiol. 26, 325-331 (1976)

Kampe, G., Crawford, E.C.: Oscillatory mechanics of the respiratory system of pigeons. Respir. Physiol. 18, 188-193 (1973)

King, A.S., Cowie, A.F.: The functional anatomy of the bronchial muscle of the bird. J. Anat. 105, 323-336 (1969)

King, J.R., Farner, D.S.: Energy metabolism, thermoregulation and body remperature. In: Biology and Comparative Physiology of Birds, Vol. II. Marshall, A.J. (ed.). New York: Academic Press, 1961, pp. 215-288

Leitner, L.-M., Roumy, M.: Vagal afferent activities related to the respiratory cycle in the duck: sensitivity to mechanical, chemical and electrical stimuli. Respir. Physiol. 22, 41-56 (1974)

Molony, V.: A study of vagal afferent activity in phase with breathing and its role in the control of breathing in Gallus domesticus. Ph.D. thesis, Univ. Liverpool, 1972

Molony, V., Graf, W., Scheid, P.: Effects of CO_2 on pulmonary air flow resistance in the duck. Respir. Physiol. 26, 333-349 (1976)

Ray, P.J., Fedde, M.R.: Responses to alterations in respiratory P_{O_2} and P_{CO_2} in the chicken. Respir. Physiol. 6, 135-143 (1969)

Scheid, P., Piiper, J.: Direct measurement of the pathways of respired gas in duck lungs. Respir. Physiol. 11, 308-314 (1971)

Zeuthen, E.: The ventilation of the respiratory tract of birds. Biol. Medd. 17, 1-51 (1942)

Origin of Carbon Dioxide in Caudal Airsacs of Birds

J. Piiper

Summary

The mechanisms responsible for the presence of CO_2 in the caudal airsacs of birds and for the O_2 concentration lower than in ambient air are analysed. Experimental evidence indicates that gas exchange through airsac walls and recirculation of lung gas via ventrobronchi ("Hazelhoff loop") play a quantitatively negligible role. Reinspiration of dead space gas, particularly when preferentially directed to the caudal airsacs, and gas exchange of neopulmonic parabronchi appear to constitute the main mechanisms. A consistent quantitative analysis based on adequate experimental data remains to be performed.

Introduction

Although important progress has been made in the analysis of gas exchange in birds, particularly relating to transfer of CO_2 and O_2 in the lungs proper (e.g., Meyer et al., 1976; Scheid et al., in press; Scheid, in press; Scheid, this volume), the gas exchange occurring in the whole complex avian respiratory system under physiological conditions is not yet satisfactorily understood. One of the important prerequisites for such an analysis is a sufficiently accurate knowledge of both the composition and the flow rate of gas entering and leaving the lung, i.e., the parabronchi, at any moment of the respiratory cycle.

In 1971 important experimental results were published from four laboratories working independently (Bouverot and Dejours, 1971; Brackenbury, 1971; Bretz and Schmidt-Nielsen, 1971; Scheid and Piiper, 1971; Scheid et al., 1972). These showed, apparently conclusively, that respiratory gas flow through the avian lung was unidirectional: from the dorsobronchi, through the parabronchi to the ventrobronchi, during both inspiration and expiration. This had been previously postulated on the basis of indirect evidence by Bethe (1925) and by Hazelhoff (1943).

Thus the "symmetrical" ventilation model proposed by Zeuthen (1942), according to which inspiration was a complete reversal of expiration in every part of the system, seemed to be refuted. Zeuthen's model, however, provided the simplest explanation for the CO_2 concentration observed in the caudal airsacs (2%-3%) amounting to about one-half the value measured in end-expired gas (5%-6%) which, according to this hypothesis, had passed the parabronchi twice. According to the experi-

Abteilung Physiologie, Max-Planck-Institut für experimentelle Medizin, D-3400 Göttingen, Federal Republic of Germany.

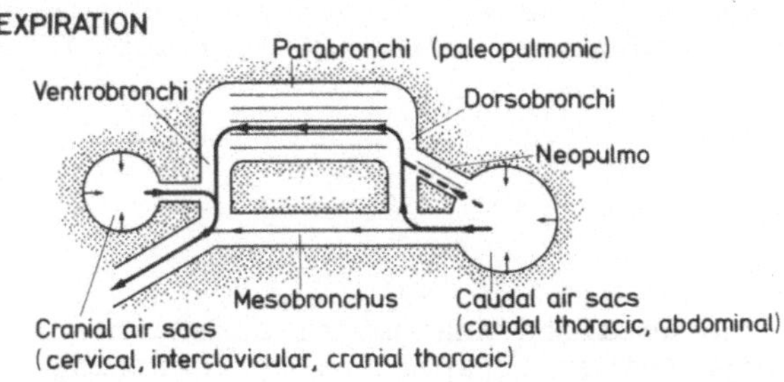

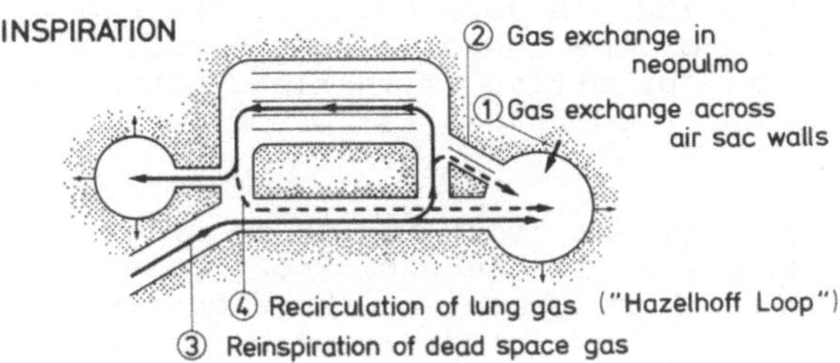

Fig. 1. Airflow direction in avian respiratory system during expiration and inspiration. The mechanisms potentially contributing to P_{CO_2} in caudal airsacs are indicated

mentally verified unidirectional flow model, part of the inspired gas enters the caudal airsacs, the rest reaches the cranial airsacs via dorsobronchi and parabronchi of the lung. How can we explain the composition of caudal airsac gas (about 2-4% CO_2 and 17-19% O_2)? The following four mechanisms will be discussed (Fig. 1):

1. Gas exchange across the airsac walls.
2. Gas exchange in the neopulmo.
3. Reinspiration of dead space gas.
4. Recirculation of lung gas ("Hazelhoff loop").

1. Gas Exchange across Airsac Walls

The great French zoologist of the early 19th century, Cuvier, assumed two types of gas exchange in birds, one pulmonary and the other, extrapulmonary, occurring in the airsacs. Sappey (1847) claimed that the scarce vascular supply of airsac walls allowed but very little gas exchange. Sappey's views were substantiated by Soum (1896) who, using the domestic duck, showed that in an abdominal airsac with blocked ostium very little CO_2 and O_2 exchange occurred and that rinsing of the airsac with CO for 15 min had no adverse effect on the bird.

A quantitative study of the gas exchange capacity of the caudal thoracic airsacs in the domestic duck was recently performed by Magnussen et al. (1976). They distinguished three routes of gas transfer: (1) with blood perfusing the airsac walls; (2) with adjoining airsacs through the airsac wall; and (3) with the lung airways through lung tissue lying on the airsac. Only the route (1) was shown to be of some importance. But the contribution of the direct airsac gas exchange to the total gas exchange of the bird amounted to a few percent only. Moreover, the difference of P_{CO_2} and P_{O_2} between inspired air and airsac, attributable to gas exchange across airsac walls, was estimated at 1.2 torr and 3.2 torr, respectively. These values represented small fractions when compared to the values observed: about 20 torr for CO_2 and 24 torr for O_2.

2. Gas Exchange in Neopulmo

Duncker (1971, 1972) has studied the detailed anatomical arrangement of the lung-airsac system in a large number of species, representing most of the major groups of birds. He distinguished between the *paleopulmo*, which corresponds to the arrangement represented in Figure 1, and the *neopulmo* which mainly consists of parabronchi connecting the caudal airsacs with the main bronchus and the dorsobronchi. The neopulmo appears to constitute up to 25% of the lungs in song birds, but was found to be wholly absent in some "primitive" flightless birds (penguins, emu). The neopulmo apparently comprises 5%-10% of duck lungs.

Anatomically, it is difficult to see how the neopulmo can be unidirectionally ventilated. The flow direction in the expiratory phase must be opposite to that in the inspiratory phase, or there must be flow during only one of the phases, the flow in the other phase passing through the main bronchial connection of the airsac. [In song birds, in which the connection from the primary bronchus to the abdominal sac is practically obliterated (Duncker, 1971, 1972), the airflow direction in neopulmonic parabronchi connected to this sac must be alternating.] It is evident that gas entering the caudal airsacs by way of the neopulmonic parabronchi during inspiration would carry CO_2 into them.

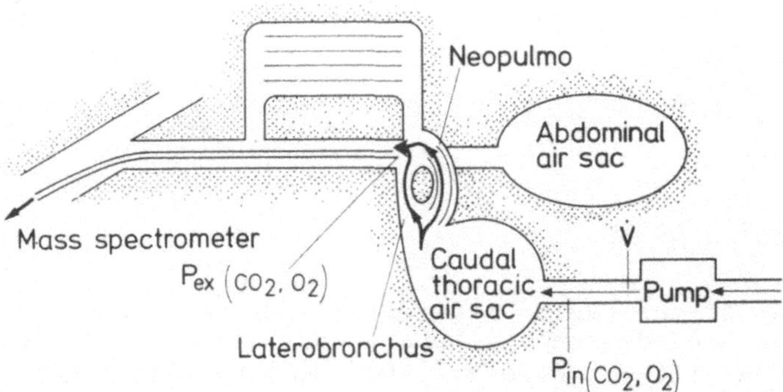

Fig. 2. Schema of experimental set-up for measurement of neopulmonic CO_2 and O_2 exchange in anesthetized duck preparations. The transfer rates are calcualted from the gas flow ($\dot{V}$) and the inflow-outflow partial pressure differences ($P_{in}-P_{ex}$)

In a preliminary series of experiments (unpublished) Magnussen and Scheid attempted to assess the role of neopulmonic gas exchange in anesthetized domestic ducks (Fig. 2). Both the left and the right caudal thoracic airsacs were cannulated and ventilated by a steady gas stream provided by gas pumps. On the experimental side, a catheter connected to a respiratory mass spectrometer was advanced into the primary bronchus till its tip lay at the opening of the laterobronchus. The CO_2 output ($\dot{M}_{CO_2}$) and O_2 uptake ($\dot{M}_{O_2}$) of the area ventilated by the gas stream (the laterobronchus and the neopulmonic parabronchi of the airsac) were calculated from the expired-inspired P_{CO_2} and P_{O_2} difference and the rate of gas flow. The $\dot{M}_{CO_2}$ and $\dot{M}_{O_2}$ values varied with the gas flow rate and the composition of gas mixture used.

From a number of measurements performed on four ducks in various conditions the following mean values were considered as the best estimate for physiological conditions: $\dot{M}_{CO_2}$ = 0.051 mmol/min, $\dot{M}_{O_2}$ = 0.053 mmol/min.

When the total neopulmonic gas exchange is estimated as three times the value for one caudal thoracic sac (i.e., 0.15 mmol/min), it appears that about 15% of the total $\dot{M}_{O_2}$ (and $\dot{M}_{CO_2}$) of the duck can be attributed to neopulmonic gas exchange. Calculating the effect of such neopulmonic CO_2 and O_2 transfer rates during inspiration upon the caudal airsacs receiving about 50% of the total ventilation of 820 ml/min (according to Scheid et al., 1974), one obtains P_{CO_2} = 7 torr and $P_{O_2} = P_{IO_2}$ - 7 torr for caudal airsacs (P_{IO_2}, O_2 pressure, in inspired gas).

These experimental results and the subsequent calculations may be challenged for quite a number of technical and physiological reasons. To explain the physiological P_{CO_2} in the caudal airsacs, measurements with reversed flow direction, i.e., flow entering the airsac after passage of neopulmonic parabronchi, should provide more direct evidence. This procedure was attempted in some experiments, but technical problems (uncontrolled leakage) could not be overcome. Therefore calculations based on these measurements, that about 7/20 ≈ 1/3 of the CO_2 in the caudal airsacs is attributable to gas exchange occurring in the neopulmo during inspiration, should be considered to represent but a very rough estimate. Although it may be inferred that this mechanism may substantially contribute to CO_2 in caudal airsacs in the duck.

In this connection it is of great interest to note that Burger et al. (1976) report that in the emu, whose lungs have been shown to be devoid of neopulmonic parabronchi, the CO_2 concentration is in the same range as reported for other birds possessing a neopulmo (duck, chicken).

3. Reinspiration of Dead Space Gas

Clearly the gas mixture high in CO_2 and low in O_2, contained in the airways from the external opening to, at least, the openings of the ventrobronchi at end-expiration, must be reinspired into caudal airsacs, lungs, and cranial airsacs at the beginning of the subsequent inspiration. There seems to be no reason why one should not estimate the value of the dead space in a bird from end-expired, mixed-expired, and inspired CO_2 concentrations by the Bohr formula devised for mammalian lungs.

On the basis of mean values of an experimental series in chickens, breathing spontaneously via tracheal cannula and a respiratory valve into a spirometer (Piiper et al., 1970), the dead space was calculated as 11.3 ml (34% of the tidal volume) and the resulting P_{CO_2} of gas entering the caudal airsacs was therefore (A) 11 torr, if it was assumed that inspiration of the dead space gas was apportioned according to flow rate to caudal and cranial airsacs, (B) 22 torr, if the entire dead space gas was assumed to be inspired into the caudal airsacs.

These values may be compared to airsac P_{CO_2} measured in similar experiments in other anesthetized chickens (Piiper et al., 1970): 15 torr in abdominal airsacs, 24 torr in caudal thoracic airsacs.

These caluclations indicate that the major part of the CO_2 in caudal airsacs of chickens, under these experimental conditions, is explainable by inspiration of dead space gas. However, a simultaneous measurement of all variables in the same individual bird is required for a more accurate assessment.

4. Recirculation of Lung Gas ("Hazelhoff Loop")

Hazelhoff (1943, 1951) has made the most outstanding experimental and theoretical contribution to the understanding of the airflow in the airways of birds. He showed that when certain structural features present in birds' lung were simulated, unidirectional flow was obtained in glass models by to-and-from movement of the piston of a syringe connected at the site of the caudal airsacs. Moreover, in a loop-like model receiving continuous flow, unidirectional parabronchial flow was also obtained. In these models he omitted an equivalent of the cranial airsacs altogether. Therefore, all the parabronchial flow had to return, i.e., recirculate, through the primary bronchus. It is relevant to observe that if sufficiently ventilated cranial airsacs are present, such a recirculation is no more a necessity.

Recently, experiments were started in our laboratory with the aim of quantitatively assessing the role of such a recirculation mechanism in gas exchange (Powell and Scheid, unpublished). In anesthetized ducks, breathing spontaneously or ventilated by pump, a specially prepared catheter connected to a respiratory mass spectrometer was introduced into the main bronchus. The spectrometer measured CO_2 and O_2 continuously at a number of specific sites from the trachea to the origin of the dorsobronchi.

Preliminary results showed that, during inspiration, inspired CO_2 and O_2 were recorded in the mesobronchus between the openings of the ventrobronchi and those of the dorsobronchi as well as in the region of the openings of the dorsobronchi as far caudally as the catheter could be advanced. Only in the region of the openings of the ventrobronchi the inspired values were not reached. This finding was attributed to diffusional or cardiogenic mixing of gas at this site with ventrobronchial gas. Its contribution in the sense of Hazelhoff loop, however, must have been very small, since at more caudal sites (opening of the dorsobronchi) the inspired gas CO_2 and O_2 values were clearly reached during inspiration.

Conclusions

The experimental evidence and the lines of reasoning presented above suggest that reinspiration of dead space gas and neopulmonic gas exchange are the major factors contributing to the relatively high CO_2

levels found in the caudal airsacs. A more accurate analysis is highly desirable particularly since the caudal airsac gas apparently represents the gas entering the parabronchi during the experitory phase and, therefore, is of particular importance in the analysis of the overall pulmonary gas exchange in birds.

References

Bethe, A.: Atmung: Allgemeines und Vergleichendes. In: Bethe, A., Bergmann, G.V., Embden, G., Ellinger, A. (eds.) Handbuch der normalen und pathologischen Physiologie, Bd. II. Berling: Springer, 1925, pp. 1-36

Bouverot, P., Dejours, P.: Pathway of respired gas in the air sacs-lungs apparatus of fowl and ducks. Respir. Physiol. 13, 330-342 (1971)

Brackenbury, J.H.: Airflow dynamics in the avian lung as determined by direct and indirect methods. Respir. Physiol. 13, 319-329 (1971)

Bretz, W.L., Schmidt-Nielsen, K.: Bird respiration: flow patterns in the duck lung. J. Exp. Biol. 54, 103-118 (1971)

Burger, R.E., Coleridge, J.C.G., Coleridge, H.M., Nye, P.C.G., Powell, F.L., Ehlers, C., Banzett, R.B.: Chemoreceptors in the paleopulmonic lung of the emu: discharge patterns during cyclic ventilation. Respir. Physiol. 28, 249-259 (1976)

Duncker, H.-R.: The lung-air sac system of birds. Ergebn. Anat. Entwickl.-Gesch. 45, 1-171 (1971)

Duncker, H.-R.: Structure of avian lungs. Respir. Physiol. 14, 44-63 (1972)

Hazelhoff, E.H.: Bouw en functie van de vogellong. Versl. gewone Vergad. Afd. Natuurk. Kon. Ned. Akad. Wet. 52, 391-400 (1943). English translation: Structure and function of the lung of birds. Poultry Sci. 30, 3-10 (1951)

Magnussen, H., Willmer, H., Scheid, P.: Gas exchange in air sacs: contribution to respiratory gas exchange in ducks. Respir. Physiol. 26, 129-146 (1976)

Meyer, M., Worth, H., Scheid, P.: Gas-blood CO_2 equilibration in parabronchial lungs of birds. J. Appl. Physiol. 41, 302-309 (1976)

Piiper, J., Drees, F., Scheid, P.: Gas exchange in the domestic fowl during spontaneous breathing and artificial ventilation. Respir. Physiol. 9, 234-245 (1970)

Sappey, Ph.C.: Recherches sur l'appareil respiratoire des oiseaux. Paris: Germer Baillière, 1847

Scheid, P.: Analysis of gas exchange between air capillaries and blood capillaries in avian lungs. Respir. Physiol. (in press)

Scheid, P., Piiper, J.: Direct measurement of the pathway of respired gas in duck lungs. Respir. Physiol. 11, 308-314 (1971)

Scheid, P., Slama, H., Piiper, J.: Mechanisms of unidirectional flow in parabronchi of avian lungs: measurements in duck lung preparations. Respir. Physiol. 14, 83-95 (1972)

Scheid, P., Slama, H., Willmer, H.: Volume and ventilation of air sacs in ducks studied by inert gas wash-out. Respir. Physiol. 21, 19-36 (1974)

Scheid, P., Worth, H., Holle, J.P., Meyer, M.: Effects of oscillating and intermittent ventilatory flow on efficacy of pulmonary O_2 transfer in the duck. Respir. Physiol. (in press)

Soum, J.M.: Recherches physiologiques sur l'appareil respiratoire des oiseaux. Paris: Masson, 1896

Zeuthen, E.: The ventilation of the respiratory tract in birds. Kgl. Danske Videnskab. Selskab. Biol. Medd. 17, 1-50 (1942)

Session IV

Control of Ventilation: Intrapulmonary Carbon Dioxide Sensitivity and Respiratory Centers

Chairman: P. Dejours

Effects of Intrapulmonary Chemoreceptors in Perfused and Non-Perfused Lungs

R. E. Burger, M. R. Barker, P. C. G. Nye, and F. L. Powell

Summary

We measured respiratory movements in 21 thoracotomized cockerels whose lungs were unidirectionally ventilated separately. The right lungs were denervated so that variations in right pulmonary P_{CO_2} or blood flow could not affect respiratory movements. P_{aCO_2} was kept at 27 torr by adjusting gas flow to perfused lungs. Amplitude of respiratory movements were the same in six cockerels with intact and in six with blocked left pulmonary artery, the latter receiving 1 l/min 28 torr P_{CO_2} in O_2 in the left lung. We injected 5 mg/kg 2,4-dinitrophenol in these, and after adjusting ventilation so that P_{aCO_2} was again 27 torr, we found that respiratory amplitude was slightly smaller in those with perfused than in those non-perfused left lungs. In three cockerels, we did not find any alteration of intrapulmonary chemoreceptor inhibition of respiratory movements in non-perfused left lungs when dinitrophenol was administered. We estimated the respiratory inhibition from calculated discharge frequency in six cockerels given three P_{CO_2} levels in left, non-perfused lungs. We calculated P_{CO_2} profiles resulting from various $C_{\bar{v}CO_2}$; higher $C_{\bar{v}CO_2}$ gave lower initial P_{CO_2} but higher P_{eCO_2}. As most receptors are within the initial P_{CO_2} region, calculated discharge increases, increasing predicted respiratory inhibition.

1. Introduction

Discharge frequencies of intrapulmonary chemoreceptors (IPC) are inversely proportional to receptive site P_{CO_2}. In chickens IPC are found only in the gas exchanging region (Nye, 1977). Respiratory control by IPC during increased metabolic rate is unknown (Scheid et al., 1974; Banzett and Burger, 1977). We have been studying the ventilatory response to increased metabolic rate following administration of 2,4-dinitrophenol (DNP). We report here: 1. DNP does not affect IPC discharge rate, 2. IPC in perfused lungs does not stimulate and may depress respiratory movements following DNP administration compared to IPC in non-perfused lungs, and 3. a model which explains the lack of ventilatory stimulation by IPC during increased metabolic rate.

Avian Sciences Department, University of California, Davis, CA 95616/USA.

2. Methods

a) General Preparation

We thoracotomized 15 cockerels. We cannulated the mesobronchi of three as described before (Burger et al., 1974). In 12, a double lumen catheter was placed in the left mesobronchus which placed a blocking ring between the mediodorsales and medioventrales. We plugged all left airsac ostia, delivering gas caudal to the block, and drew exchanged gas from cranial to the block. We kept body temperature at 41.5°C with a heating lamp and a fan. We drew arterial blood samples from a cannula in the right common carotid artery. We cut the right vagus just above the recurrent nerve and the right cardiac sympathetic nerve . Ventilation in all was adjusted to give 27 ± 1 torr P_{aCO_2}. When both lungs were perfused, each lung received one-half of the ventilation.

b) Measurements

We recorded respiratory (vertical sternal) movements (Burger et al., 1974), averaging three breaths. For content, we used a modification of the method of Linden et al. (1963) by injecting 0.02 ml blood in a "microsyringe" (Lexington Instruments) into 0.4 ml of HCl-ferricyanide solution in a 1 ml syringe with a Chaney adaptor (Hamilton). We calibrated with acid-standardized bicarbonate solutions (Nye, 1977). We measured P_{CO_2} of the acid solution and P_{aCO_2} with separate blood gas electrodes (Model 113, Instumentation Laboratories). Ventilation was measured with Model 20, Vol-O-Flo flow meters.

c) Protocols

Series I. We blocked the left pulmonary artery in three cockerels. We gave 5 mg/kg DNP i.v., adjusting the flow of O_2 through the perfused lung to give 27-torr P_{aCO_2}. We gave 28-torr P_{CO_2} in O_2 to the left lung. We measured respiratory movements and P_{aCO_2}, 15 min after DNP administration. We then released the snare around the left pulmonary artery to expose IPC to DNP-containing blood for several minutes, then snared it, and measured the variables after a further 15 min. We again gave 5 mg/kg DNP i.v. and after adjusting flow of O_2, measured respiration again.

Series II. We adjusted P_{aCO_2} to 27-torr in six cockerels with intact and in six with blocked left pulmonary circulation, giving 28-torr P_{CO_2} to the non-perfused lung and measured respiratory movements. We gave 5 mg/kg DNP to all, and after adjusting O_2 flow to the perfused lung(s), we measured respiratory movements again 15 min later.

3. Results

Series I. Initial respiratory movements were 2 mm at 110/min. After exposing the lung to DNP-containing blood, we found no change in respiration (Table 1). Respiratory movements increased 1 mm, and flow of O_2 was increased by 0.39 l/min to keep 27-torr P_{aCO_2}, following double dosage of DNP.

Table 1. Effect of 2,4-dinitrophenol on IPC discharge frequency and respiratory movements

Condition	A_R	f_R	P_{aCO_2}	$\dot{V}_g$
	mm	min^{-1}	torr	l/min
A	2.0 (0.3)	108 (24)	26.3 (1.1)	0.54 (0.05)
B	2.0 (0.3)	105 (19)	27.0 (1.3)	0.56 (0.05)
C	3.0 (0.5)	88 (20)	26.6 (1.5)	0.95 (0.02)

Three cockerels were thoractomized, the left lung ventilated with 28-torr P_{CO_2}, and the left pulmonary artery ligated. The right lung was denervated and ventilated with O_2. A_R, amplitude of respiratory movements; f_R, frequency of A_R. Flow of O_2 required to give indicated P_{aCO_2}, $\dot{V}_g$. All values given as means (S.E.).

In condition A, the measurements were made 15 min after 5 mg/kg 2,4-dinitrophenol was administered. In condition B, the left lung was exposed to flow of venous blood containing 2,4-dinitrophenol for several minutes. Measurements were made 15 min after an additional 5 mg/kg 2,4-dinitrophenol had been administered in condition C.

Series II. Both groups had 1.5-mm respiratory movements initially and required about the same flow of O_2 to give 27-torr P_{aCO_2}. After DNP administration, cockerels with blocked left pulmonary circulation increased respiratory amplitude by 1.3 mm more than those with intact pulmonary circulation (Table 2). The former required more O_2 flow and may have responded more to the administered DNP. If we correct the difference in respiratory amplitude for the difference in required O_2 flow between the two groups, we find the difference is still 0.8 mm more than for those with blocked pulmonary circulation.

Table 2. Effect of dinitrophenol-induced increased metabolism on respiratory movements

Left pulmonary artery	DNP	A_R	f_R	P_{aCO_2}	$\dot{V}_g$	T_B
	mg/kg	mm	min^{-1}	torr	l/min	°C
Intact	-	1.5 (0.4)	61 (8)	26.7 (0.6)	0.35 (0.03)	41.3 (0.2)
	5	2.3 (0.5)	37 (4)	27.1 (7)	0.74 (0.06)	41.3 (0.2)
Blocked	-	1.5 (0.2)	35 (9)	26.8 (0.7)	0.38 (0.03)	41.1 (0.1)
	5	3.6 (0.7)	35 (6)	27.0 (0.5)	0.92 (0.18)	41.2 (0.2)

There were six unidirectionally ventilated cockerels per group, measured before and after administration of 2,4-dinitrophenol. In both groups, the right lung was denervated, and the left lung was innervated. Symbols as in Table 1. T_B, body temperature.

4. Model of Respiratory Control by IPC

a) P_{CO_2} Profiles at Constant P_{aCO_2} and Varying $C_{\bar{v}CO_2}$

We used a serial CO_2-mass balance model similar to that described before (Meyer et al., 1976) except that we used 1. a formula, $C_{CO_2}^{-1}$ =

$b + m \cdot P_{\bar{C}O_2}$ (Nye, 1977), to describe the relation of C_{CO_2} and P_{CO_2}; 2. CO_2 of mixed venous blood; 3. complete oxygenation of end-capillary blood; and 4. 100 instead of 20 segments. We used 0.0303 and 0.612 for b and m, respectively and 20.8 mmol/l for normal $C_{\bar{v}CO_2}$ - typical values we have found for chickens in similar experiments - and 27-torr P_{aCO_2}. We varied $C_{\bar{v}CO_2}$ from 0.95 normal (ca. minimal value that will still give 27 torr P_{aCO_2}) to 1.26 normal [ca. 10% O_2 saturation of venous blood with a respiratory exchange ratio (R) of 0.8], adjusting ventilation-perfusion ratio to give 27 torr P_{aCO_2}.

We found that the P_{CO_2} profiles of higher than normal $C_{\bar{v}CO_2}$ increased more slowly initially, but had higher final P_{CO_2} than those of normal $C_{\bar{v}CO_2}$ (Fig. 1).

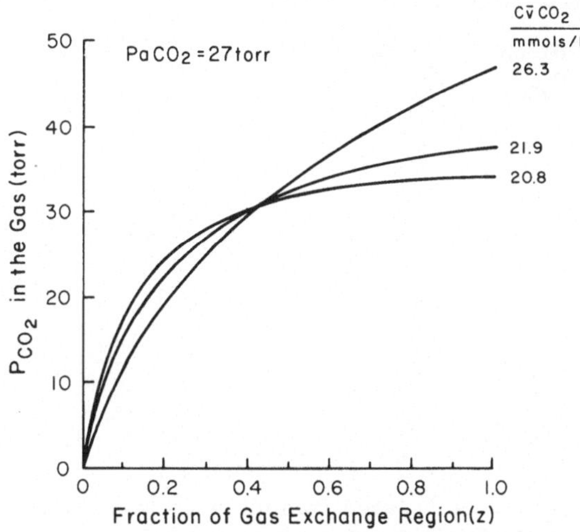

Fig. 1. Variation in P_{CO_2} along the gas exchange region ($0 \leq z \leq 1$), calculated at three different levels of $C_{\bar{v}CO_2}$. The ratio of ventilation to perfusion has been adjusted so that P_{aCO_2} is 27 torr in each case

b) Histographs of IPC Discharge Frequencies along the Gas Exchange Region

We used a sample of 46 receptors whose location (z) and sensitivity to receptive site P_{CO_2} had been estimated previously (Nye, 1977). There were a large concentration of receptors from 0.05 to 0.15 z, a smaller density at 0.35 to 0.55 z, and a few at 1.0 z. We determined discharge frequency of each receptor at the z of the P_{CO_2} of each profile, found at the location of the receptor.

We found the discharge frequencies of the receptors at 0.05 to 0.15 z were higher at high (Fig. 2B) than normal (Fig. 2A) $C_{\bar{v}CO_2}$, while those at 0.35 to 0.55 z were relatively insensitive to $C_{\bar{v}CO_2}$ (Fig. 2). Those at 1.0 z had small discharge frequencies at all $C_{\bar{v}CO_2}$.

c) Relation of IPC Discharge Frequencies to $C_{\bar{v}CO_2}$ at Constant P_{aCO_2}

We averaged discharge frequencies at all z for all receptors for each of seven $C_{\bar{v}CO_2}$ and P_{CO_2} profiles (Fig. 3). Discharge frequency varies as $6.29-17.3\ (C_{\bar{v}CO_2}-14.4)^{-1}$ ($r = 0.99995$).

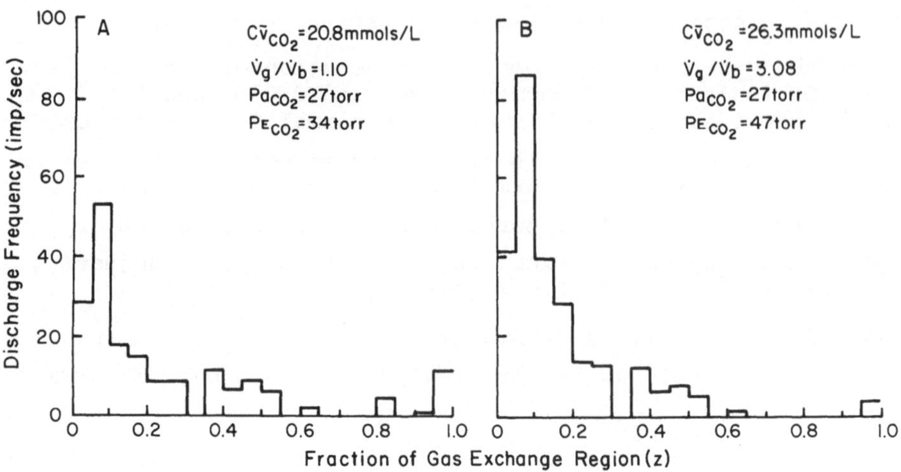

Fig. 2. Calculated variation in the sum of discharge frequencies from 46 IPC along the gas exchange region, for two levels of $C_{\bar{v}CO_2}$

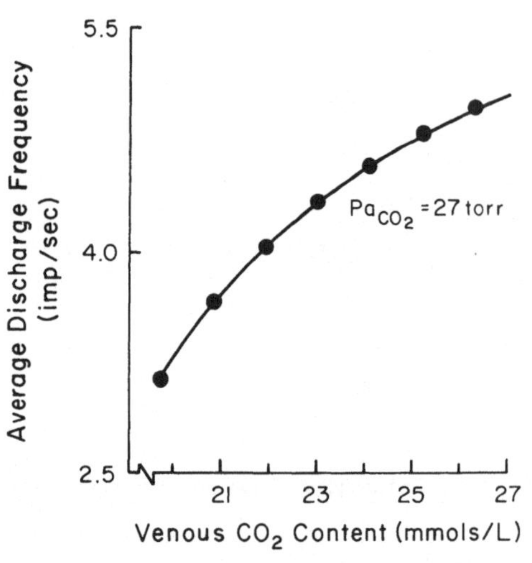

Fig. 3. Calculated variation in total discharge frequencies of 46 IPC as $C_{\bar{v}CO_2}$ varies from 19.7 mmol/l to 26.3 mmol/l. The ratio of ventilation to perfusion has been adjusted for each $C_{\bar{v}CO_2}$ so that P_{aCO_2} is 27 torr

d) Relation of Respiratory Movements to IPC Discharge Frequency

Respiratory movements of five cockerels were measured at 27-torr P_{aCO_2} and 16-, 32-, and 48-torr P_{CO_2} of gas ventilating the left non-perfused lung. The right perfused lung was denervated. We estimated averaged IPC discharge frequencies at these P_{CO_2}. We found that respiratory movements are estimated (Fig. 4) by 2.1-0.24 (discharge frequency) (r = 0.9996).

e) Relation of Respiratory Movements to $C_{\bar{v}CO_2}$ at Constant P_{aCO_2}

We substituted predicted respiratory movements for discharge frequencies into the relation in Figure 3. We found respiratory movements to be

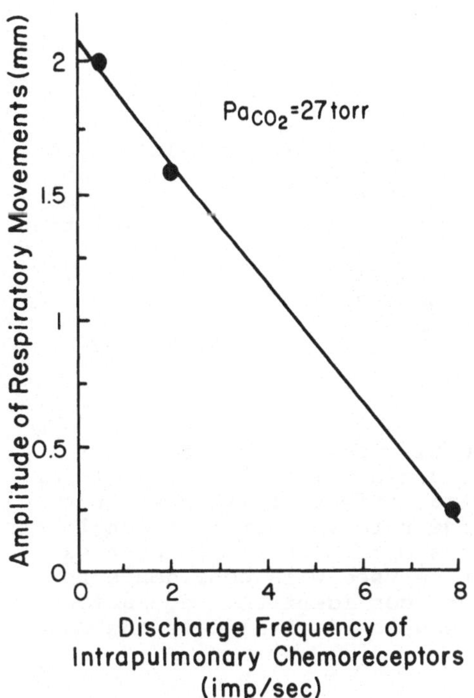

Fig. 4. Variation in amplitude of respiratory movements of six cockerels at calculated average discharge frequencies of 46 IPC at 16.1-, 31.9- and 48.0-torr P_{CO_2} of gas ventilating a non-perfused lung. Ventilation with O_2 to the perfused lung was adjusted so that P_{aCO_2} was 27 ± 1 torr. at the highest average discharge frequency (i.e., 16.1-torr P_{CO_2}), amplitude of respiratory movements is smallest

estimated by $0.59 + 4.07\ (C_{\bar{v}CO_2} - 14.4)^{-1}$ ($r = 0.99995$), that is, respiratory movements should not increase, but even decrease slightly as $C_{\bar{v}CO_2}$ increases. Figure 5 shows the relation between the predicted change in amplitude of respiratory movements when $C_{\bar{v}CO_2}$ changes from 20.8 mmol/l.

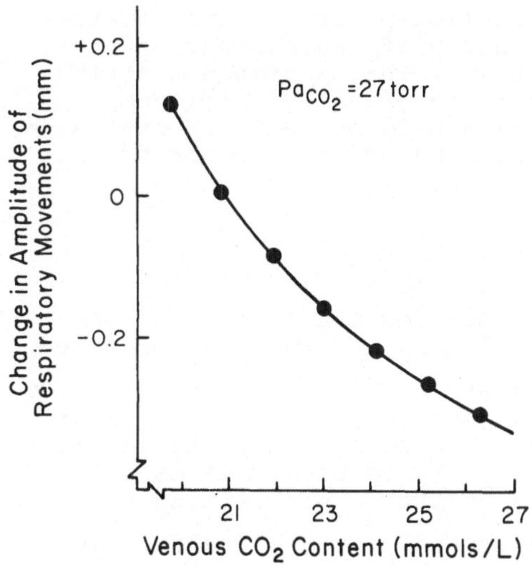

Fig. 5. Calculated variation in amplitude of respiratory movements when $C_{\bar{v}CO_2}$ varies from 19.7 mmol/l to 26.3 mmol/l, after adjusting the ventilation:to perfusion ratio to give 27-torr P_{aCO_2}

5. Discussion

a) Critique of Methods

We did not attempt to quantify CO_2 transfer so we cannot relate these results directly to the model calculations. More importantly, we do not know if the ventilation-perfusion inequalities in the innervated, perfused left lung, that may exist, were constant when the ventilation and perfusion was altered following DNP administration. To correlate results from such experiments with spontaneously breathing animals, quantitative gas transfer estimates will be essential in both ventilated and spontaneously breathing animals.

b) Critique of Model

We assumed no ventilation-perfusion inequalities, no shunts, equal density of receptors in all parabronchi, and no bias in the sample of receptors. We also assumed that the central effect of IPC discharge is related to the average of IPC discharge rate and not to a non-linear function of that discharge rate. These assumptions have not been explicitly tested so that the model cannot be used with confidence for quantitative predictions. We are, however, confident that deviations form these assumptions would have to be severe to invalidate the model for qualitative predictions though.

c) Correlation of Experimental Results with Model's Predictions

The qualitative agreement between the model's predictions and experimental observations gives us increased confidence in our conclusion that IPC do not, and in fact, cannot stimulate ventilation during increased metabolic rate if P_{aCO_2} is constant. IPC may even depress ventilation somewhat if P_{aCO_2} is held constant experimentally or by some other afferent input, although the predicted effect is small. The lack of quantitative agreement between the observed and predicted results are probably the result of altered ventilation-perfusion inequalities, before and after DNP administration, although if the central nervous system effect of high discharge rates is more than a simple contribution of those high frequencies to the average discharge frequency - such as would occur if a power law relation held between IPC discharge and respiratory movements - a similar discrepancy would occur between these predicted results and observed results.

d) Future Applications

The comparison of reflex effects with calculated IPC discharge frequencies in various profiles, including reversal of the P_{CO_2} profile by reversed flow, can help answer many questions about the role of IPC in respiratory control. For example, a major question is the form of the input-output relation between IPC discharge frequency and ventilation, a question of interest for sensory physiology as well as for respiratory control. Another is the role of IPC in stabilizing P_{aCO_2} when inhaled P_{CO_2} is varied experimentally and, through the interaction of O_2 and CO_2 exchange, in inhaled P_{O_2}. One interesting question is the relation of IPC to acid-base disturbances of metabolic origin. We will need more studies of receptor location and discharge frequencies in different forms of the P_{CO_2} profile, coupled with quantitative analysis of gas transfer to estimate ventilation-perfusion inequalities. Finally, an extension of these kinds of studies to varying P_{CO_2} and flow of ventilatory gas and their effects on discharge frequencies and respiratory movements would yield one of the most complete analyses of a major sensory system in the control of ventilation.

References

Banzett, R.B., Burger, R.E.: Response of avian intrapulmonary chemoreceptors to venous CO_2 and ventilatory gas flow. Respir. Physiol. 29, 63-72 (1977)

Burger, R.E., Osborne, J.L., Banzett, R.B.: Intrapulmonary chemoreceptors in *Gallus domesticus:* Adequate stimulus and functional localization. Respir. Physiol. 22, 87-97 (1974)

Linden, R.J., Ledsome, L.R., Norman, J.: Simple methods for the determination of the concentrations of carbon dioxide and oxygen in blood. Brit. J. Anaesth. 37, 77-88 (1965)

Meyer, M., Worth, H., Scheid, P.: Gas-blood CO_2 equilibration in parabronchial lungs of birds. J. Appl. Physiol. 41, 302-309 (1976)

Nye, P.C.G.: Stimulus-response relations and location of intrapulmonary chemoreceptors in the lung of *Gallus domesticus*. Ph.D. thesis Univ. California, Davis, CA 95616, 1977

Role of Avian Intrapulmonary Chemoreceptors in the Ventilatory Response in Inhaled Carbon Dioxide: Effects of Acetazolamide in the Duck

F. L. Powell, M. R. Fedde, R. K. Gratz, and P. Scheid

Summary

Acetazolamide (Diamox) was administered to ducks to block carbonic anhydrase (CA). Two series of experiments were performed. In Series I, ventilation was measured in spontaneously breathing, unanesthetized birds at various static levels of inspired CO_2 concentration (F_{ICO_2}). In Series II, single unit activity was recorded from intrapulmonary chemoreceptors (IPC) in anesthetized birds unidirectionally ventilated with gas mixtures containing various fractions of CO_2. Results of Series I showed that at any level of F_{ICO_2}, ventilation was lower after CA inhibition than before. Furthermore, the increment in ventilation for a given increment in F_{ICO_2} was markedly reduced after CA inhibition. In Series II, the discharge frequency of IPC was increased, and their CO_2 sensitivity attenuated, after acetazolamide. The reduced ventilation at various levels of F_{ICO_2} following acetazolamide may be explained by the increased discharge and the reduced CO_2 sensitivity of IPC.

1. Introduction

Although correlation between avian intrapulmonary chemoreceptors (IPC) and control of breathing is heavily supported, a causal relationship has not been demonstrated. If IPC activity could be controlled and directly related to ensuing ventilatory changes, evidence substantiating the role of IPC in controlling breathing would be more direct. Such an approach is similar to using NaCN to stimulate carotid body chemoreceptors and inferring their role from the changes induced in breathing. Scheid et al. (1977) reported that inhibition of carbonic anhydrase (CA) with acetazolamide increased IPC discharge frequency in lizards and decreased sensitivity to CO_2 changes. We have observed the effects of acetazolamide on IPC activity and on ventilatory sensitivity to CO_2 in the duck and correlated both results to assess the role of IPC in ventilatory control.

2. Methods

Adult domestic muscovy ducks, *Cairina moschata*, average weight 1.72 kg, were used. Carbonic anhydrase (CA) activity was blocked by intravenous

Max-Planck-Institut für experimentelle Medizin, Abteilung Physiologie, 3400 Göttingen, Federal Republic of Germany.

injection of acetazolamide (Diamox, Lederle, München, 50 mg/kg). Respired gas concentrations were measured with a respiratory mass spectrometer (Centronic, Type Q806R, Croydon, England).

a) Series I. Ventilatory Response to Inhaled CO_2

Ventilatory response to inhaled CO_2 was measured in five unanesthetized, spontaneously breathing ducks, which were only lightly restrained. Earlier, the trachea was cannulated, and catheters installed in arteries and in the right side of the heart under local anesthesia (Depot Novanest, Haury, München, FRG). The bird inhaled gases of various CO_2 concentrations (F_{ICO_2}) (inhaled O_2, F_{IO_2} = 0.3) through a low resistance, low dead space nonrebreathing valve. Respiratory frequency was measured with a pneumotachometer (Godart-Statham, type 17212 Bilthoven, Holland). After 15 min at a given level of F_{ICO_2}, expired gas volume was measured with a spirometer. CO_2-sensitivity curves were thus obtained before and after blocking CA activity with acetazolamide.

b) Series II. Carbon Dioxide Sensitivity of IPC

In this series, four anesthetized ducks were used. The experimental arrangement and the technique of searching for and recording the activity of pulmonary afferents from IPC were essentially the same as those described by Fedde et al. (1974). Briefly, ventilatory gas (F_{ICO_2}= 0.3) was insufflated at a steady rate of 1 l/min into the right caudal thoracic airsac. The right vagus was filamented for single-unit recording from IPC afferents, which were identified by their rapid response to sudden changes in F_{ICO_2}.

Activity of IPC afferents was recorded at various static levels of F_{ICO_2} before and after CA inhibition.

3. Results

a) Series I. Influence of Acetazolamide on Sensitivity to Inhaled CO_2

Both before and after blocking CA, ventilation ($\dot{V}_E$) increased when F_{ICO_2} was increased (cf. Fig. 1A). The effect of CO_2 was mainly mediated by increased tidal volume. At each level of F_{ICO_2}, ventilation was significantly diminished when CA was blocked. Thus, the ventilatory response to inhaled CO_2, expressed as the change in $\dot{V}_E$ for a given change in F_{ICO_2}, was markedly suppressed without CA activity.

Before administering acetazolamide, F_{ICO_2} could be raised to only about 0.054 without the animal struggling; after acetazolamide, this tolerance level was elevated to 0.071.

b) Series II. CO_2 Sensitivity of IPC

Before acetazolamide, all four IPC recorded in the four birds showed the inverse relation of afferent discharge frequency to F_{ICO_2} (Fig. 1B), which is typical for these receptors (see Fedde and Kuhlmann,

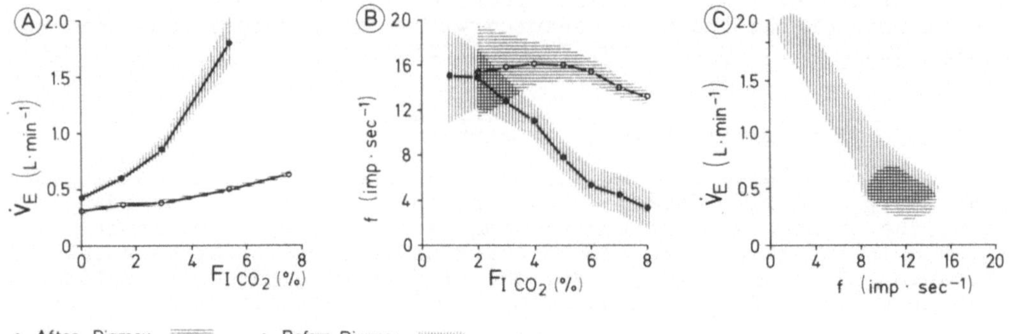

Fig. 1. Correlation between IPC discharge frequency (f) and ventilation ($\dot{V}_E$) before and after carbonic anhydrase inhibition with acetazolamide. *Shaded areas* encompass ± 1 standard error of the mean

this symposium). After acetazolamide the discharge frequency at any level of F_{ICO_2} was higher than before and, furthermore, the receptors were less sensitive to CO_2 (Fig. 1B).

4. Discussion

Effects of Acetazolamide on Ventilation

In mammals at rest, ventilation increases when CA is inhibited (cf. Maren, 1967): however, we found a significant ventilatory decrease in our ducks. Again, the ventilatory response to CO_2 in man apparently is unaffected by CA inhibition (Chiesa et al., 1969), while in our experiments a depression was marked.

Changes in IPC discharge evidently relate to changes in the output from the motoneuronal pool to the respiratory muscles (Fedde et al., 1969; Fedde and Peterson, 1970); therefore, we account for the ventilatory change after CA inhibition by the altered discharge frequency from these receptors. We have combined the results of both series of experiments from the averaged data in Figures 1A, B, so ventilation can be plotted as a function of IPC discharge; for each value of ventilation at a given F_{ICO_2} (Fig. 1A) the corresponding receptor discharge can be read from the average curve (Fig. 1B) at the same F_{ICO_2}. Such a plot is shown in Figure 1C for values before (vertical lines) and after CA is blocked with acetazolamide (horizontal lines). Before acetazolamide, there is an apparent linear increase in ventilation with decreased receptor discharge.

After acetazolamide, the responses of ventilation and IPC to changes in CO_2 concentrations are altered such that ventilation is described by the same function of IPC discharge frequency as before CA inhibition, as seen by the overlap and alignment of the areas describing both relations in Figure 1C.

The experiments thus provide direct evidence that the effect of acetazolamide on the ventilatory response of ducks to inhaled CO_2 is mainly mediated by the effect of this drug on the sensitivity of IPC and suggests that these receptors are important in controlling ventilation in birds.

Acknowledgments. This study was supported, in part, by a grant-in-aid from the American Heart Association, Kansas Affiliate, Inc. Contribution No. 78-49a, Department of Anatomy and Physiology, KAES, Kansas State University, Manhattan, Kansas 66506, USA.

References

Chiesa, A., Stretton, T.B., Massoud, A.A.E., Howell, J.B.L.: The effects of inhibition of carbonic anhydrase with dichlorphenamide on ventilatory control at rest and on exercise in normal subjects. Clin. Sci. 37, 689-706 (1969)

Fedde, M.R., Gatz, R.N., Slama, H., Scheid, P: Intrapulmonary CO_2 receptors in the duck: I. Stimulus specificity. Respir. Physiol. 22, 99-114 (1974)

Fedde, M.R., Peterson, D.F.: Intrapulmonary receptor response to changes in airway-gas composition in *Gallus domesticus*. J. Physiol. (London) 209, 609-625 (1970)

Fedde, M.R., deWet, P-D., Kitchell, R.L.: Motor unit recruitment pattern and tonic activity in respiratory muscles of *Gallus domesticus*. J. Neurophysiol. 32, 995-1004 (1969)

Maren, T.H.: Carbonic anhydrase: Chemistry, physiology, and inhibition. Physiol. Rev. 47, 595-781 (1967)

Scheid, P., Kuhlmann, W.D., Fedde, M.R.: Intrapulmonary receptors in the Tegu lizard: II. Functional characteristics and localization. Respir. Physiol. 29, 49-62 (1977)

Ventilatory Responses During Arterial Homeostasis of P_{CO_2} at Low Levels of Inspired Carbon Dioxide

J. L. Osborne and G. S. Mitchell

Summary

Ventilatory responses in anesthetized chickens were measured during arterial P_{CO_2} (P_{aCO_2}) homeostasis at low levels of inspired carbon dioxide (P_{ICO_2}). The P_{aCO_2} and hydrogen ion concentration in arterial blood ($[H^+]a$) were regulated at control levels (partial pressure of CO_2 in inspired gas, P_{ICO_2} = 0) as P_{ICO_2} was varied between 0 and 21 torr under both normoxic (Series I) and hyperoxic (Series II) conditions. During normoxia, mean expiratory minute volume ($\dot{V}_T$) increased by 53%, mean tidal volume (V_T) increased by 82%, and mean respiratory frequency (f) decreased by 25%. We conclude that CO_2-sensitive intrapulmonary chemoreceptors mediate the ventilatory response to inhalted CO_2 during arterial isocapnia and that parabronchial ventilation is the regulated variable accounting for arterial hemoestasis of P_{CO_2}.

1. Introduction

Recent experiments in our laboratory utilizing anesthetized, supine chickens have shown that during hyperoxia, arterial P_{CO_2} (P_{aCO_2}) is regulated at or below its control level as the inspired partial pressure of CO_2 (P_{ICO_2}) is either increased or decreased between 0 and 21 torr (Osborne and Mitchell, 1977). We have concluded from these experiments that the isocapnic hyperpnea associated with inhalation of CO_2 is a CO_2-coupled phenomenon and that it is mediated by CO_2-sensitive intrapulmonary chemoreceptors.

In the present studies, we have investigated whether or not P_{aCO_2} regulation occurs during inhalation of CO_2 in the upright anesthetized chicken when breathing a normoxic gas mixture. In addition, we measured the associated isocapnic hyperpnea in terms of changes in tidal volume (V_T), respiratory frequency (f), and expiratory minute volume (V_E).

The results from these studies are compared to similar studies in the supine anesthetized chicken breathing an hyperoxic gas mixture.

Department of Physiology, University of California, Irvine, Irvine, California 92717/USA.

2. Methods

Experiments were conducted on male White Leghorn chickens ranging in weight from 1.1 to 1.9 kg and 17 to 28 weeks of age. All animals were anesthetized with phenobarbital (160 mg/kg). Colonic temperature was continuously monitored and regulated at $41.0^{o} \pm 0.2^{o}C$. A tracheal cannula was inserted low in the neck and catheters were placed in the brachial vein and sciatic artery (P.E. 90).

The tracheal cannula was connected to a Fleisch pneumotachograph (type 00) to measure ventilatory flow rates. The flow signal from a differential pressure transducer (Statham, model PM 15) was electronically integrated to obtain tidal volume (V_T).

Arterial blood samples were withdrawn from the arterial cannula and analyzed for P_{CO_2}, P_{O_2}, and pH with a micro blood gas analyzer (Radiometer; model BMS 3, MK 2, and pHM 73) at 41.0°C. Blood gas and pH electrodes were calibrated before and after each measurement. The P_{CO_2} and P_{O_2} electrodes were calibrated with gases provided by a precision gas mixing pump (Radiometer, model GMA 2), and the pH electrode with standard buffer solutions.

The ventilatory gas was passed perpendicular to the pneumotachograph through a tee-connector at approximately 5 l/min. This flow rate exceed the maximum inspiratory and expiratory flow rates. Samples of the inspired gas were withdrawn at approximately 0.15 l/min and analyzed with an infrared CO_2 analyzer (Beckman, model LB-2). The CO_2 analyzer was calibrated at the beginning and periodically throughout the experiment with two known concentrations of gas: 4.97% CO_2 in O_2 and 10.24% CO_2 in N_2.

Series I. Eight chickens were placed in an upright position so as not to restrict ventilation (Fig. 1). The ventilatory gas in this series of experiments was air; CO_2 was added to the gas in accordance with the experimental protocol. In five experiments, P_{ICO_2} was increased sequentially from 0 to 28 torr and in three experiments P_{ICO_2} was changed in a varied pattern.

Series II. Three chickens placed in a supine position were ventilated with an hyperoxic gas mixture (50% O_2 in N_2). In two experiments, P_{ICO_2} was increased sequentially from 0 to 35.6 torr; in the third experiment, the pattern of CO_2 administration was varied.

At each level of P_{ICO_2}, ten min were allowed for equilibration before ventilatory parameters were measured and arterial blood samples withdrawn. Three blood samples (approximately 0.3 ml per sample) were withdrawn at each level of P_{ICO_2} and immediately analyzed.

3. Results

a) Series I: Upright Chicken Breathing Normoxic Gas

Figure 2 shows the changes in P_{aCO_2} and $[H^+]_a$ for six experiments in which P_{aCO_2} was regulated at its control level as P_{ICO_2} was varied

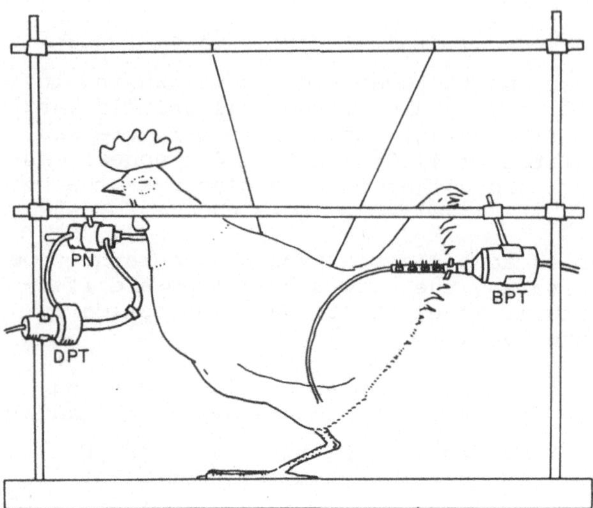

Fig. 1. Animal preparation. Chickens were suspended from a supporting frame in an upright position. *BPT*, blood pressure transducer; *DPT*, differential pressure transducer; *PN*, pneumotachograph. Four 3-way valves were placed in series with the blood pressure transducer and arterial catheter for withdrawl of arterial blood samples into heparinized 0.5 ml glass syringes

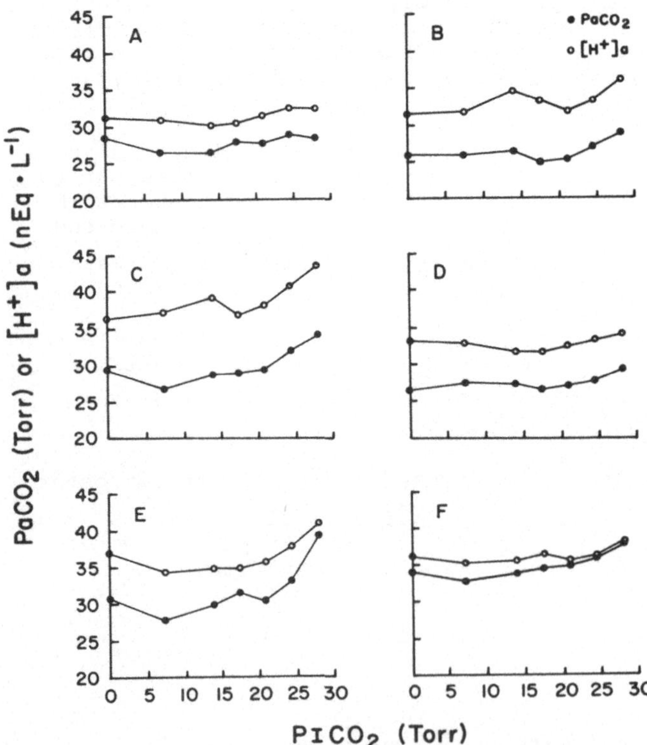

Fig. 2. Changes in P_{aCO_2} and $[H^+]_a$ during inhalation of CO_2 in six anesthetized, upright chickens under normoxic conditions. Corresponding ventilatory changes are shown in Figure 3

from 0 to 21 torr. In two experiments, not shown, P_{aCO_2} was regulated between 0 and 14 torr but not at 21 torr.

Mean values for the six experiments shown in Figure 2 are listed in Table 1. At 24.4 torr P_{ICO_2}, both P_{aCO_2} ($P < 0.01$) and $[H^+]_a$ ($P < 0.10$) are significantly greater than control levels.

Table 1. Blood gas and ventilatory values during inhalation of CO_2 in upright, anesthetized chickens (normoxia)

P_{ICO_2} (torr)[a]	P_{aCO_2} (torr)[b]	$[H^+]_a$ (n Eq/l)[c]	P_{aO_2} (torr)[d]	V_T (ml)[e]	f (min^{-1})[f]	$\dot{V}_E$ (ml/min)[g]
0	29.2 ±1.2	34.24 ±1.08	91 ±3	12.2 ±2.7	55.8 ±17.7	474 ±54
7.2	27.8 ±1.0	33.68 ±0.96	98 ±3	12.9 ±2.5	50.4 ±15.4	483 ±53
14.0	28.7 ±1.5	34.34 ±1.31	101 ±2	17.4 ±3.8	45.6 ±13.5	563 ±40
17.5	29.0 ±1.4	33.89 ±1.07	105 ±2	19.2 ±3.9	44.5 ±11.9	658 ±58
21.0	29.1 ±1.4	34.20 ±1.08	109 ±1	22.2 ±4.0	41.8 ±11.4	727 ±55
24.4	30.7 ±1.4	35.60 ±1.34	112 ±1	26.4 ±4.7	37.9 ±8.5	821 ±64
28.1	32.8 ±1.9	37.59 ±1.72	111 ±3	31.5 ±4.4	36.6 ±8.9	982 ±97

Values are means from six experiments ± S.E.

[a] P_{ICO_2}, partial pressure of CO_2 in inspired gas; [b] P_{aCO_2}, CO_2 tesnion of arterial blood; [c] $[H^+]_a$, hydrogen ion concentration in arterial blood; [d] P_{aO_2}, O_2 tension of arterial blood; [e] V_T, tidal volume; [f] f, respiratory frequency; [g] $\dot{V}_E$, expiratory minute volume.

Ventilatory changes for each of the six experiments are illustrated in Figure 3 and mean values are listed in Table 1. Two general types of breathing patterns were observed. In two experiments (Fig. 3A, B) respiratory frequency (f), at $P_{ICO_2} = 0$, was high and tidal volume was low; as P_{ICO_2} was elevated, respiratory frequency markedly decreased and tidal volume increased. In the four remaining experiments, respiratory frequency at $P_{ICO_2} = 0$ was significantly lower and remained relatively constant or only slightly decreased as P_{ICO_2} was raised to 21 torr. Tidal volume always increased.

Through the range of isocapnia (0 to 21 torr P_{ICO_2}), the increase in mean minute volume was 53% (range, 38%-85%), the increase in mean tidal volume was 82% (range, 43%-168%) and the decrease in mean respiratory frequency was 25% (range, 0-48%).

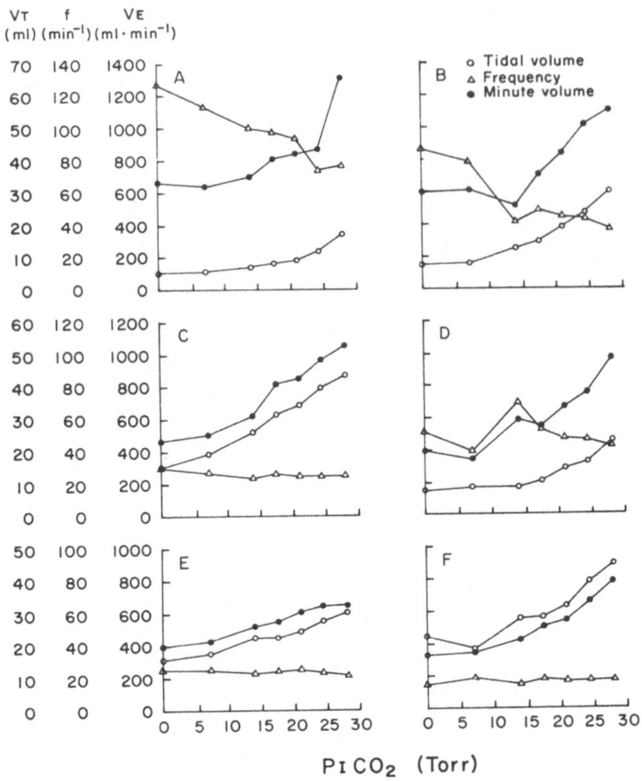

Fig. 3. Changes in ventilatory parameters during inhalation of CO in six anesthetized, upright chickens under normoxic conditions. V_T, tidal volume; f, respiratory frequency; $\dot{V}_E$, expiratory minute volume

b) Series II: Supine Chickens Breathing Hyperoxic Gas

Mean values from this set of experiments are listed in Table 2. Changes in P_{aCO_2} and $[H^+]_a$ are similar to those changes observed in upright animals breathing air (Table 1). At 21 torr P_{ICO_2}, there was no significant difference in either P_{aCO_2} or $[H^+]_a$; however, at 28.4 torr P_{ICO_2}, both P_{aCO_2} ($P < 0.1$) and $[H^+]_a$ ($P < 0.1$) significantly increased. These observed changes in P_{aCO_2} and $[H^+]_a$ are similar to our earlier study in which ventilatory parameters were not simultaneously measured (Osborne and Mitchell, 1977).

Ventilatory changes during the isocapnic range were similar to the corresponding changes observed during normoxia in the upright chicken (Table 1) although the control level of respiratory frequency was significantly lower and remained relatively constant as P_{ICO_2} was altered.

Table 2. Blood gas and ventilatory values during inhalation of CO_2 in supine, anesthetized chickens (hyperoxia)

P_{ICO_2} (torr)	P_{aCO_2} (torr)	$[H^+]_a$ (n Eq/l)	P_{aO_2} (torr)	V_T (ml)	f (min^{-1})	$\dot{V}_I$ (ml/min)
0	28.7 ±1.4	37.70 ±1.85	190 ±13	11.5 ±2.4	34.9 ±0.6	400 ±76
7.4	28.8 ±1.3	38.74 ±2.89	193 ±16	13.0 ±2.9	32.1 ±5.1	531 ±94
14.4	28.2 ±1.0	40.04 ±3.52	198 ±15	15.7 ±2.5	39.7 ±2.7	620 ±91
20.9	29.1 ±1.4	38.59 ±3.26	195 ±16	19.5 ±2.3	38.5 ±2.8	750 ±100
28.4	31.8 ±1.4	40.75 ±2.84	196 ±15	27.6 ±2.9	36.1 ±0.8	981 ±82
35.6	35.3 ±1.8	47.08 ±2.97	200 ±14	33.7 ±3.1	34.9 ±1.4	1168 ±65

Values are means from three experiments ± S.E. (see Table 1 for definition of terms). $\dot{V}_I$, inspiratory minute volume.

4. Discussion

It has been suggested that the amplitude of breathing is significantly greater in an anesthetized bird placed in an upright position than in the supine position (King and Payne, 1964). In light of these observations, we have chosen to study ventilatory changes during P_{aCO_2} regulation in both upright and supine positions.

In these studies, we have noticed relatively little difference in tidal volume, V_T, between the upright and supine chicken. Our mean values for V_T in the control upright and supine chickens (12.4 and 11.5 ml respectively) are considerably smaller than the 33.0 ml reported by Piiper et al. (1970) for unanesthetized White Leghorn hens, but resemble the values reported by Weiss et al. (1963) in similar animals: 15.4 ml. It is difficult to reconcile the differences, but we attribute them primarily to the effects of the anesthesia as well as to the age and size of our birds. In the first series of experiments, three birds were 26-30 weeks of age and weighed 1.9 kg each and three birds were 17-18 weeks of age and weighed 1.4, 1.1, and 1.1 kg. At P_{ICO_2} = 0, the older birds characteristically breathed slower (f = 25, 30, and 15 min^{-1}) and deeper (V_T = 16, 15, and 22 ml) than the three younger, lighter birds (f = 86, 51, and 128 min^{-1}; V_T = 7, 8, and 5 ml).

Isocapnic hyperpnea during inhalation of CO_2 is evident in these studies as minute volume, $\dot{V}_E$, increased without a significant increase in either P_{aCO_2} or $[H^+]_a$. Through the range of P_{aCO_2} regulation in the first series of experiments, average $\dot{V}_E$ at 21 torr P_{ICO_2} increased by 53% of the control value. Chickens with the highest f and lowest V_T (Fig. 2A, B) showed the smallest increase in $\dot{V}_E$ through the range of P_{aCO_2}

regulation. This most likely reflects an increase in the parabronchial to dead space ventilatory ratio. Since resporatory frequency decreases and tidal volume increases, one would expect the parabronchial to dead space ventilatory ratio to increase if little or no change occurred in the physiological dead space volume.

Under both normoxic and hyperoxic conditions, P_{aCO_2} was regulated equally well at its control level as P_{ICO_2} was varied between 0 and 21 torr. This observation suggests that the carotid body chemoreceptors probably are not necessary for the regulation of P_{aCO_2} since hyperoxia depresses carotid body chemoreceptor activity (Bouverot and Leitner, 1972) and the ventilatory response to hypercapnia mediated by the carotid body chemoreceptors (Jones and Purves, 1970).

The existence of CO_2-sensitive intrapulmonary chemoreceptors in birds has been well documented (Fedde and Peterson, 1970; Osborne and Burger, 1974) and these receptors are thought to play a major role in the control of breathing during normocapnic and hypocapnic conditions (Osborne et al., 1977). We believe that this receptor system is responsible for the increase in ventilation and thus the regulation of P_{aCO_2} during inhalation of CO_2 (Osborne and Mitchell, 1977). This hypothesis is further supported by the data presented here; and the variation in ventilatory patterns that we have observed (Fig. 2) strongly suggests that parabronchial ventilation and not minute volume is most likely the regulated variable.

Acknowledgments. The authors wish to thank Mr. Michael Salazar and William Satmary for their excellent technical assistance. This research was supported in part by the National Institutes of Health, Grant No. HL-18685.

References

Bouverot, P., Leitner, L.-M.: Arterial chemoreceptors in the domestic fowl. Respir. Physiol. 15, 310-320 (1972)

Fedde, M.R., Peterson, D.F.: Intrapulmonary receptor response to changes in airway-gas composition in *Gallus domesticus*. J. Physiol. (London) 209, 609-625 (1970)

Jones, D.R., Purves, M.J.: The effect of carotid body denervation upon the respiratory response to hypoxia and hypercapnia in the duck. J. Physiol. (London) 211, 295-309 (1970)

King, A.S., Payne, D.C.: Normal breathing and the effects of posture in *Gallus domesticus*. J. Physiol. (London) 174, 340-347 (1964)

Osborne, J.L., Burger, R.E.: Intrapulmonary chemoreceptors in *Gallus domesticus*. Respir. Physiol. 22, 77-85 (1974)

Osborne, J.L., Mitchell, G.S.: Regulation of arterial P_{CO_2} during airway CO_2 loading in chickens. Respir. Physiol. 31, 357-364 (1978)

Osborne, J.L., Mitchell, G.S., Powell, F.: Ventilatory responses to CO_2 in the chicken: intrapulmonary and systemic chemoreceptors. Respir. Physiol. 30, 369-382 (1977)

Piiper, J., Drees, F., Scheid, P.: Gas exchange in the domestic fowl during spontaneous breathing and artifical ventilation. Respir. Physiol. 9, 234-245 (1970)

Weiss, H.S., Frankel, H., Hollands, K.G.: The effect of extended exposure to a hot environment on the response of the chicken to hyperthermia. Can. J. Biochem. Physiol. 41, 805-815 (1963)

The Effect of Middle Cardiac Nerve Stimulation Upon the Respiratory Response to Pa_{CO_2} in the Chicken

J. A. Estavillo and M. L. Youther

Summary

Respiratory movements in 13 thoracotomized cockerels were recorded. The pulmonary circulation of the right lung was blocked, and the left perfused lung was denervated in order to determine the effects on respiratory movements of intrapulmonary CO_2 tension, P_{iCO_2}, arterial CO_2 tension (P_{aCO_2}) and electrical stimulation of the left middle carciac nerve. At low P_{iCO_2} respiratory movements changed little with varying P_{aCO_2}; conversely, at low P_{aCO_2} respiratory movements changed little with varying P_{aCO_2}. At 70-torr P_{aCO_2}, respiratory movements were maximal at P_{iCO_2} of 33 torr or higher. Stimulation of the middle cardiac nerve (MCN) produced a reduction in respiratory movements at most P_{CO_2} levels but showed greater sensitivity at low intrapulmonary P_{iCO_2} and P_{aCO_2} levels. Receptors with axons in the MCN may be important in the control of avian respiration.

1. Introduction

Chickens have CO_2-sensitive receptors in their lungs called intrapulmonary chemoreceptors which control ventilation if arterial CO_2 tension, P_{aCO_2}, is high. Additionally, the ventricles of chickens also contain receptors which are sensitive to P_{aCO_2}, as well as CO_2 pressure (Estavillo and Burger, 1973b).

The interaction of the intrapulmonary chemoreceptors with the other receptor systems sensitive to P_{aCO_2} is not known. However, when chicken lungs are overventilated with N_2 gas, the low intrapulmonary P_{CO_2} causes apnea from intrapulmonary chemoreceptor activity despite the lethal effect of the simultaneously induced hypoxia (Ray and Fedde, 1969).

Evidence shows that carotid bodies in chickens do not appear to play a role in respiratory control, for as Jones reported (1976), ventilatory sensitivity to inhaled CO_2 is not greatly affected by hypoxia.

In recent work attempting to define the functional role of thoracic vagal rami in the chicken, Peterson and Nightingale (1976), using 1-min

---

Department of Basic Medical Education, School of Medicine, Southern Illinois University, Carbondale, IL 62901/USA.

electrical stimulation of the various rami, concluded that thoracic receptors important in regulation of respiration are confined primarily to the lungs and that these receptors do not play a direct role in regulation of cardiovascular function.

There is a strong possibility that ventricular end-net receptors, which are in a unique anatomical position to monitor CO_2 exchange with the environment, by altering their activity in response to cardiac output as a function of P_{aCO_2} from ventilation-perfusion, may interact with that activity from intrapulmonary chemoreceptors to produce a central response.

This study was conducted to determine the interactive effects on respiratory movements of the receptor systems sensing P_{CO_2} in an isolated lung and of P_{aCO_2} levels, and also the respiratory effect of electrical stimulation of the middle cardiac nerve afferents at various P_{aCO_2} and intrapulmonary P_{CO_2} levels.

2. Methods

a) General Preparation

We used 13 White Leghorn cockerels, 16-20 weeks of age, weighing 1.4 to 2.0 kg. They were anesthetized with intravenously administered sodium pentobarbital (35 mg/kg) and restrained on their backs. We opened the thoracic cavity and ventilated the lungs as described by Banzett and Burger (1977). We always used 2 l/min of gas, total flow, to ventilate each lung. In both series O_2 and N_2 were heated to 41.5°C and humidified by passing through a hot water spray and CO_2 added as desired, adjusting N_2 flow. A snare was placed, loosely, around the right pulmonary artery.

We cannulated the left carotid artery for recording blood pressure, and the left brachial vein for drug infusion. We cut both cardiac sympathetic nerves as well as right and left vagal rami as previously described (Estavillo et al., 1977).

In the series using electrical stimulation, the left middle cardiac nerve (MCN) was dissected free and cut at its most peripheral end, laid across bipolar silver electrodes and bathed in warmed parafin oil. The right MCN was also cut.

b) Measurements

We measured sternal movements as described before (Ray and Fedde, 1969). Sternal movements are highly correlated to tidal volume in the closed chest animal (Kuhlman and Fedde, 1976).

We recorded temperature from the large intestine maintaining normal body temperature. We recorded sternal deflections, heart rate for ECG, blood pressure, and mean blood pressure on a pen writer (Model R411, Beckman).

Electrical stimulus parameters selected maximized the responses; 15 volts, 30 pulses/s, 0.5 ms pulse duration.

c) Interaction of Intrapulmonary and Arterial Carbon Dioxide Pressures

In this series, we changed P_{CO_2} of the gas ventilating the non-perfused right lung from 7 to 105 torr in 6 steps at P_{aCO_2} of 70 torr. After completing a sequence of P_{CO_2} changes, we released the snare around the right pulmonary artery and altered P_{CO_2} to both lungs to establish a new level of P_{aCO_2}. After 5 min, we snared the right pulmonary artery and recorded sternal deflections after an additional 5 min before changing P_{CO_2}.

d) Influence of Carbon Dioxide Pressures on Middle Cardiac Nerve Response

In this series we altered the gas tensions in each lung in the same manner as the first series above. Gas tensions were 7, 14, 21, 33, 42 torr left lung and 14, 21, 33 torr in the vascularly isolated innervated right lung. After each change in gas tension we waited 5 min and then recorded sternal deflections for 30 s prior to stimulating for 20 s. Recording continued until all responses returned to base levels.

3. Results

a) Ventilating Gas P_{CO_2} and P_{aCO_2} Correlation

Earlier work by Estavillo et al. (1977) found that P_{aCO_2} could be described by 3.9 + 0.98 P_{CO_2} ($r > 0.99$) where P_{CO_2} was the partial pressure of CO_2 in the gas ventilating the denervated perfused left lung. Since P_{aCO_2} and P_{CO_2} were closely correlated, the predicted levels of P_{aCO_2} are given instead of the level of the gas ventilating the left lung.

b) Interaction of Intrapulmonary and Arterial Carbon Dioxide Pressures

Respiratory amplitude changed directly with the level of P_{CO_2} in the innervated vascularly isolated lung and was directly correlated to the level of P_{aCO_2} (Fig. 1). At 7-torr P_{aCO_2}, respiratory amplitude was little affected by P_{iCO_2}. Respiratory movements were affected over 2 times more by variations in P_{iCO_2} at 14 torr than at 7 torr P_{aCO_2}. At 70-torr P_{aCO_2}, respiratory amplitude was nearly maximized at 33-torr P_{iCO_2}.

At 7-torr P_{iCO_2} the effect of a change from 7 to 70 torr P_{aCO_2} on respiratory amplitude is small, but increased significantly at 33 torr and higher (Fig. 1).

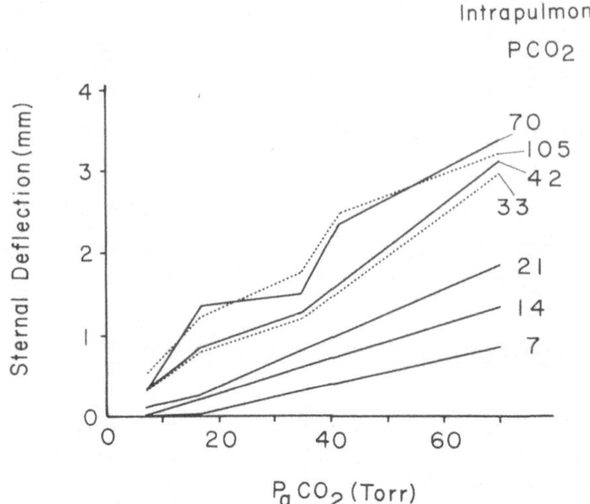

Fig. 1. Respiratory movement sensitivity to P_{iCO_2} and P_{aCO_2}. P_{aCO_2} level was maintained by the F_{CO_2} of gas overventilating the denervated left lung. Intrapulmonary P_{CO_2} was varied in the right lung with innervation intact and pulmonary circulation blocked. Sternal deflections were measured from the ventral edge of six thoracotomized cockerels. Note the maximum sensitivity to P_{iCO_2} is when P_{aCO_2} is 70 torr with P_{iCO_2} level of 33 torr producing near maximum sternal deflections

c) Effect of Middle Cardiac Nerve

The degree to which stimulation of MCN reduced respiratory movements depended on the level of P_{iCO_2} and P_{aCO_2}. At a P_{aCO_2} of 33 torr and P_{iCO_2} of 21 torr, respiratory movements fell during stimulation (Fig. 2B). When the levels of CO_2 are reversed the decrease in respiratory movements was as great (Fig. 2A). However, the prestimulus amplitudes were not equal with isolated lung: P_{CO_2} 33 torr produced a larger prestimulus drive (Fig. 2A).

As P_{aCO_2} was varied from 14 to 42 torr while P_{iCO_2} was kept constant at 14 torr, respiratory movements showed a decreasing response to the stimulus (Fig. 3C). In addition, stimulating the MCN show a decreased response of the respiratory movement as P_{iCO_2} was increased to 33 and 21 torr (Fig. 3A, B). On average, stimulation of the MCN caused an immediate decrease in amplitude within the next breath.

4. Discussion

Mechanoreceptors, whose axons are carried in the middle cardiac nerve with endings in the ventricles and mediastinal tissue, have been identified with a very small number of endings in tissue beyond the aortic valves and presumably a few aortic chemoreceptors (Estavillo and Burger, 1973a).

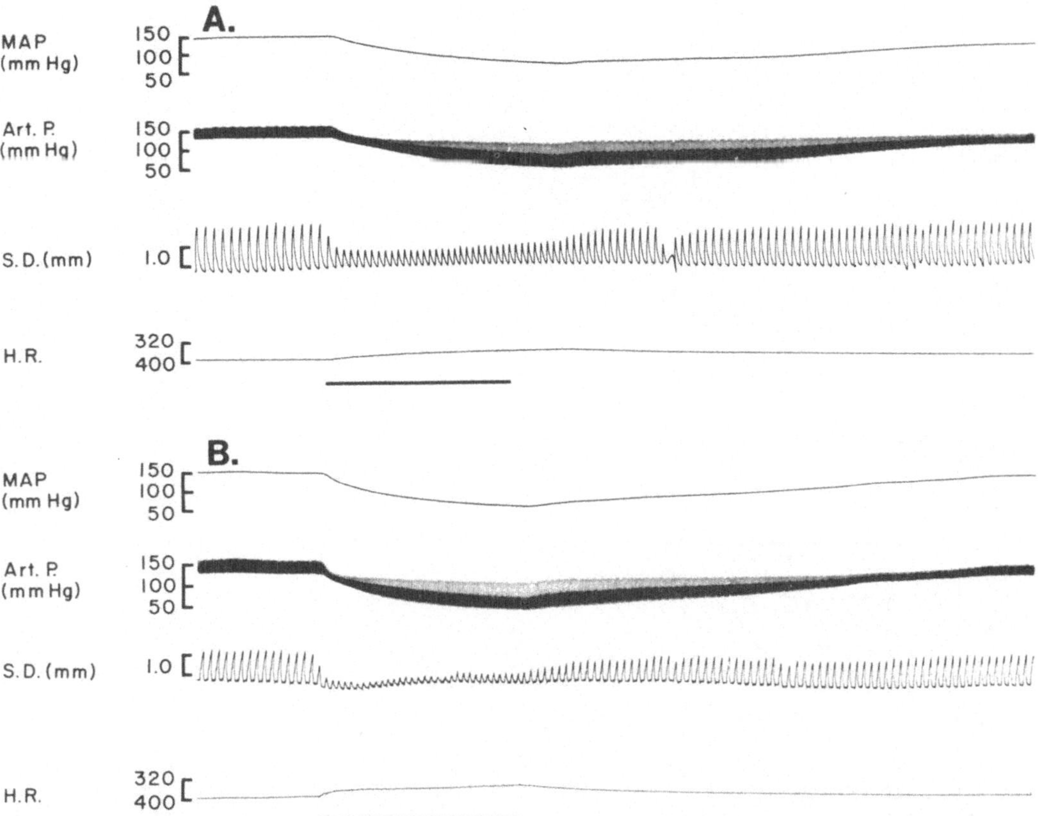

Fig. 2. Effect of P_{aCO_2} and MCN stimulation on sternal deflection in one animal. In A P_{iCO_2} is 33 torr, P_{aCO_2} is 21 torr. In B P_{aCO_2} is 33 torr and intrapulmonary P_{iCO_2} is 21 torr. The *bar* below (A), and (B), indicates time of 20-s stimulation of MCN. Note that the decrease in sternal movements is nearly proportional. *MAP*, mean arterial pressure; *Art. P.*, arterial pressure; *SD*, sternal deflection; *HR.*, heart rate

No reflex effects produced by mediastinal receptors are known nor have large effects of P_{aCO_2} on the activity of the few tested been observed. Similarly, we have no quantitative data to suggest that the few aortic arch baroreceptors tested are influenced in their activity by variations in P_{aCO_2}. However, in the chicken, the activity of most ventricular receptors is altered over a wide range by varying P_{aCO_2}. This activity is carried by the axons of the middle cardiac nerve which are predominantly afferent (Estavillo, 1977) with fibers of ventricular receptors comprising a significant portion of the total number of axons in the middle cardiac nerve. To date activity from ventricular receptors has not been identified in any other vagal rami.

Information supporting any functional description for ventricular receptors is lacking. However, recent work has shown that inflating a balloon in the left ventricle produces inhibition of spontaneous breathing in closed chest dogs on cardiac by-pass (Kostreva et al., 1977).

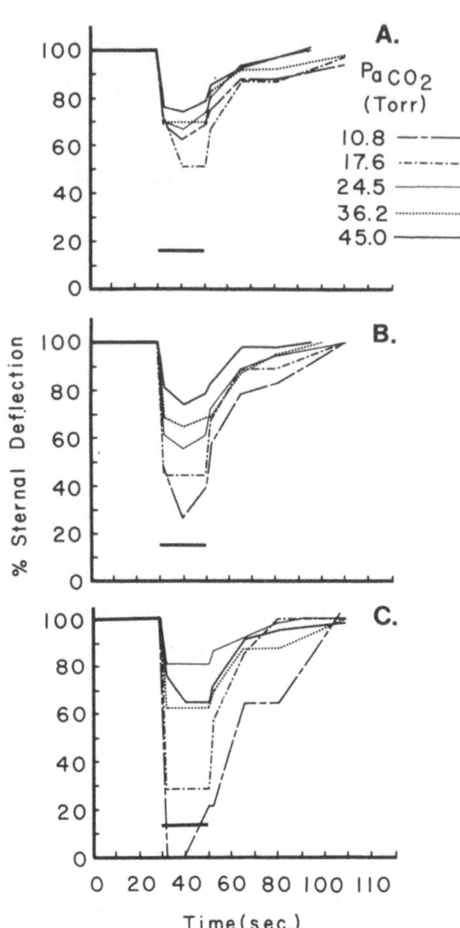

Fig. 3. Percent depression of sternal deflection by stimulation of MCN at three levels of intrapulmonary P_{CO_2}. Intrapulmonary P_{CO_2} in A, 14 torr; B, 21 torr; C, 33 torr. *Bar* in each graph indicates time of stimulus. P_{aCO_2} levels shown in *top right*

intrapulmonary chemoreceptors. Elucidation of these interactions will require a greater understanding of the correlation of discharge frequency to sternal deflection as well as a correlation of discharge frequencies to sternal deflection at various levels of blood pressure and blood volume at different P_{CO_2} levels.

Although no direct comparison of discharge frequency to sternal deflections over the varied range of P_{CO_2} used in previous studies is possible, we know that amplitude of sternal deflections is, roughly, inversely related to ventricular receptor discharge frequency determined in our earlier work. In this study as P_{CO_2} increased, sternal deflections increased in amplitude.

Stimulation of the MCN in these experiments consistantly produced a decrease in respiratory movements similar to those seen in the intact MCN animal at low P_{aCO_2} (Fig. 3). Although these results are contrary to the conclusions of Peterson and Nightingale (1976) their work fails to show either in the diagram or discussion any evidence for stimulation of the MCN.

In recently completed work (Estavillo et al., 1977) transient increases in blood pressure, which increase the rate of receptor activity, also produce a decrease in respiratory movements for the time of pressure increase. As P_{aCO_2} and P_{iCO_2} are increased the effective stimulus response decreases for receptor activity produced by blood pressure or electrical stimulation of the MCN. Hence there are probably complex interactions between the central effects of ventricular receptors and

Acknoewledgments. This study was supported, in part, by USPHS NIH HL 19149 and the Illinois Heart Association.

References

Banzett, R.B., Burger, R.E.: Response of avian intrapulmonary chemoreceptors to venous CO_2 and ventialtory gas flow. Respir. Physiol. 29, 63-72 (1977)

Burger, R.E., Osborne, J.L., Banzett, R.B.: Intrapulmonary chemoreceptors in *Gallus domesticus*: adequate stimulus and functional localization. Respir. Physiol. 22, 87-97 (1974)

Estavillo, J.A.: Fiber size and sensory endings of the middle cardiac nerve of the domestic fowl *Gallus domesticus*. Acta Anatomica in press (1977)

Estavillo, J.A., Burger, R.E.: Cardiac afferent activity in the depressor nerve of the chicken. J. Appl. Physiol. 225, 1063-1066 (1973a)

Estavillo, J.A., Burger, R.E.: Avain cardiac receptors: activity changes by blood pressure, carbon dioxide and pH. J. Appl. Physiol. 225, 1067-1071 (1973b)

Estavillo, J.A., Youther, M.L., Burger, R.E.: Respiratory control by the middle cardiac nerve in the chicken: effects of blood pressure. Respir. Physiol. submitted (1977)

Jones, D.R.: The control of breathing in birds with particular reference to the initiation and maintenance of diving apnea. Federation Proc. 35, 1975-1982 (1976)

Kostreva, D.R., Czarnecki, D., Zuperku, E.J., Cook, R.L., Kampine, J.P.: Cardiac vagal afferent reflex alteration of respiratory drive. Federation Proc. 36, 489 (1977)

Kuhlman, W.D., Fedde, M.R.: Upper respiratory dead space in the chicken, its fraction of tidal volume. Comp. Biochem. and Physiol. 54A, 409-411 (1976)

Peterson, D.F., Nightingale, T.E.: Functional significance of thoracic vagal branches in the chickens. Respir. Physiol. 27, 267-275 (1976)

Ray, P.J., Fedde, M.R.: Responses to alterations in respiratory P_{O_2} and P_{CO_2} in the chicken. Respir. Physiol. 6, 135-143 (1969)

Effect of $F_{L_{CO_2}}$ Dynamics on T_i and T_{tot} in Spontaneously Breathing Birds

A. L. Kunz and R. D. Tallman, Jr.

Summary

This study investigates whether the model of the respiratory pacemaker devised by Kunz and Miller (1974) is responsible for respiratory frequency in awake spontaneously breathing birds. This model hypothesizes that each inspiration causes an oscillation in fractional concentration of CO_2 in the lung (F_{LCO_2}) and that the timing of an event in this F_{LCO_2} cycle sets the time of the next inspiration. These experiments used awake chickens which were spontaneously breathing from a gas flow line in which the fractional concentration of CO_2 (F_{ICO_2}) could be rapidly changed. During the first portion of the inspiratory period, F_{ICO_2} was kept at a value near that of end expiratory F_{CO_2} (0.05). Then τ seconds after the beginning of inspiration, the F_{ICO_2} was decreased to 0.0 for 0.6 s and then returned to 0.05. Ventilatory movements were monitored by a whole body plethysmograph. When τ was varied within the range 0.1 to 0.6 s the inspiratory period, T_i, and total respiratory period, T_{tot}, were found to change proportionately. Sinusoidal and step forcing of τ showed an immediate response of T_i and T_{tot} with almost no hysteresis. These findings are believed to support the model.

1. Introduction

When oscillations are forced in the fractional concentration of CO_2 in the lungs (F_{LCO_2}) of awake unidirectionally ventilated chickens, ventilatory movements mirror the frequency of the F_{LCO_2} oscillations (Kunz and Miller, 1971; Kunz et al., 1973; Kunz and Miller, 1974). The bird's ventilatory frequency, and hence F_{LCO_2} oscillations, can be increased or decreased over a range of about two octaves (1/2× to 2×).

Kunz and Miller (1974) proposed a model which stated that during normal breathing the oscillation in F_{LCO_2} caused by the bird's breathing triggers the next breath. That is, as the bird breathes, a fluctuation of F_{LCO_2} is induced which affects the length of that breath. This reciprocal interaction constitutes an oscillator that paces ventilation. Admittedly, there is also a central neural oscillator. But the central oscillator normally serves only as a back-up oscillator to pace ven-

Department of Physiology, The Ohio State University, Columbus, Ohio/USA.

tilation when there are no F_{LCO_2} oscillations. Ordinarily, the period of ventilation (T_{tot}) is determined by the period of the peripheral oscillator.

Molony (Edinburgh, unpublished communication) devised an experiment to test this model. He increased the delay time between the beginning of inspiration and when F_{LCO_2} was lowered, by adding measured lengths of tubing to the trachea, increasing the dead space. His expectation was that increasing the delay would increase the ventilatory period by a like amount. Molony was disappointed in that the increases in tubing length did not slow respiration (unpublished communication). He offered the explanation that mean F_{LCO_2} may also have risen in his experiments, but found no evidence that the pacing phenomenon played a role in normal ventilation. He also suggested that the phenomenon may be an artifact of unidirectional ventilation, because in the unidirectionally ventilated chicken the wave of low F_{LCO_2} reached all areas of the lung at once, simultaneously firing all intrapulmonary CO_2 receptors (IPCs): while in normal spontaneous breathing, some receptors fire during inspiration and others fire during expiration.

2. Strategy and Methods

We designed an analogous experiment which allowed us to manipulate electronically the delay between the time of inspiration (t_i) and the time F_{LCO_2} went down, (t_d). The only surgery performed was a tracheostomy. The trachea was then connected by way of a T-tube to an airflow system in which the CO_2 concentration in the inspired gas, F_{ICO_2}, could be rapidly changed by a gas mixer which was controlled by an analog computer (Fig. 1).

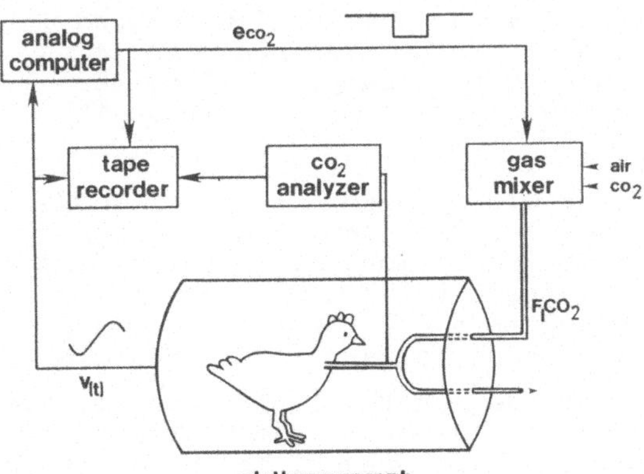

Fig. 1. Schematic of experimental setup

The strategy was to keep the fraction of CO_2 in the airflow line (F_{ICO_2}) the same as the end-expiratory F_{CO_2} (about 0.05) during the first portion of the inspiratory period and then to change F_{ICO_2} to 0.0 after a length of time, τ. In these experiments, τ served as the independent variable and was systematically altered. After the beginning of inspiration the IPCs are first exposed to the F_{CO_2} of the gas that was in the tracheal deadspace and then to F_{ICO_2} of about the same value until the 0.0 F_{ICO_2} gas reached the lung. We anticipated that increasing τ would increase the period of ventilation. Results using this technique were similar to those of Molony's. Incremental increases in the delay, τ, did not cause similar incremented increases in ventilatory period.

In the meantime, further experiments using the unidirectionally ventilated chicken showed that the event in the F_{LCO_2} cycle which triggered inspiration was not when F_{LCO_2} went down, t_d, but when F_{LCO_2} went back up, t_u, (Miller and Kunz, 1977). This explained the lack of results in the experiments in which the trachea was lengthened either physically (by adding tubing) or virtually (by computer manipulation). Those experiments delayed t_d but not t_u.

After realizing the importance of delaying t_u, we modified the strategy so that we lowered F_{ICO_2} as before but kept F_{ICO_2} down for only 0.6 s so that the rise in F_{ICO_2} still occurred during inspiration. That is, the F_{ICO_2} was kept at 0.05 except for a pulse of 0.0 F_{ICO_2} which started τ seconds after the beginning of inspiration and lasted 0.6 s.

3. Results

Finally we increased the delay, τ, which in turn lengthened the inspiratory period, T_i, and total ventilatory period, T_{tot}. Figure 2 shows how changing the delay, τ, effected the ventilatory cycle. In the lower tracing τ was 0.12 s. As τ was increased to 0.33 and then to 0.60 s both the T_i and T_{tot} increased progressively. Figure 3 is a recording of the results of dynamically forcing the delay, τ. The first one-third of the figure is run at a fast paper speed to show the nature of the tracings. The remainder of the record is run at a much slower speed to emphasize the contours of the envelopes of the tracings. The *top tracing* is the recording from the CO_2 analyzer of samplings from the trachea. During inspiration, this tracing records F_{ICO_2} while during expiration it records F_{ECO_2}. End-expiratory F_{CO_2} stayed relatively constant during this experiment at about 0.06. The *second tracing* shows ventilation as measured by a whole body plethysmograph (inspiration is up). The lower four tracings are ramps which monitor specific time intervals in the ventilatory cycle. The tracing labeled τ monitors the interval t_i-t_d, the delay between the beginning of inspiration and when F_{ICO_2} starts down. It is independently forced to vary in a sinusoidal pattern. The inspiratory period, T_i, recorded on the forth tracing, has about the same amplitude of oscillation as that of τ. The ventilatory period recorded on the *next to the bottom tracing*, however, has an amplitude of oscillation about three times as large as that of τ.

The *bottom tracing*, K, is the interval t_u-t_e; the time between when F_{ICO_2} starts up and the beginning of expiration. The minor variations in K are seen to show little of the periodicity of the sinusoidal oscillations of τ.

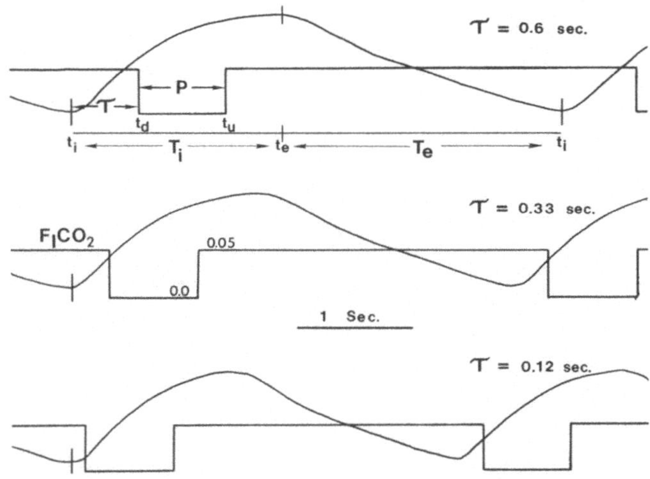

Fig. 2. Three ventilatory cycles at different delay times

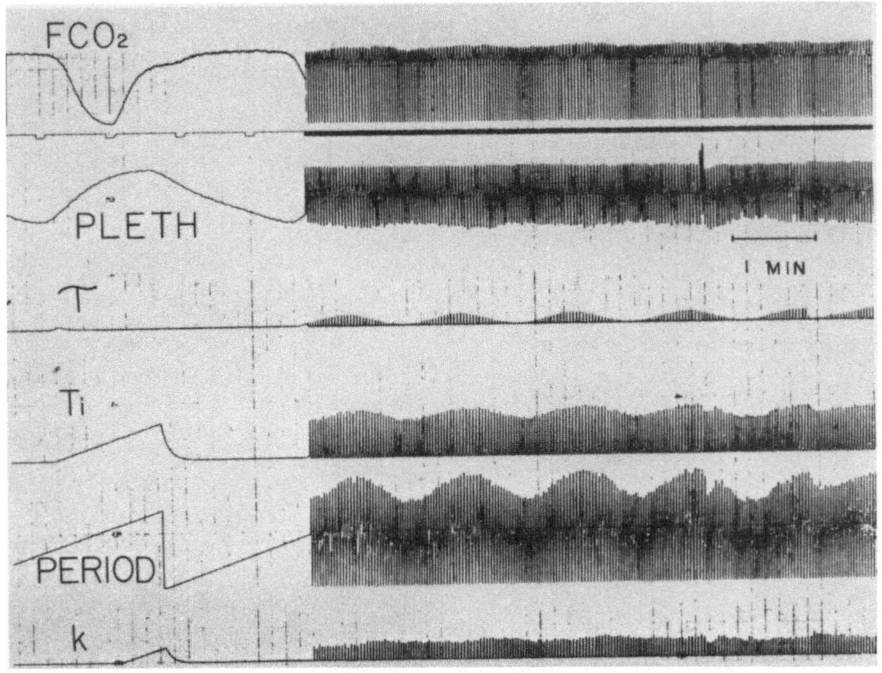

Fig. 3. Dynamic response of T_i and period to sinusoidal forcing of τ

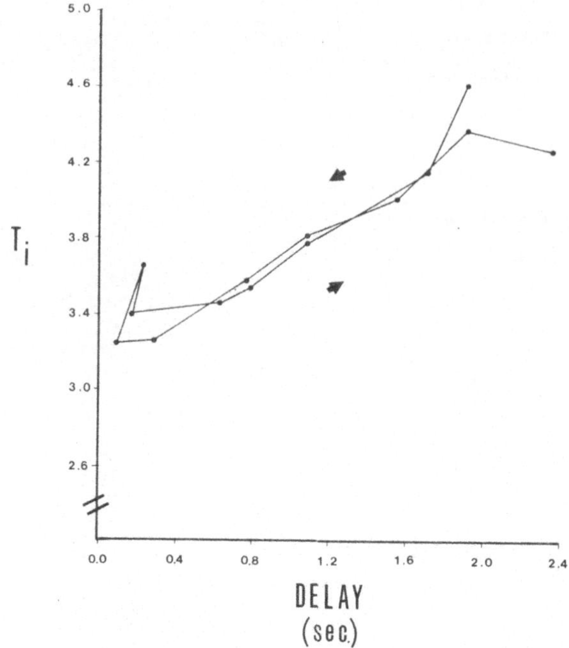

Fig. 4. Plot of T_i vs τ for one complete cycle of sinusoidal forcing of τ

EFFECT OF A STEP CHANGE IN DELAY ON T_i

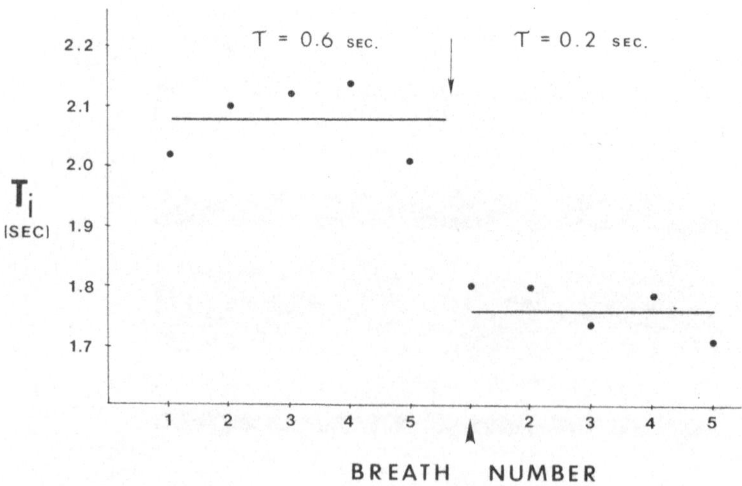

Fig. 5. Response of T_i to step change in τ

Figure 4 is a plot of τ vs. T_i on a breath-by-breath basis. The absence of a hysteresis loop indicates that the two oscillations are in phase. Figure 5 shows the response of inspiratory period to a step increase in τ. Inspiratory period usually followed a step change in delay on the first breath.

4. Discussion

The birds in these experiments are not only awake and standing upright, they are also breathing spontaneously and the air is presumably passing through the lungs in a normal pattern so that the criticisms of unidirectional ventilation do not apply.

The absence of a hysteresis loop during sinusoidal forcing in conjunction with the rapid response to a step change in delay infers that changes in F_{LCO_2} during a breath result in change in the T_i and T_{tot} of that same breath.

We believe these results support and clarify the model of Kunz and Miller; that a breath lowers F_{LCO_2} and when F_{LCO_2} again begins to rise it triggers the beginning of the next expiration K seconds later, and sets the time of the beginning of the next inspiration. This inspiration again lowers F_{LCO_2} and the cycle repeats. The effect of increasing delay on total period, T_{tot}, is also linear (Tallman and Kunz, 1976, 1977) but is not of unity gain. That is, it does not behave as if the time of the delay is simply being added to the ventilatory period. This relationship is more complex and is being investigated further.

Acknowledgments. This study was supported in part by National Heart and Lung Institute Grant HL 14870-02 and Office of Naval Research Grant NR. 101-733.

References

Kunz, A.L., Miller, D.A.: Pacing respiratory movements with alveolar carbon dioxide oscillations. Federation Proc. 30, 270 (abstract) (1971)

Kunz, A.L., Miller, D.A.: Pacing of avian respiration with CO_2 oscillation. Respir. Physiol. 22, 167-177 (1974)

Kunz, A.L., Miller, D.A., Weissberg, R.M.: The avian respiratory pacemaker: An analysis of the effect of CO_2 dynamics. In: Iberall, A.S., Guyton, A.C. (eds.). Regulation and Control in Physiological Systems. Int. Federation Automatic Control, Rochester, N.Y. (1973) p. 300-303

Miller, D.A., Kunz, A.L.: Evidence that a cyclic rise in pulmonary CO_2 triggers inspiration in birds. Respir. Physiol. 31, 193-202 (1977)

Tallman, R.D., Kunz, A.L.: The effect of changing the temporal pattern of F_{CO_2} in the inspiratory gas flow of birds. Physiologist 19, 385 (abstract) (1976)

Tallman, R.D., Jr., Kunz, A.L.: Changes in inspiratory duration as a function of when F_{CO_2} is increased in the inspiratory gas flow of birds. Proc. Inter. Union of Physiol. Sciences 13, 742 (1977)

Effect of Stretch on the Respiratory Pattern of a Chicken

D. A. Miller

Summary

This study has examined the effects of stretch on the respiratory pattern of the chicken. When intrapulmonary pressure was increased at the output tubes of a unidirectionally ventilated preparation, it resulted in a decrease in tidal volume nearly equal to the stretch-induced thoracic volume change. This behavior appeared to be dependent on intact vagus nerves. Period between breaths was relatively unaffected by stretch with some animals showing small increases (10-20%) and others showing no detectable changes. The most effective mechanical stimulus was pressure in the airsacs, suggesting these or associated structures as the localization of mechanoreceptors.

1. Introduction

Recent avian respiratory studies have characterized receptors as mechano and CO_2 types based on their vagal afferent fiber discharge rate in response to various stimuli (Burger et al., 1974; Fedde et al., 1974; Molony, 1974). Studies of the interaction between pulmonary CO_2 and the central respiratory organization have led to the formulation of well-defined and critically tested hypotheses regarding the underlying mechanisms of the respiratory oscillator (Kunz and Miller, 1974; Miller and Kunz, 1977). It was the hope of the present study that, by examining the effect of another stimulus modality (stretch) on the respiratory pattern, additional hypotheses would be suggested regarding respiratory rhythmicity and control.

Mechanoreceptors which respond to changes in intrapulmonary pressure are slowly adapting, relatively insensitive to CO_2, and fewer in number ($\sim$ 20% of total) than the CO_2 receptors (Fedde et al., 1974; Molony, 1974). The stretch receptors increase their discharge frequency in response to increased volume or pressure in the lung-airsac system. When the intrapulmonary pressure is increased in a unidirectionally ventilated duck, a nearly linear relationship between pressure and discharge frequency is obtained (Fedde et al., 1974). When the respiratory system is inflated from a syringe, the relationship is not nearly as clear (Molony, 1974), i.e., the volume changes seem to correlate better with maximal frequency discharge during the inflation. This latter uncertainty may be due to the volume distribution and transient nature of the inflation to the various parts of the airways and airsacs.

Department of Physiology, Medical College of Georgia, Augusta, Georgia 30902/ USA.

Past studies (Eaton et al., 1971; Leitner, 1972) examining the effect of mechanical distortion of the lung-airsac system on the respiratory pattern were complicated by the use of an anesthetized animal and by the lack of control of pulmonary CO_2. The respiratory pattern is exquisitely sensitive to airway CO_2 thus, this variable needs to be maintained constant during a stretch perturbation. The unidirectionally ventilated avian preparation is ideally suited for this study since pulmonary CO_2 and stretch can be controlled independently and the experiments can be performed in the awake animal.

2. Methods

In this preparation (Fig. 1) air can be forced in through a tracheostomy, passed sequentially through the lungs, and then through implanted exit tubes to the outside. This allows pulmonary CO_2 concentration to be easily controlled experimentally. The pulmonary gas is warmed and humidified. Respiratory rate and tidal volume are measured by a whole body plethysmograph. The CO_2 concentration of the inflow gas is generated by an electropneumatic transducer modulated by an electrical signal generator. The CO_2 concentration is recorded by diverting part of the input gas flow through a Beckman Medical Gas Analyzer.

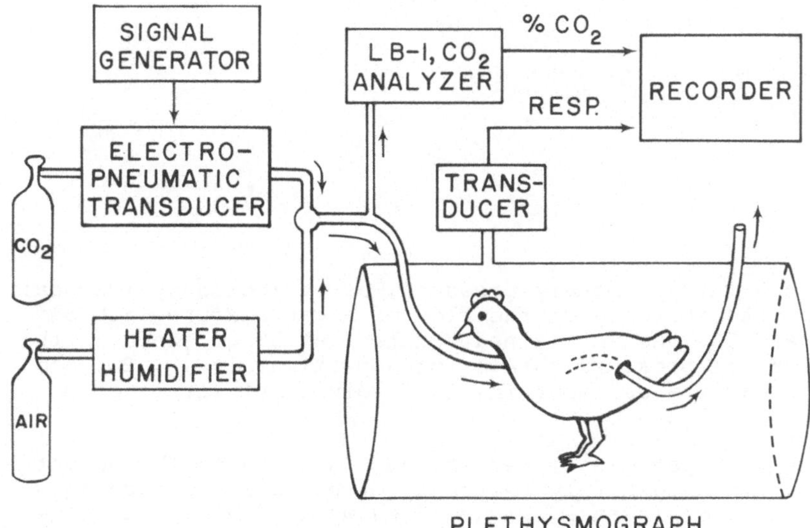

Fig. 1. Experimental set-up

Mechanical distortion of the lungs and airsac membranes is accomplished by increasing the intrapulmonary pressure. This can be done either by increasing the flow of gas through the animal (normally 5 l/min) or by increasing the resistance at the output tubes which provide an exit for the gas from the airsacs.

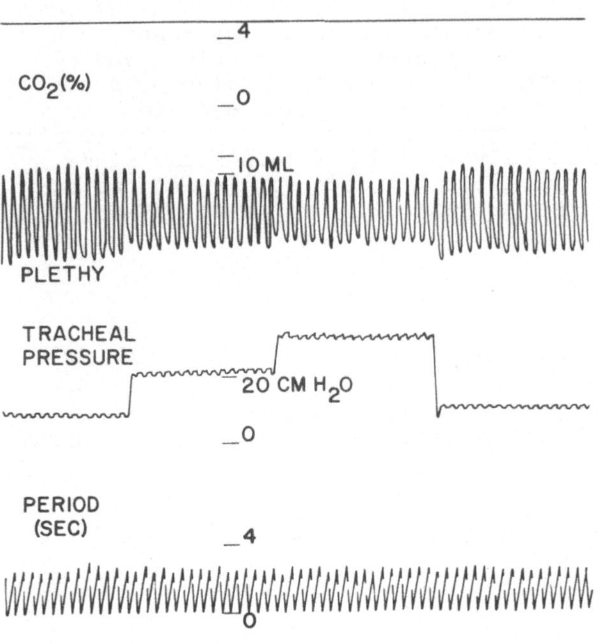

Fig. 2. Effect of increasing intrapulmonary pressure by increasing unidirectional flow

3. Results

Figure 2 records how intrapulmonary pressure, as reflected by tracheal pressure, went up with increase in the flow of a gas mixture with 5% CO_2. Thus, pulmonary CO_2 was maintained constant at 5% as shown in the first tracing. Tracheal pressure changes of approximately 10, 20, and 30 cmH_2O appear to cause a decrease in tidal volume but little or no effect on period.

In the next series of experiments, resistance was added to the output tubes by a valve which alternately switched between a resistance and the atmosphere or zero resistance. Figure 3 shows several cylces alternately adding and removing the output tube resistance. The effect of this resistance is seen in the plethysmographic record as an increase in the mean thoracic volume. For this particular thoracic stretch, tidal volume decreased from 56 ml to 39 ml.

The effect of the output tube resistance is also observed in the mean tracheal pressure which is elevated by an amount equal to the output tube pressure. The change in the period between breaths due to the thoracic stretch was small in this record amounting to about 0.3 s. Five other chickens showed similar small increases or no detectable changes in period. An interesting finding using the output tube resistances was that a pressure change of several cmH_2O had as much effect on tidal volume as did the 10, 20, and 30 cmH_2O tracheal pressure changes. In fact when tidal volume was plotted (Fig. 4) against output tube pressure for several different tracheal pressure ranges the points appear to fall on the same line indicating that the mechanical stimulus is much more closely related to output tube pressure than tracheal pressure.

Since the respiratory movements are measured in a plethysmograph, it was possible to measure the thoracic volume change that had been caused

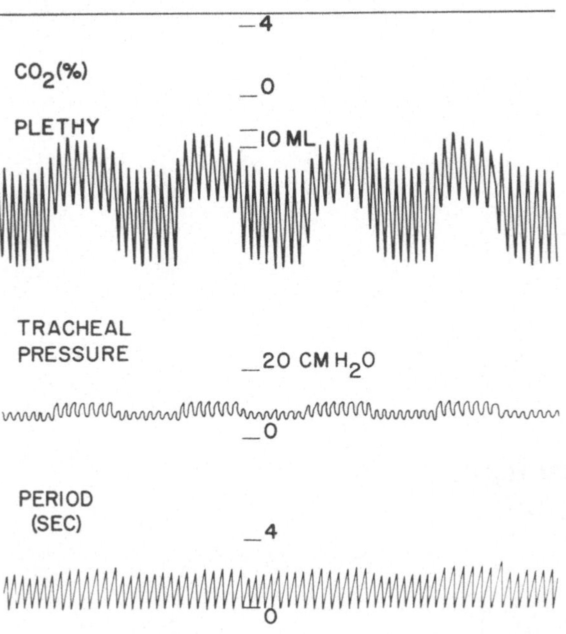

Fig. 3. Effect of increasing intrapulmonary pressure by adding resistance to the output tubes

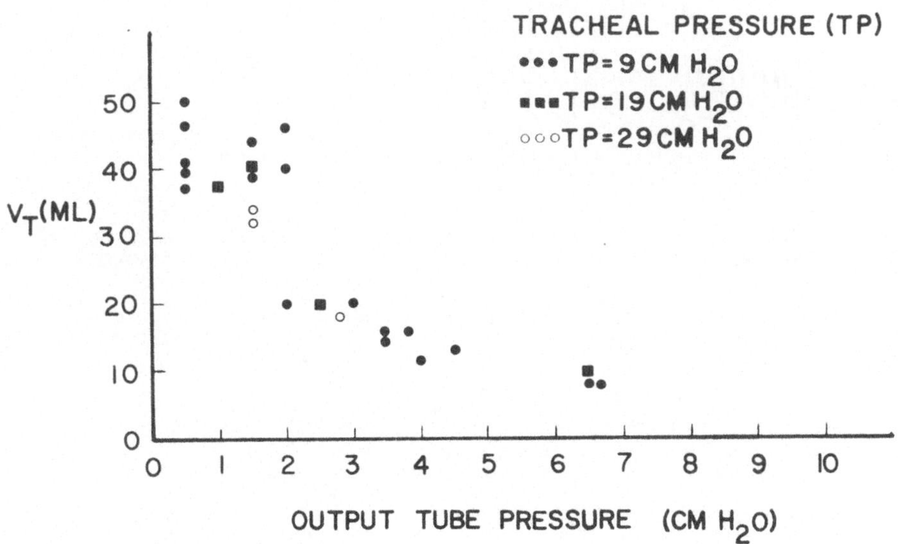

Fig. 4. Relation between tidal volume and output tube pressure for three different tracheal pressure ranges

by adding resistance to the output tubes. This experimentally produced increase in thoracic volume could then be compared to the decrease in tidal volume. Figure 5 shows how the thoracic volume change induced by the output tube resistance was measured on the plethysmographic record. After running several cycles with and without the stretch stimulus, the pulmonary CO_2 level is lowered till the animal is apneic, at which

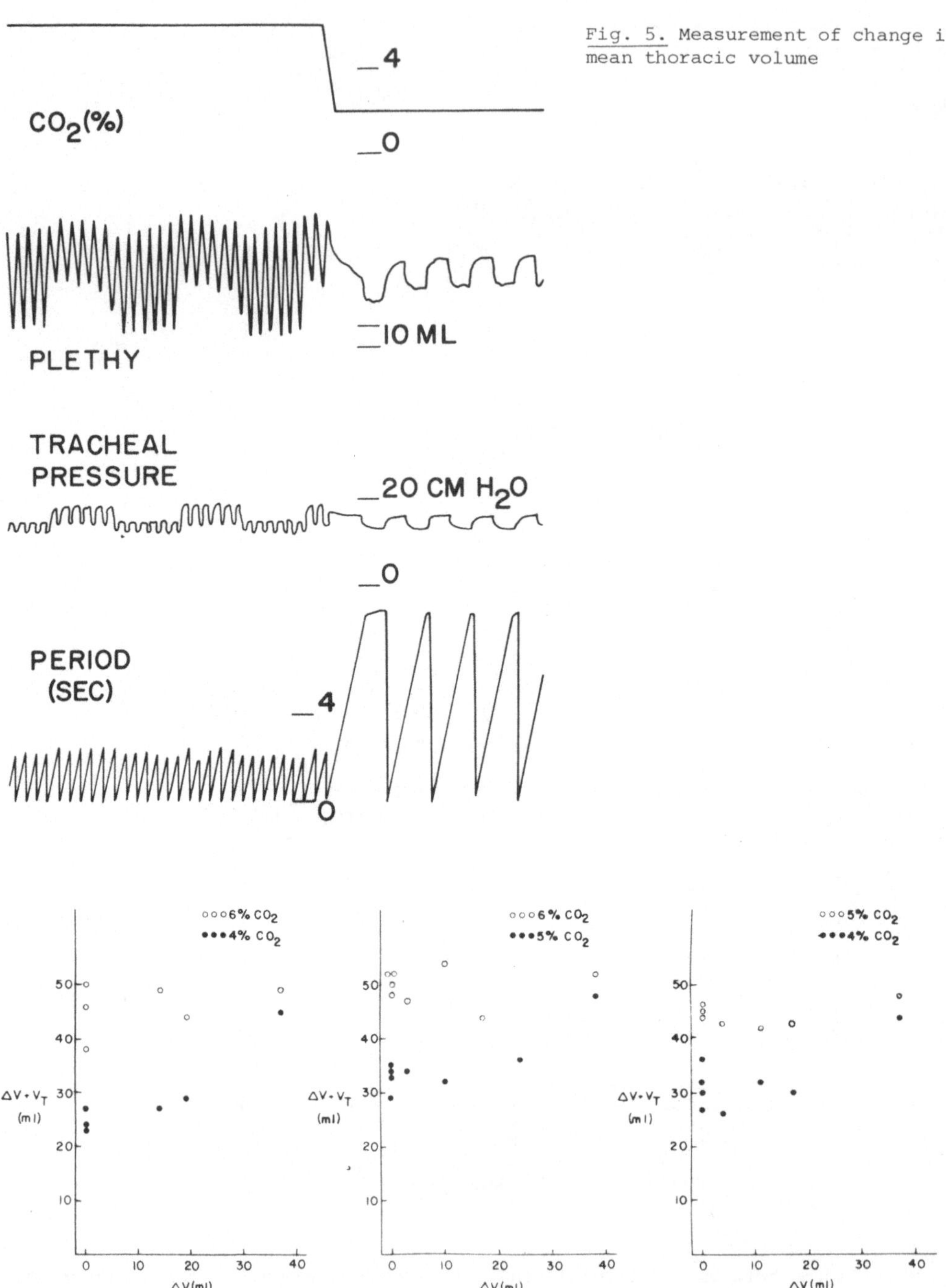

Fig. 5. Measurement of change in mean thoracic volume

Fig. 6. Graph of the sum of the change in mean thoracic volume (ΔV) and tidal volume (V_T) vs. ΔV at two different pulmonary CO_2 levels in three chickens

point the plethysmographic record shows only the thoracic volume change being caused by the output tube resistance. When the output tube resistance is altered, the thoracic volume change can be made larger or smaller so that the effects of stretch on tidal volume can be evaluated over a range of thoracic volume changes. The increased thoracic volume was never greater than 60 ml since this was considered to be the upper extreme of the physiological range. It was also found that animals would not generally tolerate inflations much greater than this amount. On initial measurement the decreases in tidal volume appeared to be nearly identical to the increases in thoracic volume due to stretch. A convenient graphical way (Fig. 6) of illustrating the relationship of decreased tidal volume in response to increased mean thoracic volume is to plot the sum of the tidal volume (V_T) and the mean thoracic volume change (ΔV). If the sum $\Delta V + V_T$ remains constant, it indicates that tidal volume has decreased by exactly the same amount that mean thoracic volume has increased. The sum of $\Delta V + V_T$ in three animals each run at two different mean CO_2 levels are shown in Figure 6. In the six animals tested this sum remains nearly constant as long as the thoracic volume increase does not exceed the original tidal volume. The result holds at different mean CO_2 levels ruling out any type of mechanical stop or obstruction causing the results. It seemed very curious and quite a coincidence that the increase in thoracic volume would nearly equal the decrease in tidal volume.

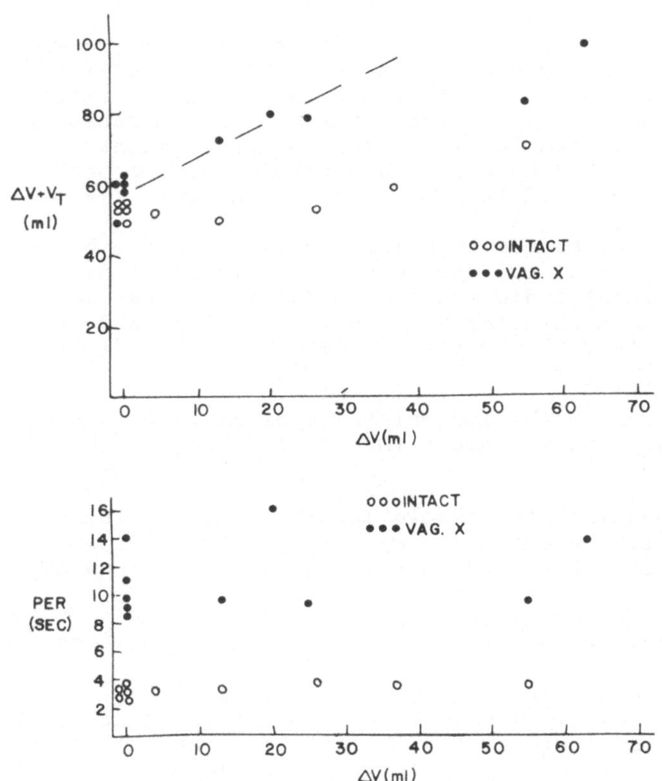

Fig. 7. Relation between ($\Delta V + V_T$) and ΔV in a chicken with intact and sectioned (VAG.X) vagus nerves

In an effort to rule out some of the possible explanations, vagotomies were performed on three chickens. The strategy here was that mechanoreceptor afferents are known to travel in the vagus nerves. If the observed behavior, i.e., constancy of $\Delta V + V_T$, is highly dependent on an intact vagus, it would imply involvement of pulmonary mechanoreceptor. The results of a typical experiment are shown in Figure 7. With an intact vagus, $\Delta V + V_T$ remains relatively constant for ΔV increases less than the original tidal volume. Period between breaths is relatively unaffected by increase in ΔV. In the vagotomized state at zero stretch, the typical increased tidal volume and period are observed. The upper graph shows that with the vagotomy the constant relationship of $\Delta V + V_T$ no longer appears to hold. A line of identity shows that the tidal volume does not decrease at all for small increases in thoracic volume. In nine experiments in six chickens with intact vagus nerves the average slope of $(\Delta V + V_T)$ vs. ΔV was 0.13 over a range where V_T was less than ΔV. In three vagotomized chickens the average slope of $(\Delta V + V_T)$ vs. ΔV was 0.67. These results strongly implicate the participation of mechanoreceptors in the maintenance of a near constant $\Delta V + V_T$.

4. Discussion

The most significant effect of stretch on the respiratory pattern was a decrease in tidal volume. This decrease was much more sensitive to increases in output tube pressure than increases in tracheal pressure. Molony (1974) has suggested the paired bronchoperitoneal membranes or inter-airsac membranes as possible locations for pulmonary mechanoreceptors. Afferent nerve endings have been described in the bronchoperitoneal membranes by Bennett and Malmfors (1970) and Groth (1972).

Decreases in tidal volume in response to pulmonary stretch were quantitated by plotting $(\Delta V + V_T)$ vs. ΔV. The very small slope of this line (0.13 in nine experiments) indicated that tidal volume was decreased almost the same amount that the mean thoracic volume had increased due to the stretch. Possible explanations for this finding include:

1. Pulmonary mechanoreceptors in the lung-airsac system: when these receptors are stretched they might inhibit the motor discharge that establishes tidal volume.

2. Abdominal or thoracic muscles: the muscle length-tension relationship or muscle-spindle reflex behavior (De Wet et al., 1971; De Anda and Rebollo, 1968) might alter the force generation at an increased length.

3. Respiratory load: at an increased thoracic volume the compliance might be decreased.

The slope of the $(\Delta V + V_T)$ vs. ΔV line increased to 0.67 in three vagotomized chickens implying that somehow the intact vagus influences the relationship. Since mechanoreceptor afferents travel in the vagus nerve, these receptors may account for the present findings.

Graphs of output tube pressure vs. ΔV were linear over the range of thoracic volumes examined so that decreases in compliance do not account for the effect of pulmonary stretch on tidal volume.

Therefore, in terms of the normal operation of the respiratory apparatus, the response of tidal volume to stretch is interpreted as a manifestation of a reflex mechanism which normally allows for some adjustment to varying respiratory loads. Different loads might be encountered due to changing posture or activity of the animal.

The effects of pulmonary stretch on tidal volume found in this study are similar to findings in mammals (McClelland et al., 1972). There the behavior has been referred to as ventilatory stability.

Pulmonary stretch produced a small increase in the period between breaths in some birds and no detectable change in others. These findings are consistant with the findings of Eaton et al. (1971) who found a small increase in perido when chickens were inflated with a gas containing 5%-8% CO_2. No detectable change in period was found in some birds in the present study probably because range of thoracic volume changes examined was small compared to the range studied by Eaton et al. (1971). Leitner (1972) found increases in the ratio of time of expiration of the first breath after inlfation with 5% CO_2:the time of expiration before inflation ranging from 1.2 to 2.2. However, these results were obtained over a volume range (90-185 ml) beyond that examined in this study.

Acknowledgments. I would like to thank Millie Reeves for her fine technical support on this study. This work was supported in part by General Research Support Grant NIH 5 SO1 - RR 05365 - 15.

References

Bennett, T., Malmfors, T.: The adrenergic nervous system of the domestic fowl (*Gallus domesticus*). Z. Zellforsch. Mikrosk. Anat. 106, 22-50

Burger, R.E., Osborne, J.L., Banzett, R.B.: Intrapulmonary chemoreceptors in *Gallus domesticus*: adequate stimulus and functional localization. Respir. Physiol. 22, 87-97 (1974)

DeAndra, G., Rebollo, M.A.: The neuromuscular spindles in the adult chicken I. Morphology. Acta Anat. (Basel) 67, 437-451 (1968)

DeWet, P.D., Farrell, P.R., Fedde, M.R.: Number and morphology of muscle spindles in the transversus abdominis muscle of the chicken. Poult. Sci. 50, 1349-1359 (1971)

Eaton, J.A., Jr., Fedde, M.R., Burger, R.E.: Sensitivity to inflation of the respiratory system in the chicken. Respir. Physiol. 11, 167-177 (1971)

Fedde, R.M., Gatz, R.N., Slama, H., Scheid, P.: Intrapulmonary CO_2 receptors in the duck: II. Comparison with mechano-receptors. Respir. Physiol. 22, 115-121 (1974)

Groth, H.P.: Licht- und fluoreszenmikroskopische Untersuchungen zur Innervation des Luft-sacksystems der Vögel. Z. Zellforsch. Mikrosk. Anat. 127, 87-115

Kunz, A.L., Miller, D.A.: Pacing of avian respiration with CO_2 oscillations. Respir. Physiol. 22, 167-177 (1974)

Leitner, L.M.: Pulmonary mechanoreceptor fibres in the vagus of the domestic fowl. Respir. Physiol. 16, 232-244 (1972)

McClelland, A.R., Sproule, B.J., Lynne-Davies, P.: Functional importance of the Bruer-Hering reflex. Respir. Physiol. 15, 125-130 (1972)

Miller, D.A., Kunz, A.L.: Evidence that a cyclic rise in avian pulmonary CO_2 triggers the next inspiration. Respir. Physiol. in press (1977)

Molony, V.: Classification of vagal afferents firing in phase with breathing in *Gallus domesticus*. Respir. Physiol. 22, 57-76 (1974)

Central Nervous Mechanisms Regulating Thermal Panting

S. A. Richards and P. Avery

Summary

Recent experimental evidence on the regulation of thermal panting in birds (in particular the fowl) is briefly reviewed in relation to present knowledge of the neural organisation of polypnea in mammals. A pacemaker mechanism in the midbrain or rostral pons is the dominant frequency controller in panting, although the fowl is unusual in that vagal reflexes also play a crucial role. Thermosensitive elements in the spinal cord can initiate and inhibit panting and those in the hypothalamus and midbrain are also involved. The "paradoxical" effects of CO_2 during polypnea may be understood in terms of the release of central receptors from vagal control.

1. Introduction

The polypneic pattern of breathing seen in many birds and mammals under heat stress is characterised by its high frequency and low tidal volume, a change that must result from a specific modification to the mechanisms normally governing eupnea. The purpose of this article is to examine experimental evidence on the neural processes involved in panting which has appeared since two previous reviews (21, 22). Emphasis is placed on the avian work, no attempt having been made comprehensively to survey the mammalian literature. Nevertheless, since ideas about the genesis of respiratory rhythm in birds are still derived very largely by analogy from mammalian data, it remains inevitable that these be considered in any but the most superficial analysis.

2. Results on Birds

a) Ablation and Electrical Stimulation

Nothing has yet been published which might provide insight into the organisation of panting at the neuronal level and there is still very little work using the techniques of stimulation and ablation. A "panting centre" was first described in the midbrain of the pigeon by Saalfeld (26), and later mapped in stereotaxic coordinates for this species and for the fowl by Richards (24). It was shown (Table 1) that stimulation of a discrete area in the rostral midbrain would elicit an increase in respiratory frequency of up to 20-fold; a similar response has since been confirmed by others (10, 15).

Wye College, University of London, Ashford, Kent TN25 5AH, U.K.

Table 1. Summary of preliminary data on the effects of stimulation and ablation on polypnea in the fowl

Site	Technique	Conditions	Effects on respiratory frequency[a]	Remarks	References[b]
Hypothalamus	Ablation	Anesthetized; hyperthermic	↓	Raised thermal threshold	23
	Electrical stimulation	Anesthetized	↑↑		7
	Cooling	Conscious; hyperthermic	↓		*
	Heating	Conscious	↑		*
Midbrain	Ablation	Anesthetized; hyperthermic	↓↓↓	Abolition in all conditions	23
	Electrical stimulation	Anesthetized and conscious	↑↑↑	Sufficient in normothermia and after vagotomy	23
	Cooling	Conscious; hyperthermic	↓		*
	Heating	Conscious	-		*
Spinal cord	Cooling	Anesthetized and conscious; hyperthermic	↓		*
	Heating	Anesthetized and conscious	↑↑		*
Vagus nerve	Bilateral section	Anesthetized; hyperthermic	↓↓↓	Partial recovery in a feq preparations	18
	Unilateral stimulation (after bilateral section)	Anesthetized; hyperthermic	↑	Necessary but not sufficient without heat	18 20

[a] The arrows provide a qualitative indication of the strength of the induced effects: one, weak or irregular; two, moderately pronounced; three, very strong.
[b] * Previously unpublished results of Avery and Richards.

Transection of the brain stem also provided information on the function of structures at various levels (Table 1). Sections through the rostral hypothalamus had no effect on heat-induced panting and those caudal to the anterior commissure, while at first appearing to eliminate the response, did not in fact do so provided there was a 2-4 h period of recovery from surgical trauma and survival to rectal temperature 45°-46°C. Similar results were obtained after eliminating the entire hypothalamus, but sections caudal to the posterior commissure caused apnea and death in normothermic fowls, although it was subsequently found that if the section passed vertically only to about the level of the red nucleus, eupneic breathing continued but the polypneic response to heat was abolished.

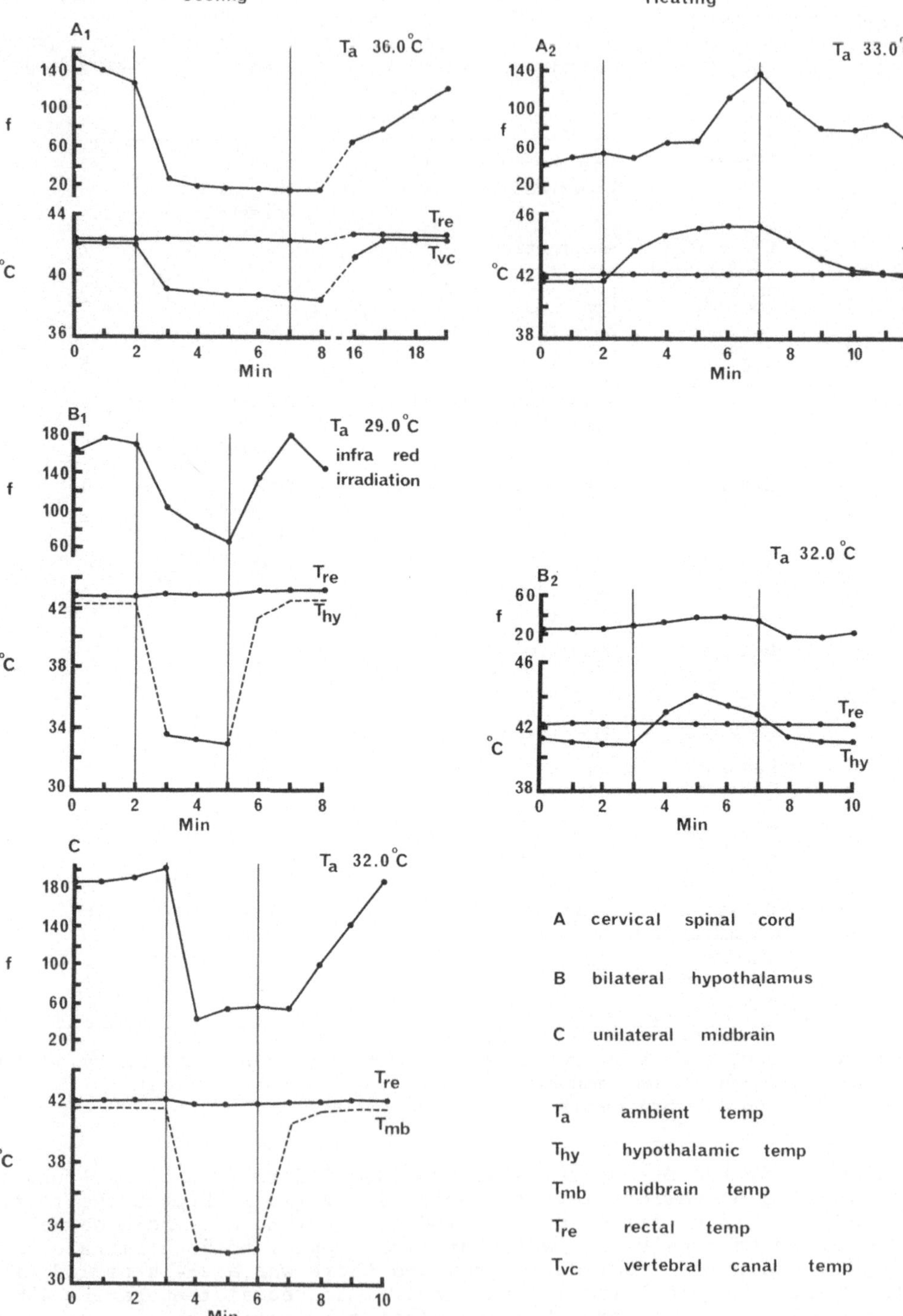

Cooling
Heating
A1
Ta 36.0°C
A2
Ta 33.0°C
f
°C
Tre
Tvc
Min
B1
Ta 29.0°C
infra red
irradiation
Thy
B2
Ta 32.0°C
C
Ta 32.0°C
Tmb
A cervical spinal cord
B bilateral hypothalamus
C unilateral midbrain
Ta ambient temp
Thy hypothalamic temp
Tmb midbrain temp
Tre rectal temp
Tvc vertebral canal temp

Bilateral cervical vagotomy abolishes panting in the fowl (although not in the pigeon) (20). However, of considerable interest was the finding that midbrain stimulation could still elicit polypnea in the vagotomized preparation, despite a reduced respiratory frequency and raised current threshold. Following total hypothalamic ablation and bilateral vagotomy, weak electrical stimulation of one vagus would often support a relatively normal respiratory rhythm in normothermia and permit the establishment of a panting response to rising body temperature, even when the stimulus parameters were themselves unchanged ("tonic"). Furthermore, in a minority of preparations polypnea, once established in this way, would then continue after withdrawal of the vagal stimulation, suggesting that even in the fowl, in which vagal drive is normally indispensable, some independent mobilization of central mechanisms may be possible.

b) Central Thermal Stimulation

Initiation or inhibition of panting by locally raising or lowering the temperature of the hypothalamus and spinal cord has been widely performed in mammals and has led to the hypothesis that thermosensitive neurons in these areas are capable of exerting a controlling influence on the lower brain stem respiratory centres. Enough work has now been performed with birds to conclude that there are many similarities between the two classes, although here the role of the spinal cord appears to be dominant (19, 25). Changes in hypothalamic temperature do influence polypnea in the fowl (23, 28), house sparrow (12), and pigeon (19), although in the latter species the effect may be opposite to that expected when the stimulus is applied during intermittent panting (27).

The results of some typical experiments performed recently on the fowl are summarized in Figure 1 (see also Table 1). Little or no effect on respiratory frequency resulted from unilateral changes in hypothalamic temperature under a wide variety of conditions, although bilateral heating caused a modest acceleration, and bilateral cooling a marked inhibition. In the midbrain polypnea could not be elicited by heating but, again, inhibition of heat-induced panting was achieved by cooling; the midbrain implantations were all unilateral owing to the difficulties of achieving viable chronic preparations after surgery in this area. Spinal heating was particularly effective at eliciting polypnea, and inhibiton by cooling was usually possible.

These results give general support to the extensive studies of Rautenberg and his colleagues with the pigeon (see 19). However, direct comparisons between the effects of thermal stimuli at the spine and in the brain must always be made with caution owing to the widely different surface areas presented by the two types of thermode and the inevitable ambiguity of the stimulus.

◀ Fig. 1. Effects of thermal stimulation of the central nervous system on respiratory frequency in the unanesthetized fowl. (A1) Cooling the spinal cord inhibited panting induced by high ambient temperature. (A2) Heating the spinal cord elicited panting. (B1) Cooling the hypothalamus reduced a high respiratory frequency resulting from infra-red irradiation. (B2) Heating the hypothalamus accelerated respiratory frequency slightly but did not cause panting. (C) Cooling the midbrain inhibited panting in moderate ambient heat. Note the interrupted time axis in (A1). *Broken lines* in B1 and C represent estimated temperatures

c) The Influence of Transmitter Substances

In mammals a considerable amount of work has been carried out with the purpose of elucidating motor pathways from the rostral brainstem by using transmitter materials to interfere with impulse activity or synaptic function at this level (5). Observations on birds show that intrahypothalamic or intramuscular noradrenaline elevates respiratory frequency by 3-10 fold in fowls and pigeons at thermoneutrality or in moderate heat (11, 17); 5-hydroxytryptamine has no effect on panting in pigeons exposed to high temperatures (8). These results do not conform to the synaptic model proposed for mammals (see 5).

3. The Organisation of Respiratory Rhythm in Polypnea

a) Neurophysiological Evidence

The genesis of the normal breathing pattern (in mammals) is a function of the nervous system that is perhaps better understood than most in terms of the activity of single cells, yet our knowledge of how these mechanisms are modified in polypnea remains poor. Recent expressions of the classical hypothesis of rhythmicity, as for example presented by Cohen et al. (3), explain the inspiratory-expiratory switching characteristics of the system as the result of interactions between medullary and pontine networks and assign a crucial role to the pneumotaxic centre. An alternative hypothesis advanced by Wyman (29) questions the whole concept of reciprocal inhibition between inspiratory and expiratory neurons and argues for a medullary "central pattern generator" which can, in the absence of all extrinsic information from sense receptors, drive a relatively normal breathing pattern; the pontine centres are seen as having no more than a modulating influence.

It is not yet clear whether published data on neuronal activity in panting have relevance in the context of these hypotheses of fundamental rhythmicity; it seems that there may be radical, and as yet unexplained, differences between the control mechanisms in eupnea and polypnea. For example, experiments on the cat have shown that an inversion of the discharge pattern in phrenic motoneurons (6, 14) and medullary inspiratory neurons (13) occurs during polypnea as compared to eupnea, while there are conflicting reports of a decline (13) and a pronounced increase (16) in the activity of expiratory neurons.

There is agreement, however, that while neurons at all levels of the brain stem may be involved in the organisation of polypnea, the response is primarily mediated by way of a region in the midbrain or rostral pons. Lesions in this area in the cat and dog (see 21), pigeon (26), and fowl (24) eliminate panting altogether, whereas vagotomy is most commonly without significant effect.

b) Effects of Carbon Dioxide

Another unresolved question is that concerning the "paradoxical" response of the panting animal to alveolar CO_2. In the intact animal breathing normally, increased CO_2 accelerates respiratory frequency, an effect that is abolished by vagotomy. In polypnea however, raised CO_2 causes a fall in frequency (1, 2) which is unaffected by vagotomy (4); the paradoxical effect thus appears to be exerted directly on the brain stem mechanisms.

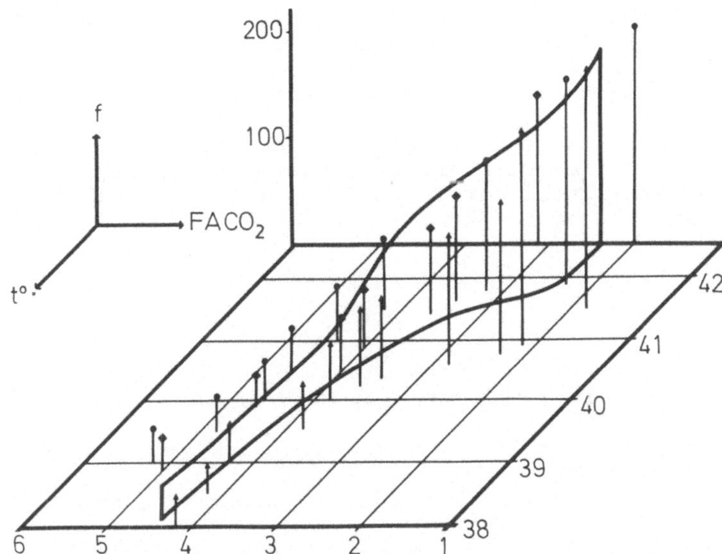

Fig. 2. The relationship between respiratory frequency (*f*, breaths/min), rectal temperature (t^oC) and alveolar carbon dioxide (F_{ACO_2}%) during the onset of thermal polypnea in three naturally breathing cats. When rectal temperature was below 40°C respiratory frequency increased very slowly and F_{ACO_2} decreased slightly. At higher temperatures respiratory frequency and F_{ACO_2} changed rapidly, reaching 200 breaths/min and 2% respectively at 42.5°C. Reproduced from Monteau et al. (14) by kind permission of the authors and Masson & Cie, Paris

This reversal of the effect of CO_2 could provide the necessary positive feedback that initiates full panting (Fig. 2) and it is possible that the neural substrates involved are common to the various types of polypnea, including thermal and decerebrate panting (2, 4, 24), hypocapnic polypnea (2, 6), and the rapid respiratory oscillations associated with vocalization (15). Polypnea could then be construed as a non-specific net effect of disturbances in the cerebral mechanisms (e.g. hypocapnic vasoconstriction and hypoxia) that are triggered, but not necessarily controlled, by the thermoregulatory system (9).

In any event this is of interest in relation to the observations that hypocapnia can cause polypnea even in normothermic animals, but in normocapnia hyperthermia may be ineffective (Fig. 2) (14). Hypocapnic polypnea occurs in vagotomized animals (2) and it may be that the simplest explanation of the "paradoxical" effect of CO_2 in panting is that these animals are, as it were, "functionally vagotomized". For example, in the panting cat the Hering-Breuer reflex is intact when stimulated by inflation (14), but the normal activity of the pulmonary stretch receptors is insignificant (7). Therefore, under these conditions the basic response of the central receptors might be "released" from control that is otherwise exerted by the peripheral CO_2 and proprioceptive reflexes. Raised CO_2 would then lengthen the respiratory cycle and a reduction would promote acceleration.

But this problem is particularly complex for while alveolar CO_2 may act as a controller of respiratory frequency, the frequency-depressing mechanism in thermal polypnea has itself a part to play in the regulation of arterial P_{CO_2}. The variations in frequency and tidal volume in the cat (4) and the somewhat similar compound pattern in the pigeon (18) would suggest that very shallow rapid panting might allow the accumulation of CO_2, which would retard the frequency and increase the amplitude. The deeper breathing would then increase alveolar ventilation, reduce the CO_2, and promote a return to the more rapid rhythm.

Clearly a great deal more work is needed and the avian preparation in which CO_2 levels can be controlled with great precision by unidirectional ventilation might lend itself particularly well to the analysis of this intriguing problem.

References

1. Bucher, K., Baettig, P.: Some special respiratory effects of CO_2 in pigeons. Agents and Actions 5, 378-382 (1975)
2. Cohen, M.I.: Respiratory periodicity in the paralyzed, vagotomized cat: hypocapnic polypnea. Am. J. Physiol. 206, 845-854 (1964)
3. Cohen, M.I., Piercey, M.F., Gootman, P.M., Wolotsky, P.: Respiratory rhythmicity in the cat. Federation Proc. 35, 1967-1974 (1976)
4. Euler, C. von, Herrero, F., Wexler, I.: Central mechanisms determining rate and depth of respiratory movements. Respir. Physiol. 10, 93-108 (1970)
5. Hellon, R.F.: Monoamines, pyrogens and cations: their actions on central control of body temperature. Pharmacol. Rev. 26, 289-321 (1975)
6. Hilaire, G., Monteau, R.: Activité des motoneurones phréniques au cours de la polypnée thermique ou hypocapnique. J. Physiol. (Paris) 68, 193-203 (1974)
7. Hilaire, G., Moneau, R.: Activité en polypnée des récepteurs pulmonaires d'inflation. J. Physiol. (Paris) 70, 219-238 (1975)
8. Hissa, R., Rautenberg, W.: Thermoregulatory effects of intrahypothalamic injections of neurotransmitters and their inhibitors in pigeons. Comp. Biochem. Physiol. 51A, 319-326 (1975)
9. Karczewski, W.A.: Thermal polypnea. A non-specific phenomenon. In: Temperature Regulation and Drug Action. Lomax, P., Schonbaum, E., Jacob, J. (eds.). Basel: Karger, 1975, pp. 348-351
10. Kotilainen, P.V., Putkonen, P.T.S.: Respiratory and cardiovascular responses to electrical stimulation of the avian brain with emphasis on inhibitory mechanisms. Acta Physiol. Scand. 90, 358-369 (1974)
11. Marley, E., Nisticò, G.: Central effects of clonidine 2- (2, 6- dichlorophenylamine) -2-imidazoline hydrochloride in fowls. Brit. J. Pharmacol. 55, 459-473 (1975)
12. Mills, S.H., Heath, J.E.: Responses to thermal stimulation of the preoptic area in the house sparrow, *Passer domesticus*. Am. J. Physiol. 222, 914-919 (1972)
13. Monteau, R., Hilaire, G.: Modification d'activité des neurones respiratoires bulbaires lors de l'installation de la polypnée thermique. J. Physiol. (Paris) 72, 205-231 (1976)
14. Monteau, R., Hilaire, G., Ouedraogo, C.: Contribution à l'étude de la fonction ventilatoire ou course de la polypnée thermique. J. Physiol. (Paris) 68, 97-120 (1974)
15. Peek, F.W., Phillips, R.E.: Repetitive vocalizations evoked by local electrical stimulation of avian brain. Brain, Behav. Evol. 4, 417-438 (1971)
16. Pleschka, K., Wang, S.C.: The activity of respiratory neurons before and during panting in the cat. Pflüg. Arch. 353, 303-315 (1975)
17. Pyörnilä, A., Hissa, R., Saarela, S.; Effects of adrenergic blocking agents on peripheral noradrenaline induced thermoregulatory responses in the pigeon at heat. Pflüg. Arch. 365, 167-170 (1976)
18. Ramirez, J.M., Bernstein, M.H.: Compound ventilation during thermal panting in pigeons: a possible mechanism for minimizing hypocapnic alkalosis. Federation Proc. 35, 2562-2565 (1976)
19. Rautenberg, W., Necker, R., May, B.: Thermoregulatory responses of the pigeon to changes of the brain and spinal cord temperatures. Pflüg. Arch. 338, 31-42 (1972)

20. Richards, S.A.: Vagal control of thermal panting in mammals and birds. J. Physiol. (London) 199, 89-101 (1968)
21. Richards, S.A.: The biology and comparative physiology of thermal panting. Biol. Rev. 45, 223-264 (1970a)
22. Richards, S.A.: Physiology of thermal panting in birds. Ann. Biol. Anim. Bioch. Biophys. 10, 151-168 &1970b)
23. Richards, S.A.: The role of hypothalamic temperature in the control of panting in the chicken exposed to heat. J. Physiol. (London) 211, 341-358 (1970c)
24. Richards, S.A.: Brain stem control of polypnea in the chicken and pigeon. Respir. Physiol. 11, 315-326 (1971)
25. Richards, S.A.: Thermal homeostasis in birds. In: Avian Physiology. Peaker, M. (ed.). Sympl. Zool. Soc. Lond. 35, 65-96 (1975)
26. Saalfeld, F.E. von: Untersuchungen über das Hacheln bei Tauben. Z. vergl. Physiol. 23, 727-743 (1936)
27. Schmidt, I.: Paradoxical changes of respiratory rate elicited by altering rostral brain stem temperatures in the pigeon. Pflüg. Arch. 367, 111-113 (1976)
28. Scott, N.R., van Tienhoven, A.: Thermoregulatory responses of poultry to local heating and cooling of the hypothalamus. Paper 74 - 5512. Am. Soc. Agric. Eng. (1974)
29. Wyman, R.J.: Neural generation of the breathing rhythm. Ann. Rev. Physiol. 39, 417-448 (1977)

Control of Panting by Thermosensitive Spinal Neurons in Birds

W. Rautenberg, B. May, R. Necker, and G. Rosner

Summary

Selective heating of the spinal cord evoked panting in pigeons exposed to neutral ambient temperature. Cooling the spinal cord stopped panting which was elicited by peripheral heat (ambient temperature 38°C).

Electrophysiological recordings from ascending pathways of the spinal cord showed the existence of warm-sensitive neurons that were excited by heating the spinal cord. There was a nearly linear increase of impulse frequency in the range 35°-45°C spinal cord temperature.

Recordings in the preoptic area of the hypothalamus showed warm-responsive neurons that were excited by heating the spinal cord. There was generally a steep non-linear increase of impulse frequency in a temperature range where panting was observed (42°-45°C).

1. Introduction

The thermal panting of endothermic animals increases the evaporative heat loss by which the body temperature can be stabilized during heat load. Birds have no sweat glands and it is only by altering the rate of ventilation that they can regulate their evaporative heat loss. That means, however, that the effector mechanism, which was developed to regulate the blood gases by controlling the rate of lung ventilation, is simultaneously used to avoid an excessive hyperthermic state. Thus, one and the same effector is used in two feedback loops: 1. in the regulation of the blood gases and 2. in the control of the body temperature. The interaction or competition of the two control systems have hardly been investigated in birds: the following study deals with temperature-sensitive structures in the central nervous system of pigeons which effect the rate of respiration under conditions of selective heating.

2. Methods

Domestic pigeons (*Columba livia*) weighting about 500 g were used for the experiments. Thermodes were chronically implanted into the vertebral canal and in the rostral brain stem, as previously described (Rautenberg et al., 1972). The unanaesthetized birds were placed in a climatic

Institut für Tierphysiologie, Ruhr-Universität Bochum, D-4630 Bochum, Federal Republic of Germany.

chamber at an ambient temperature of 30°C (neutral) or 38°C (heat load). The chamber was continuously ventilated (1.2 l/min) with dry air. The spinal cord and the brain stem could be cooled or heated selectively by circulating water through the thermodes. The rate of respiration and several body temperatures were recorded continuously.

For the electrophysiological studies the pigeons were anaesthetized (1.5-2 g/kg urethane, i.p.) and artificially respirated. Recordings were made in the lateral funiculus of the spinal cord after cutting the cord at C_4 (Necker, 1975). To avoid a direct thermal influence on the recording site recordings were made cranial to the end of the implanted vertebral thermode. Further electrophysiological experiments dealt with temperature-responsive neurons at higher levels of the CNS. For these sterotaxic recordings were made in the preoptic area of pigeons during thermal stimulations of the spinal cord. In both studies steel needle electrodes were used and the recording site was histologically verified by Prussian blue staining. The stimulus temperature in the vertebral canal and the impulse frequency were recorded on an oscilloscope and were stored on a magnetic tape for computer analysis. The body temperatures outside the selectively stimulated part of the spinal cord were generally held at the normal level of about 41°C by an electric pad.

3. Results

If conscious pigeons are exposed to an ambient temperature of about 38°C, the deep body temperature rises progressively and the rate of ventilation also increases exponentially with body temperature. The rate of breathing reaches about 600 c/min when the deep body temperature

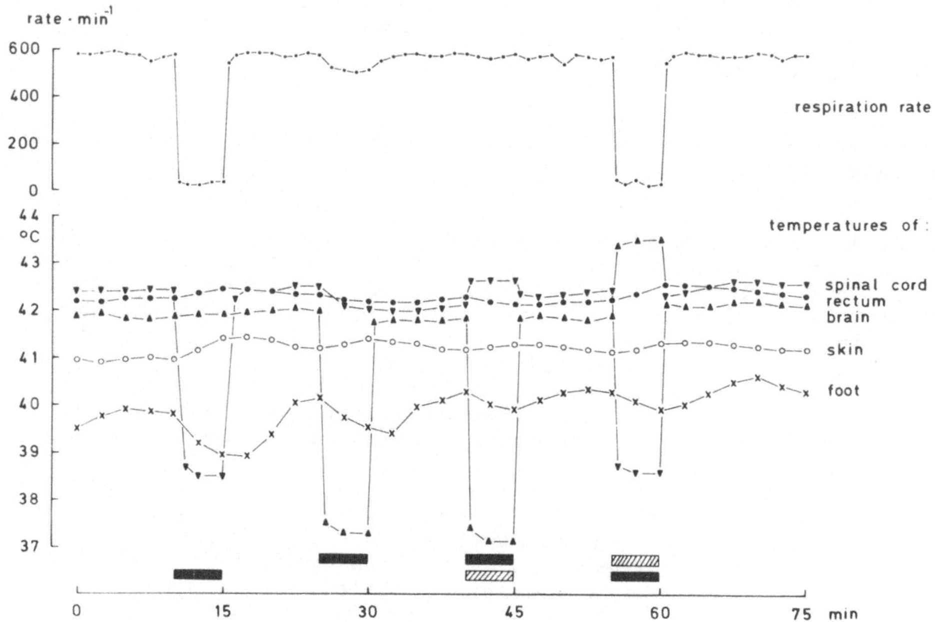

Fig. 1. Effect of cooling the brain stem and the spinal cord on thermal panting in a pigeon. Air temperature 38°C. *Bars* mark stimulus periods

exceeds about 42.5°C (Fig. 1). Under these conditions selective cooling of the spinal cord to about 38.5°C reduces the rate of respiration to about 20 c/min. However, a cooling of the brain stem to about 37°C has only a small effect on respiration which can be compensated by a slight warming of the cord by 0.5°C, as it is shown in the third stimulus period of Figure 1. The last stimulus period involves both a cooling of the vertebral canal and a heating of the brain. Here, the same effect on respiration results as is shown in the first period where the spinal cord is cooled alone.

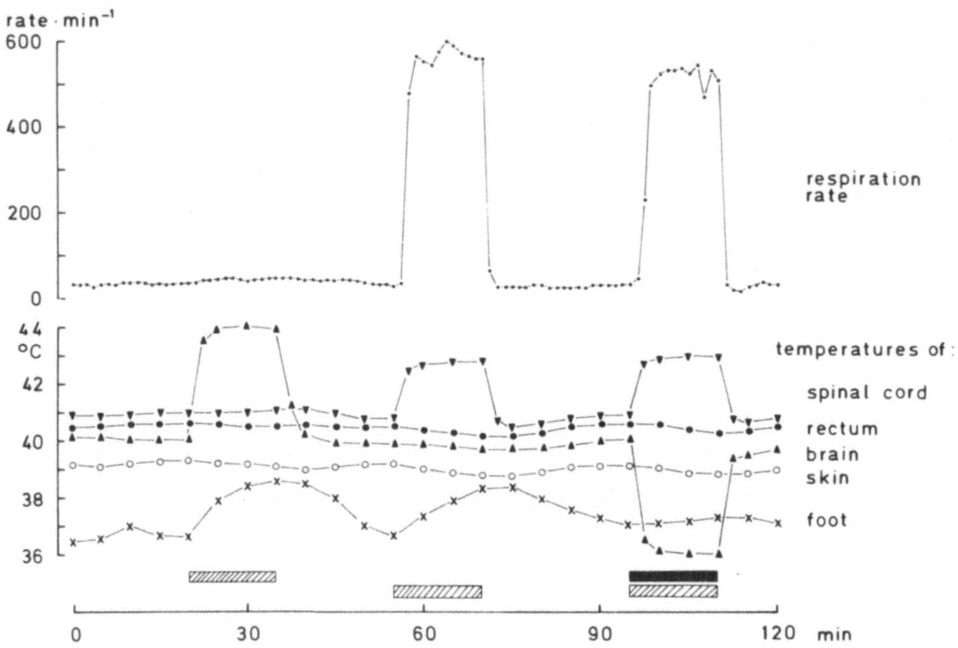

Fig. 2. Effect of heating the brain stem and the spinal cord on the rate of respiration in a pigeon. Air temperature 30°C. *Bars* mark stimulus periods

Figure 2 shows an experiment where a pigeon was exposed to a neutral ambient temperature of 30°C. The rate of breathing was about 18 c/min. Heating the brain stem to about 44°C was without an effect on respiration, whereas heating of the vertebral canal to 43°C drives the rate of ventilation up to the level of panting (Fig. 2). This effect of spinal heating on respiration could not be influenced by simultaneous cooling of the brain stem. The two examples are representative for numerous trials performed under similar conditions (Rautenberg, 1969; Rautenberg et al., 1972). The selective heating of the spinal cord drives both the ventilation of the lung and the gular fluttering, which is mode of respiratory heat loss from a dead space ventilation of the upper part of the respiratory tract (Richards, 1970). The gular fluttering, which was measured in some experiments, only occurs if the rate of lung ventilation exceeds 500 c/min. Its frequency is similar or a little higher than that of the lung ventilation.

The threshold temperature at which pigeons start to pant when the spinal cord is selectively heated depends on several factors, e.g., other body temperatures, the humidity, the windspeed, the time of day when investigations are carried out, etc. Apart from the other body temperatures, these factors were held constant in the following study which investigates the temperature threshold of panting by heating the spinal cord at different skin and core temperatures. Figure 3 shows that there

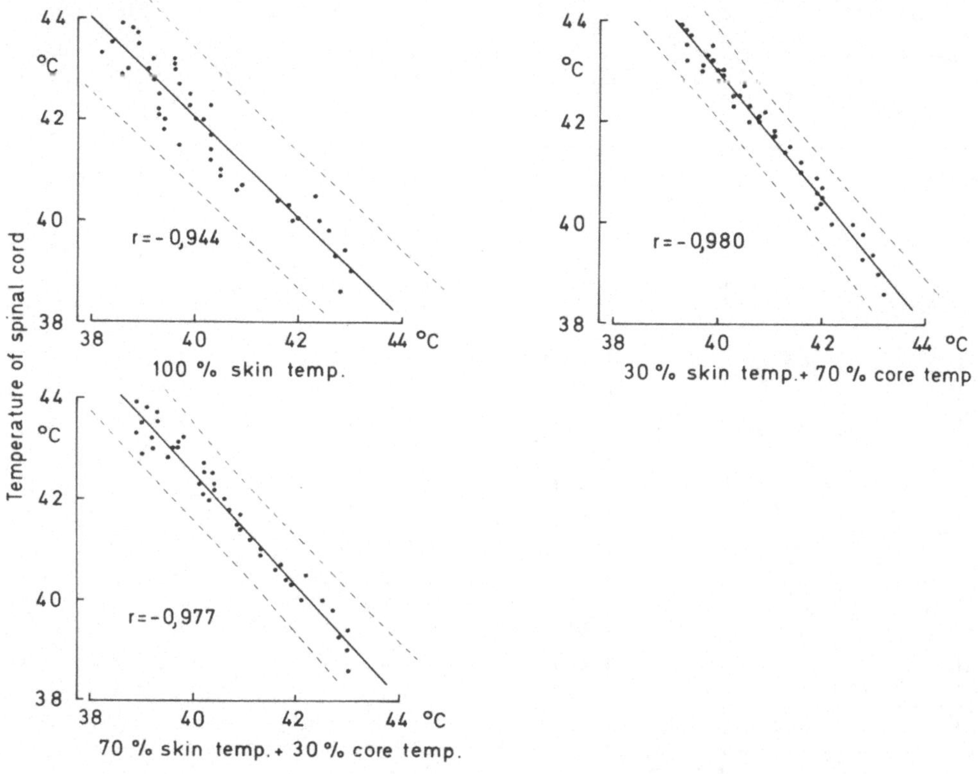

Fig. 3. Threshold of thermal panting of 15 pigeons. Relationship between artificially warmed spinal cord temperature (*ordinates*) and the other body temperatures calculated for different ratios of skin and core temperature. Tolerances of regressions (*dashed lines*) are calculated by $= \bar{y} + b_{yx}(x - \bar{x}) \pm t_{2\alpha}s_T$

is an interaction of spinal temperature and the other body temperatures with regard to the threshold of panting. The ordinates always show the spinal cord temperature at which panting starts (180-200 c/min), where the temperature of the vertebral thermode is continuously raised by 0.42°C/min. The abscissae indicate the mean skin temperature or temperatures calculated from different portions of skin and core temperatures. A linear relationship exists for the threshold of panting between the temperature of the spinal cord and the other body temperatures, within a physiological range. The dashed lines present the tolerance limits of the regression functions and may indicate the special role of the deep body temperature in regulating the evaporative heat loss (smallest tolerance limit at 70% core temperature).

The results presented above led to the suggestion that temperature-sensitive neurons should exist in the spinal cord of pigeons and that it should be possible to record the activity of these neurons with electrophysiological methods. In electrophysiological experiments, the thermal stimulus was applied to the same spinal segments where the thermoregulatory responses described above could be elicited. The recording site was rostral to the thermode, i.e., the neuronal activity was recorded from the ascending pathways in the lateral funiculus of the spinal cord.

Most of the spontaneously firing neurons in the lateral funiculus could not be influenced by thermal stimulation of the spinal cord: only a few neurons showed responses to thermal stimulation (Necker, 1975). Most of the temperature-sensitive neurons were excited by warming. In Figure 4b the response of a warm-sensitive neuron to heating of the

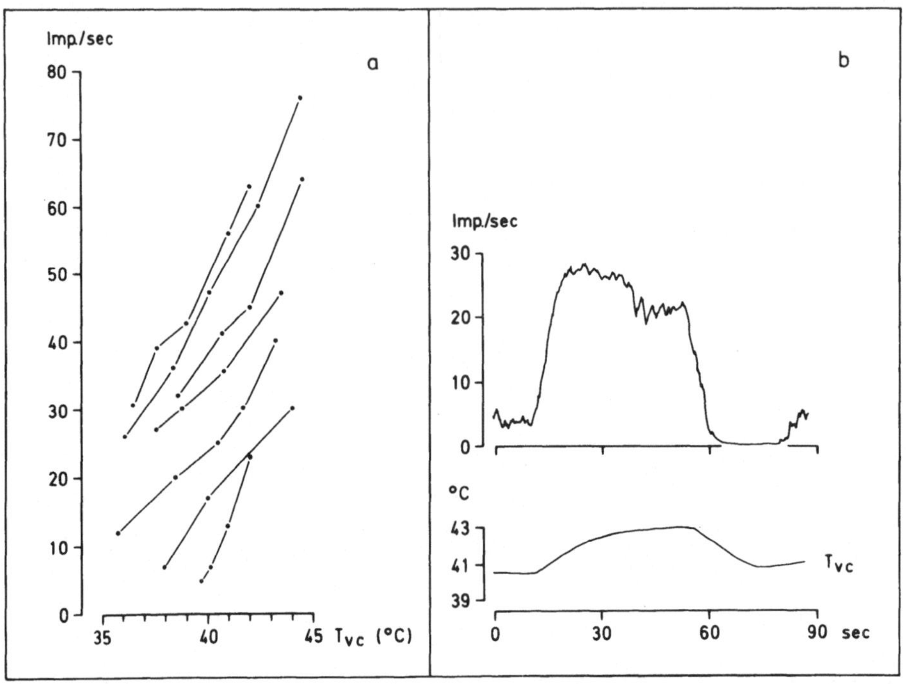

Fig. 4a, b. Response of spinal warm-sensitive neurons to thermal stimulation of the spinal cord in pigeons. (a) static curves of seven neurons (*points* are means of 1 min after 3 min of adaptation). (b) response to dynamic thermal stimulation; *upper curve*: impulse frequency, *lower curve*: spinal cord temperature (T_{vc})

cord from about 40.5°-43°C is shown. There is a strong excitation with a moderate adaptation of the impulse frequency at the higher temperature. A fall in spinal temperature is followed by transient inhibition of firing (Fig. 4b). Such phasic responses to dynamic temperature changes were found only in part of the temperature-sensitive neurons. Most warm-sensitive spinal neurons showed a proportional change in impulse frequency without any phasic responses.

During a maintained change in spinal temperature there was usually a maintained change in impulse frequency. In Figure 4a static curves of seven warm-sensitive neurons are drawn. These static curves show the relationship between impulse frequency and spinal temperature after adaptation. There is a nearly linear increase of impulse frequency in the range of 35°-45°C. This linear relationship over a wide range of spinal temperature suggests that these neurons might be sensory neurons detecting the temperature of the spinal cord. It is an interesting finding that the range of sensitivity of the neurons corresponds to the temperature range where effects of warming or cooling the spinal cord on respiration occur.

It has been shown that the integration of temperature signals in the temperature control system probably occurs in the anterior hypothalamus (Hammel, 1968). Thus one would expect to find neurons in this region of the brain whose activity is influenced by thermal stimulation of the spinal cord via the activity of the ascending spinal neurons described above. In electrophysiological experiments such neurons could be shown to exist in the preoptic area of the anterior hypothalamus of the pigeon (Rosner, 1977).

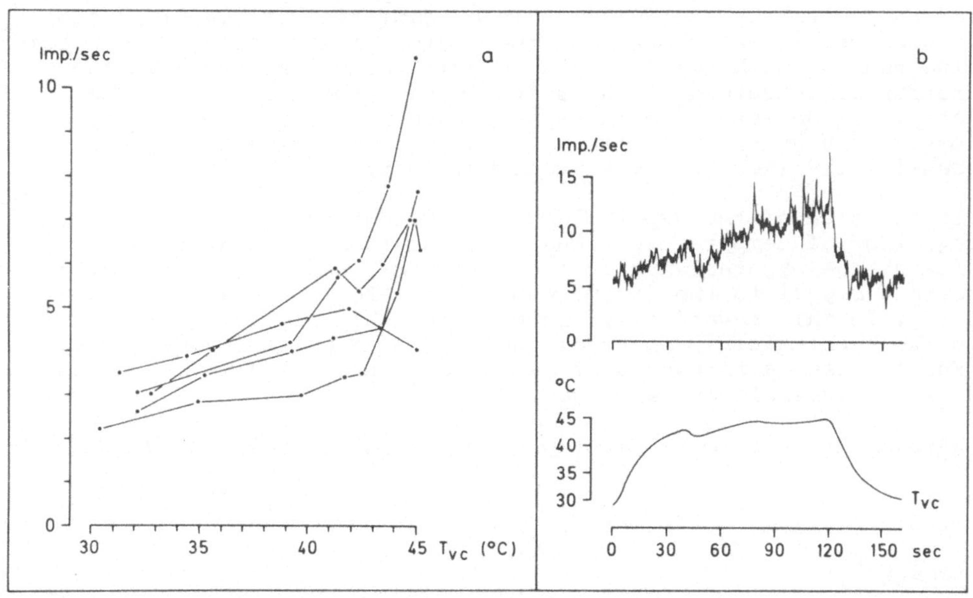

Fig. 5a, b. Response of warm-responsive neurons in the preoptic area of the hypothalamus to thermal stimulation of the spinal cord in pigeons. (a) static curves of five neurons. (b) response to dynamic thermal stimulation (note the response to small variations of temperature at high temperature values at about 45 s of the recording). Further explanations see Figure 4

Figure 5 shows the response of warm-responsive hypothalamic neurons to dynamic temperature changes of the spinal cord as well as static curves. Heating the spinal cord from about 31° to 45°C is followed by an increase in impulse frequency that is more pronounced at higher temperatures. Usually no phasic responses such as excitatory overshoot or transient inhibition of impulse frequency were observed. In contrast to the warm-sensitive neurons in the spinal cord, the static curves of the warm-responsive neurons in the hypothalamus usually were nonlinear (Fig. 5a). While there is a moderate increase of impulse frequency at spinal temperatures below about 40°C there is a steep increase in the range 42°-45°C. Such a nonlinear characteristic can be interpreted as the response of interneurons (Eisenman, 1972) which, in this case, receive an input from the temperature-sensitive neurons of the spinal cord. Only after the input increases beyond a threshold is there a strong activation of the hypothalamic interneurons. The range of maximal sensitivity corresponds well to the range of spinal cord temperatures where panting occurs. Thus one might speculate that these neurons are involved in the initiation and graduation of evaporative heat loss by elevated respiratory frequency.

4. Conclusions

There is ample evidence that the spinal cord of endotherms represents a thermosensitive area within the temperature control system (Simon, 1974). The results presented above show that changes in the rate of respiration serving to control heat loss can be elicited or influenced by thermal stimulation of the spinal cord in pigeons. The neuronal basis of the thermal sensitivity could be confirmed by electrophysiological recordings. It was shown that there are, in the spinal cord, warm-sensitive neurons with response characteristics suggesting that they may function as detectors of the spinal cord temperature. The idea that there are thermodetectors in the spinal cord has been supported by experiments with spinally deafferented pigeons which retained spinal thermosensitivity (Necker and Rautenberg, 1975).

The preoptic area of the hypothalamus is assumed to represent a centre of integration of temperature signals in the thermoregulatory system. In addition, the hypothalamus is thought to be the site of initiation of an output signal to the thermoregulatory effector mechanisms such as panting. In this regard it is interesting that recordings from neurons in the hypothalamus show that there is an input from the spinal cord. Furthermore, a threshold of response exists which is well within the range of threshold of panting.

Acknowledgment. This work was supported by the Deutsche Forschungsgemeinschaft (SFB 114).

References

Eisenman, J.E.: Unit activity studies on thermoresponsive neurons. In: Essays on Temperature Regulation. Bligh, J., Moore, R.E. (eds.). Amsterdam-London: North Holland, 1972, pp. 55-69

Hammel, H.T.: Regulation of internal body temperature. Ann. Rev. Physiol. 30, 641-710 (1968)

Necker, R.: Temperature-sensitive ascending neurons in the spinal cord of pigeons. Pflüg. Arch. 353, 275-286 (1975)

Necker, R., Rautenberg, W.: Effect of spinal deafferentation on temperature regulation and spinal thermosensitivity in pigeons. Pflüg. Arch. 360, 287-299 (1975)

Rautenberg, W.: Die Bedeutung der zentralnervösen Thermosensitivität für die Temperaturregulation der Taube. Z. vergl. Physiol. 62, 235-266 (1969)

Rautenberg, W., Necker, R., May, B.: Thermoregulatory responses of the pigeon to changes of the brain and the spinal cord temperatures. Pflüg. Arch. 338, 31-42 (1972)

Richards, S.A.: Physiology of thermal panting in birds. Ann. Biol. Anim. Bioch. Biophys. 10, 151-168 (1970)

Rosner, G.: Response of hypothalamic neurons to thermal stimulation of spinal cord and skin in pigeons. Pflüg. Arch. 368, Suppl. R 29 (1977)

Simon, R.: Temperature regulation: The spinal cord as a site of extrahypothalamic thermoregulatory functions. Rev. Physiol. Biochem. Pharmacol. 71, 1-76 (1974)

Session V

Respiration of the Embryo: Eggshell, Embryonic Membranes, Circulation, and Blood

Chairman: J. Metcalfe

The Avian Egg: In vivo Conductances to Oxygen, Carbon Dioxide, and Water Vapor in Late Development

C. V. Paganelli, R. A. Ackerman, and H. Rahn

Summary

In vivo measurements of both O_2 and water vapor conductances (G_{O_2} and G_{H_2O}) were made at 38°C on the same hen's eggs between days 12 and 18 of incubation. Neither G_{O_2} or G_{H_2O} changed systematically during this phase of incubation. Average G_{O_2} (cm^3/day/torr) was 12.60 ± 0.30 (S.D.); average G_{H_2O}, 15.26 ± 0.51; G_{O_2}/G_{H_2O} = 0.83. In vivo determination of G_{CO_2} and G_{H_2O} in a second group of eggs gave values of 9.22 ± 1.54 and 14.60 ± 0.65, respectively; G_{CO_2}/G_{H_2O} = 0.64. The conductance ratios agree closely with the binary diffusivity ratios (in air or N_2) of the gases in question. We conclude that O_2, CO_2, and water vapor diffuse between environment and air cell of the chicken egg via the same diffusion path, whose chief resistance lies in the shell itself. The outer shell membrane offers little resistance in vivo to passage of gas molecules.

1. Introduction

Chemical communication between the internal and external environments of the developing avian egg occurs only by gaseous exchange through the microscopic pores of the eggshell and the underlying shell membranes. A basic assumption underlying much of the work on avian embryonic gas exchange done in recent years both in our laboratory and at other institutions is that this process is diffusion-controlled, in marked contrast to the convection-controlled gas exchange of the hatchling or adult bird (Wangensteen et al., 1970/71; Kutchai and Steen, 1971; Ar et al., 1974; Lomholt, 1976; Tullett and Board, 1976). A corollary of this assumption is that rates of penetration per unit driving force for various gases should be proportional to their diffusivities, provided that all gases use the same diffusion path through the shell and membranes.

A mathematical description of the diffusive flux of a gas (e.g., O_2, CO_2, water vapor) through the pores of the eggshell is provided by a modified form of Fick's first law of diffusion (Wangensteen et al., 1970/71; Paganelli et al., 1975):

$$\dot{M} = (\mathcal{D}/RT)\ (Ap/L)\ \Delta P \qquad (1)$$

Department of Physiology, School of Medicine, State University of New York at Buffalo, Buffalo, New York 14214/U.S.A.

where $\dot{M}$ = gas flux, cm^3 (STPD)/s
$\mathcal{D}$ = binary diffusion coefficient, cm^2/s
A_p = effective pore area of shell, cm^2
L = length of diffusion path (equivalent to shell thickness), cm
ΔP = gas partial pressure difference across shell, torr

$\mathcal{D}$, the binary diffusivity, applies when only two gases are present in the diffusion path. RT, the product of the gas constant and the absolute temperature, has units of cm^3 (STPD)/cm^3/torr, and serves to adjust units on the two sides of the equation. The first two terms of Equation (1) can be conveniently combined into the single term, G, the diffusive conductance, with units of cm^3/s/torr (Ar et al., 1974). Thus:

$$\dot{M} = G \cdot \Delta P \qquad (2)$$

where $G = (\mathcal{D}/RT)\ (A_p/L)$. *Conductance* defined in this manner is equivalent to *diffusing capacity*. Both express the flux of a gas per unit time and pressure difference, whether applied to the egg or to the vertebrate lung.

Conductance, as we have generally measured it, represents the flux of gas per unit time and driving pressure difference between atmosphere and air cell. Two structures form this conductance: the shell itself, penetrated by pores of microscopic radius, and the outer shell membrane, a network of fibrous protein whose interstices are gas-filled. O_2, CO_2, and water vapor must traverse these structures in their passage between the environment and gas space of the air cell. The fundamental question to which we address ourselves in the present studies is whether these three gases do in fact share a common diffusion path through the shell and outer shell membrane.

To answer this question we developed methods to measure both O_2 and H_2O conductances in the same live fertile eggs at five times during incubation. In separate experiments, in vivo CO_2 and H_2O conductances were compared in the same eggs at four different times during incubation. When ratios of both O_2 and H_2O conductances and CO_2 and H_2O conductances were compared with the ratios of the corresponding binary diffusivities of these gases in air, a reasonably close agreement was found. Our principal conclusion is that within our ability to measure conductances and evaluate tabulated values of diffusion coefficients, O_2, CO_2, and H_2O share the same diffusion path through the shell and outer shell membrane of the hen's egg.

2. Methods

a) Comparison of O_2 and H_2O Conductances

Fertile White Leghorn eggs, before being set in an incubator, were placed in desiccators at 25°C and weighed on three consecutive days to establish their rate of water loss. Water vapor conductance (G_{H_2O}) for each egg was determined as described previously (Ar et al., 1974). After this treatment, during which each egg lost about 1 g of H_2O, the eggs were placed in a commercial forced-draft incubator at 38°C and incubation was begun. At appropriate stages each egg was placed in a closed vessel containing KOH as a CO_2 absorber, and the O_2 uptake measured by a manometric method. A sensitive pressure transducer recorded the pressure changes between periodic refilling of the vessels with pure O_2 by calibrated syringe. When a steady O_2 consumption was obtained, the egg was removed from the respirometer and submerged in a jar containing warm, acidified water. A small hole was made in the blunt end of the egg and a sample of air cell gas aspirated into a syringe. This gas was then analyzed for O_2 and CO_2 in a Scholander gas analyzer. This method has been previously described (Wangensteen and Rahn, 1970/71; Rahn et al., 1974). From the air cell O_2 fraction, the O_2 difference (ΔP_{O_2}) across the eggshell and outer membrane can be obtained. When this value is divided into the O_2 consumption, one obtains the O_2 conductance (G_{O_2}) which for convenience we express as cm^3 O_2 (STPD)/day/torr.

b) Comparison of CO_2 and H_2O Conductances

In this group of experiments G_{H_2O} was again determined prior to incubation as described above. Incubation was begun at 38°C in a forced-draft incubator, and eggs were removed at intervals for measurement of CO_2 production. This was accomplished by determining with a mass spectrometer the increase in CO_2 concentration of a stream of air passed through a closed container at 38°C containing five or six eggs. Knowledge of the flow rate and the difference in CO_2 concentration between gas entering and leaving permitted calculation of $\dot{M}_{CO_2}$. ΔP_{CO_2} from air cell to environment was determined by sampling air cell gas as described above for O_2 measurements.

3. Results

Mean values of O_2 conductance, G_{O_2}, and water vapor conductance, G_{H_2O}, in chicken eggs are shown in Table 1 for incubation times of 12, 14,

Table 1. Comparison of O_2 and water vapor conductances in air. Standard deviations in brackets

Age days	n[b]	G_{H_2O} - 25°C mg/day/torr	G_{H_2O} - 38°C cm^3/day/torr	G_{O_2} - 38°C cm^3/day/torr	G_{O_2}/G_{H_2O} at 38°C
12[a]	5	11.33 (1.44)	14.40 (1.83)	12.32 (2.48)	0.86
14[a]	10	12.28 (1.11)	15.61 (1.41)	12.80 (3.53)	0.82
16	8	11.99 (0.41)	15.24 (0.52)	12.6 (1.5)	0.83
17	8	12.31 (1.81)	15.65 (2.30)	13.0 (3.4)	0.83
18	3	12.10 (0.58)	15.38 (0.74)	12.3 (0.7)	0.80
Mean		12.00	15.26	12.60	0.83
S.D.		(0.40)	(0.51)	(0.30)	(0.02)
n		5	5	5	5

[a] Data from Ackerman and Rahn (1978). [b] Number of eggs.

Table 2. Comparison of CO_2 and water vapor conductances in air. Standard deviations in brackets

Age days	n	G_{H_2O} - 25°C mg/day/torr	G_{H_2O} - 38°C cm³/day/torr	G_{CO_2} - 38°C cm³/day/torr	G_{CO_2}/G_{H_2O} at 38°C
12	5	12.48 (2.84)	15.86 (3.61)	8.17	0.52
12	5	11.28 (1.78)	14.34 (2.26)	10.39	0.72
14	6	11.13 (0.87)	14.15 (1.11)	10.97	0.78
14	6	11.32 (1.47)	14.39 (1.87)	6.90	0.48
17	6	11.57 (3.00)	14.71 (3.81)	8.80	0.60
18	6	11.11 (2.83)	14.12 (3.60)	10.10	0.72
Mean		11.48	14.60	9.22	0.64
S.D.		(0.52)	(0.65)	(1.54)	(0.12)
n		6	6	6	6

16, 17, and 18 days. In Table 2 mean CO_2 and H_2O conductances are shown for eggs incubated for 12, 14, 17, and 18 days. Water vapor conductances are first given in Tables 1 and 2 as directly calculated from weight loss measurements, in units of mg/day/torr at 25°C (Ar et al., 1974). In order to compare G_{H_2O} with G_{O_2} or G_{CO_2}, the units of G_{H_2O} must first be changed to cm³ (STPD)/day/torr. The ratio of the molar volume of an ideal gas to the molecular weight of water (22.414/18.016) provides the conversion factor.

Because G_{H_2O} is routinely measured at 25°C, it must also be corrected to 38°C if comparison with G_{O_2} or G_{CO_2} at 38°C is to be made. There are two considerations here. Since G expressed in volume units is equal to $(\mathcal{D}/RT)(A_p/L)$, it must vary as 1/T. However, $\mathcal{D}$ also varies explicitly with temperature. The Chapman-Enskog equation gives this variation as $\mathcal{D} \sim (T)^{1.5}$ (Reid and Sherwood, 1966). Thus G will vary as $T^{1/2}$. G at 25°C is then multiplied by the factor $(311/298)^{1/2}$ or 1.022 to convert it to G at 38°C. [Reid and Sherwood (1966) give a range of values from 1.5 to 2.3 for the exponent of T, depending on the gases involved. However, the temperature difference 25°-38°C on the absolute scale is small enough so that the correction factor is not particularly sensitive to the exact value of the exponent chosen.]

Notice that the values of G for H_2O, O_2, and CO_2 in Tables 1 and 2 show no systematic variation with incubation age for eggs from day 12 through day 18, within the limits imposed by sample variation and measurement technique. That is, the conductance of the shell and outer shell membrane to these gases remains constant during this phase of incubation. The changes in conductance of hen's eggs as a function of incubation age have been shown to take place on the fourth or fifth day of incubation, well before the measurements reported here were made

(Kutchai and Steen, 1971; Lomholt, 1976; Tullett and Board, 1976). For any given incubation age, the same eggs were used for both G_{H_2O} and G_{O_2} determinations; this was also true for the comparison of G_{H_2O} and G_{CO_2}. However, there was no direct comparison of all three gases in the same eggs. Ackerman and Rahn (1978) recently made conductance measurements in early incubation and found that combined shell and outer shell membrane permeabilities to O_2 and H_2O were also constant from day 5 through day 14 of incubation. In the absence of variation with age, it is reasonable to average the G_{O_2}/G_{H_2O} ratios over all incubation ages, and to treat the G_{CO_2}/G_{H_2O} ratios in the same fashion, in order to arrive at a best estimate of the relative conductances of the shell and outer shell membrane to the gases involved. The average ratios are shown in Tables 1 and 2.

4. Discussion

It has been previously been assumed (Wangensteen et al., 1970/71; Ar et al., 1974) that eggshell conductance to a gas is proportional to the diffusivity of that gas. Thus $G_{H_2O}/G_{O_2} = \mathcal{D}_{H_2O}/\mathcal{D}_{O_2}$; if the conductance for one gas is known, the conductance for other gases can be calculated. However, one must first obtain from the literature values of the appropriate binary diffusivities at the correct temperature if one is to make this comparison. We present such values in Table 3. The

Table 3. Binary diffusion coefficients of water vapor, O_2, and CO_2 in N_2 or air determined experimentally at a given temperature and corrected to 38°C. Values at 1 atm

Gas pair	$\mathcal{D}$, cm^2/s	T, °C	Ref.	$\mathcal{D}$, cm^2/s corrected to 38°C
H_2O-N_2	0.253	25	4	0.27
H_2O-N_2	0.256	34	6	0.26
H_2O-N_2	0.303	55	6	0.28
			Mean	0.27
CO_2-N_2	0.144	0	1	0.18
CO_2-N_2	0.167	25	7	0.18
CO_2-air	0.177	44	5	0.17
			Mean	0.18
O_2-air	0.177	0	2	0.22
O_2-N_2	0.181	0	1	0.22
O_2-N_2	0.22	20	3	0.24
			Mean	0.23

References: (1) Cited by Chapman and Cowling (1970); (2) Cited by Cussler (1976); (3) Cited by Hirschfelder et al. (1954); (4) Paganelli and Kurata (1977); (5) Cited by Reid and Sherwood (1966); (6) Schwertz and Brow (1951); (7) Walker and Westenberg (1958).

diffusivities are first given with the temperatures at which they were measured, as cited in the references in Table 3. They were then adjusted to 38°C using the factor $T^{1.5}$ as described above. Because of uncertainty as to the exact value of the exponent of T for different gases, we also used $T^{1.7}$ and T^2 as correction factors. Again, all of these corrections gave results which were quite similar. One can also use values of $\mathcal{D}$ calculated for specific gases from the Chapman-Enskog equation at different temperatures as correction factors. Because there are other temperature-dependent terms in that equation, as well as the explicit dependence of $\mathcal{D}$ on $T^{1.5}$, one might expect a somewhat different temperature correction. However, adjusting values of $\mathcal{D}$ by this technique gave results which could not be distinguished from the $T^{1.5}$ correction, at least for the degree of precision which we required.

When the temperature corrections were applied, we obtained values of $\mathcal{D}$ for H_2O - N_2, CO_2 - N_2, and O_2 - N_2 at 38°C which are shown in Table 3. It should be clear from the foregoing that the process of temperature correction is an approximate one; thus the values of $\mathcal{D}$ are given to two decimals only.

Table 4 presents a comparison of the ratios of conductances and diffusivities calculated for the gas pairs O_2 - N_2 (or air) and CO_2 - N_2 (or air), both compared with H_2O - N_2. (For our purposes, binary dif-

Table 4. Ratios of diffusion coefficients and conductances in air at 38°C

	O_2 : H_2O		CO_2 : H_2O		CO_2 : O_2	
	$\frac{G_{O_2}}{G_{H_2O}}$	$\frac{\mathcal{D}_{O_2,N_2}}{\mathcal{D}_{H_2O,N_2}}$	$\frac{G_{CO_2}}{G_{H_2O}}$	$\frac{\mathcal{D}_{CO_2,N_2}}{\mathcal{D}_{H_2O,N_2}}$	$\frac{G_{CO_2}}{G_{O_2}}$	$\frac{\mathcal{D}_{CO_2,N_2}}{\mathcal{D}_{O_2,N_2}}$
Mean	0.83	0.85	0.64	0.67	0.77	0.78
S.D.	0.02		0.12		0.15[b]	
n[a]	5		6			

[a] n refers to the numbers of groups of data from Tables 1 and 2.
[b] Calculated by combining the standard deviations for G_{CO_2}/G_{H_2O} and G_{O_2}/G_{H_2O}, according to standard rules for the propagation of error.

fusion of a gas in N_2 is identical to its diffusion in air.) One additional comparison is calculable from the data: that directly between CO_2 - N_2 and O_2 - N_2, also shown in Table 4. The agreement between the ratios of G values and $\mathcal{D}$ values is satisfactory in all cases. Within the limits of accuracy of our techniques, we conclude that O_2, CO_2, and water vapor share a common diffusion path through the shell and outer shell membrane. Parenthetically, the agreement between G and $\mathcal{D}$ ratios also supports our basic assumption of a diffusion-limited gas exchange in incubating chicken eggs. It would be difficult to conceive of an important convective component in this process of gas exchange, since convection by its nature does not transport one gas species in preference to another, as we have shown to be the case here.

There is one further conclusion which we are able to draw from the present studies. The shell membranes, both outer and inner, play virtually no role in limiting diffusion of water vapor from the egg. The shell itself is by far the major resistance to passage of this substance (Paganelli et al., 1973). Thus, the outer shell membrane, which is in the diffusion path between atmosphere and air cell for water vapor, O_2, and CO_2 must by the same token offer very little resistance to the passage of O_2 and CO_2, at least from the 12th day of incubation onward. During this phase of incubation, air cell gas tensions will be determined almost entirely by the pore area and thickness of the shell, in relation to the metabolic rate of the embryo.

Acknowledgments. This study was supported in part by NIH Grant 5 P01 HL 14414, by NSF Grant No. PCM76-20947, and by NIH Grant R01 HL 18022. R.A. Ackerman held National Institutes of Health Postdoctoral Fellowship Award No. 5 F32 HL 05197.

References

Ackerman, R.A., Rahn, H.: In vivo O_2 and water vapor conductance during development. Respir. Physiol. to be submitted (1978)

Ar, A., Paganelli, C.V., Reeves, R.B., Greene, D.G., Rahn, H.: The avian egg: water vapor conductance, shell thickness, and functional pore area. Condor 76, 153-158 (1974)

Chapman, S., Cowling, T.G.: The Mathematical Theory of Non-Uniform Gases, 3rd ed. New York: Cambridge Univ. Press, 1970, p. 263

Cussler, E.L.: Multicomponent Diffusion. Amsterdam: Elsevier, 1976, p. 13

Hirschfelder, J.O., Curtiss, C.F., Bird, R.B.: Molecular Theory of Gases and Liquids. New York: John Wiley & Sons, 1954, p. 579

Kutchai, H., Steen, J.B.: Permeability of the shell and shell membranes of hens' eggs during development. Respir. Physiol. 11, 265-278 (1971)

Lomholt, J.P.: The development of the oxygen permeability of the avian egg shell and its membranes during incubation. J. Exp. Zool. 198, 177-184 (1976)

Paganelli, C.V., Ar, A., Rahn, H.: Relative water vapor permeabilities of the shell and shell membranes of the hen's egg. Physiologist 16, 415 (1973)

Paganelli, C.V., Kurata, F.K.: Diffusion of water vapor in binary and ternary gas mixtures at increased pressures. Respir. Physiol. 30, 15-26 (1977)

Paganelli, C.V., Ar, A., Rahn, H., Wangensteen, O.D.: Diffusion in the gas phase: the effects of ambient pressure and gas composition. Respir. Physiol. 25, 247-258 (1975)

Rahn, H., Paganelli, C.V., Ar, A.: The avian egg: air-cell gas tension, metabolism, and incubation time. Respir. Physiol. 22, 297-309 (1974)

Reid, R.C., Sherwood, T.K.: The Properties of Gases and Liquids, 2nd ed. New York: McGraw-Hill, 1966, p. 534

Schwertz, F.A., Brow, J.E.: Diffusivity of water vapor in some common gases. J. Chem. Phys. 19, 640-646 (1951)

Tullett, S.G., Board, R.G.: Oxygen flux across the integument of the avian egg during incubation. Brit. Poultry Sci. 17, 441-450 (1976)

Walker, R.E., Westenberg, A.A.: Molecular diffusion studies in gases at high temperature. I. The "point source" technique. J. Chem. Phys. 29, 1139-1146 (1958)

Wangensteen, O.D., Rahn, H.: Respiratory gas exchange by the avian embryo. Respir. Physiol. 11, 31-45 (1970/71)

Wangensteen, O.D., Wilson, D., Rahn, H.: Diffusion of gases across the shell of the hen's egg. Respir. Physiol. 11, 16-30 (1970/71)

Pore Size Versus Pore Number in Avian Eggshells

S. G. Tullett

Summary

The "porosity" of avian eggshells is analysed and discussed. Pore geometry has been shown to be related to egg weight yet experimental evidence shows that the "porosity" of eggshells can change independently of egg weight. Some ideas on the means by which such alterations occur are presented.

1. General Structure of the Avian Eggshell

The avian eggshell forms a mediating boundary between the developing embryo and the nest or incubator environment. It protects the egg contents from physical insults and microbial infection, supplies the embryo with inorganic salts, and permits the exchange of respiratory gases while at the same time preventing an excessive loss of water from the egg. The "true shell" is composed essentially of calcium carbonate built on an organic matrix and anchored to the outer of two shell membranes which surround the egg contents. Externally, the shell may be covered by an organic cuticle or an inorganic cover (Tullett et al., 1976). A full account of the chemical composition and fine structure of shells has been given by Tyler (1969) and Simons (1971).

2. Gaseous Exchange Across the Shell

Gaseous exchange across the shell is brought about by diffusion (Wangensteen et al., 1970/71). Cuticle appears to offer little resistance to the diffusion of gases (Tullett and Board, 1976) and after approximately the first quarter of the incubation period, in the species thus far studied, the inner shell membrane loses its initially high resistance to gaseous exchange (Kutchai and Steen, 1971; Lomholt, 1976; Tullett and Board, 1976). Therefore, for the period of incubation when embryonic respiration is significant, the geometry of the pores which traverse the shell becomes the limiting factor in the exchange of respiratory gases.

University of Bath, Bath, Somerset, England.

3. Pores and Shell Porosity

The variety in the form of pores found in avian eggshells is illustrated in Figure 1. In the majority of species, however, the pores traverse the shell radially without branching and are funnel-shaped, the wider orifice outermost.

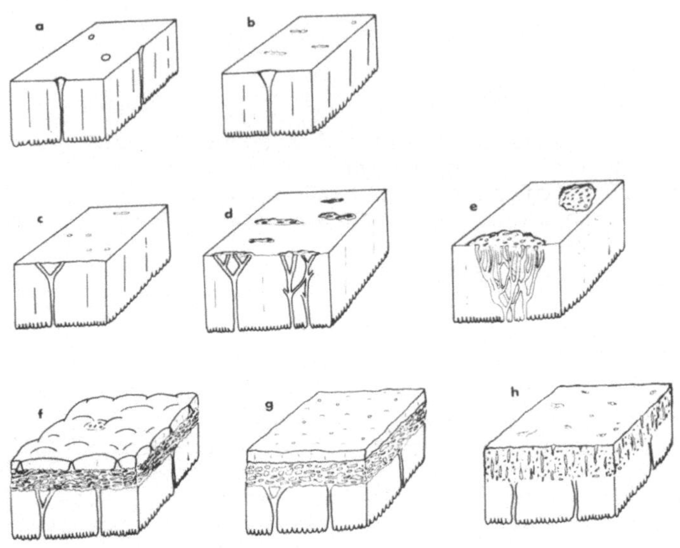

Fig. 1. The variety of pore form in avian eggshells. Diagram represents blocks of shell with shell membranes and cuticle/cover, if originally present, removed. (a) Simple, unbranched, funnel-shaped pore characteristic of the majority of eggshells. (b) Lanceolate outer orifice orientated along longitudinal axis of egg. Usually found in association with type (a) pores, e.g. Black Swan. Wandering Albatross. (c) Simple branching pores usually found in association with type (a) and (b) pores. Branching along longitudinal axis of egg; e.g., Emperor Penguin. Mute and Whooper Swans. (d) More extensively branched pores venting into grooves in external shell surface. Branching along longitudinal axis of egg, e.g., Rhea and Moa* (simple branching as on *left*) Aepyornis* (simple branching as shown on left but also more complex systems as shown on *right*). (e) Extensive branching in all planes. Ostrich. (f-h) Reticulate pore system. Pores traverse the inner shell and, with or without branching, open into a plexus of tubules running through the outer shell which vents independently via short canals to the external shell surface. (f) Emu (g) Cassowary (h) Storks and certain birds of prey (e.g., Buzzard, Osprey, Sparrowhawk). More details of these pore systems are given by Tullett (1976) and Board et al. (1977)

---

* Extinct.

A simple way of quantifying shell *porosity* is to monitor the loss of water from eggs stored under conditions of known absolute humidity (Ar et al., 1974). This technique gives an estimate of the conductance of the shell to water vapour (G_{H_2O}, in mg/day/torr water vapour pressure difference across the shell) and was used by Ar et al. (1974) in establishing the following equation which describes a relationship between shell conductance and fresh egg weight (W, in g):

$$G_{H_2O} = 0.432\ W^{0.780} \quad (1)$$

Ar et al. (1974) went on to show that shell conductance was proportional to the *total functional pore area* (A_p, in cm^2) and inversely proportional to pore length (L, in cm) or shell thickness and they predicted how these two parameters were related to egg weight. Rahn et al. (1976) have applied a correction to the diffusion coefficient of water vapour in air used by Ar et al. (1974). When this temperature-compensated factor is used the relationship between total functional pore area (A_p, in cm^2) and egg weight (W, in g) given by Ar et al. (1974) becomes:

$$A_p = 9.89 \times 10^{-5}\ W^{1.236} \quad (2)$$

The term A_p is the product of the total number of pores per egg (N) and the cross-sectional area of each pore (A_p, in cm^2) which are independently related to egg weight (W, in g):

$$N = 1449\ W^{0.420} \quad (3)$$

$$A = 5.425 \times 10^{-8}\ W^{0.804} \quad (4)$$

[Equations (3) and (4) from Tullett and Board (1977)]

The product of these two equations gives A_p such that:

$$A_p = 7.86 \times 10^{-5}\ W^{1.224} \quad (5)$$

which is in good agreement with Equation (2) derived by Ar et al. (1974) since the individual pore area measurements were made in the narrowest part of the pore canal and would thus be expected to give a lower coefficient (Tullett and Board, 1977).

Pore length (L, in cm) or shell thickness has been shown (Ar et al., 1974) to be related to egg weight (W, in g) according with the equation:

$$L = 5.126 \times 10^{-3}\ W^{0.456} \quad (6)$$

As pointed out by Ar et al. (1974), the ratio A_p/L describes the pore geometry for all gases which exchange across the shell and it is not surprising therefore that it bears some relation to the *metabolic size* of the egg (i.e., it is very nearly proportional to egg weight raised to the power 3/4). Why pore number, pore area, and pore length are independently related to egg weight in the way they are, however, is not clear.

It has been shown (Rahn et al., 1976) that any change in the environment necessitating a change in the incubation pattern would also necessitate a change in the ratio A_p/L. On the basis of the equations presented above it may be supposed that change in the conductance of birds eggs occurs by changes in egg weight. This is supported by data obtained in a comparison of eggs of domestic fowl at sea level with those of their progeny housed for 15 years at 3800 m (Wangensteen et al., 1974).

Shell thickness remained unchanged while the functional pore area in the shells of eggs laid by the high-altitude birds was reduced compared to their sea-level counterparts; presumably this restricted the increased loss of water at altitude (brought about by the increase in the diffusion coefficient of water vapour in air as a result of the decrease in barometric pressure). The data for these eggs are given in Table 1A and it can be seen that the reduction in functional pore area seems to have been achieved by a reduction of egg weight in accordance with Equation (2). No data are available for the independent changes in pore number and individual pore area.

In two other cases the supposition that changes in egg weight effect changes in the water conductance of the shell appears unjustified. The

Table 1. Data used in an examination of the simple eggshell model
A. Domestic fowl at high altitude (Wangensteen et al., 1974)

	Egg weight[a] (g)	G_{H_2O} ± S.D. (cm^3/day/torr)	$A_p(HA)/A_p(SL)$ [b]	$A_p(HA)/A_p(SL)$ [c]
High-altitude eggs (HA)	40	24.9 ± 2.2	0.68	0.72
Sea-level eggs (SL)	52	36.5 ± 5.9		

[a] Calculated from shell areas given by Wangensteen et al. (1974) according to the formula presented by Besch et al. (1968).
[b] Calculated as ratio of water vapour conductances.
[c] A_p calculated from egg weight according to Equation (2).

B. Common loon (*Gavia immer*)

Egg weight = 154 g; G_{H_2O} = 98 mg/day/torr (Rahn, pers. comm.).

This water vapour conductance would, according to Equation (1), be expected of an egg weighing 1048 g.

Expected number of pores per egg	≃ 12,000	(Eq. 3)
Actual	≃ 42,000	(Tullett and Board, 1977)
Expected area of each pore	≃ 300 μm^2	(Eq. 4)
Actual	≃ 1000-6000 μm^2	(Tullett and Board, 1977)

C. Barn swallows (*Hirundo rustica*)

	Altitude of nesting site 1000 m	3000 m	% decrease
Water vapour conductance (μg/day/torr)[d]	424	144	66.0
Conductance per cm^2 of shell[d]	57.58	22.38	61.1
Shell area (cm^2)	7.36	6.43	12.6
Egg weight (g)[e]	1.89	1.54	18.5
Predicted weights to give 66% decrease in G_{H_2O} using Eq. (1)	1.89	0.47	66.0

[d] Calculated from equations given by Packard et al. (1977); [e] Calculated from shell area according to formula given by Paganelli et al. (1974).

first involves eggs of the Common Loon (*Gavia immer*) which show that in relation to egg weight the shell has a very high water vapour conductance (Rahn, pers. comm.; Table 1B). The reason for the high conductance is probably high water vapour pressure in the nest, which is constructed near water and is often wet (Campbell and Ferguson-Lees, 1972), and the apparent need to lose a specific amount of water from the egg in order to produce an air cell of a size which will permit the initiation of lung ventilation and allow the movements necessary for the escape of the chick from the egg at hatching (Rahn et al., 1976). If the empirically determined water vapour conductance were to have been achieved by an increase in egg weight according to Equation (1) it would have required the construction of a 1048-g egg - clearly an impossible task for a bird the size of a loon. The solution has been to form many more pores than normal for eggs of this weight with some having cross-sectional areas up to 20 times the value expected from Equation (4) (Tullett and Board, 1977). [The relationship between cone frequency and pore frequency defined by Tullett (1975) does not apply, therefore, although the number of cones per unit area is normal for eggs of this weight (Tullett, 1976).]

The second instance concerns the observations by Packard et al. (1977) of an adaptive reduction in the conductance of eggs of the Barn Swallow (*Hirundo rustica*) with increasing altitude of the nesting site. It can be calculated (Table 1C) that between the heights of 1000 m and 3000 m there was a decrease of 66% in the water conductance of the eggs - apparently without a significant change in egg weight. However, to produce a 66% reduction in the water vapour conductance of the shell by egg weight changes in accordance with Equation (4) would entail the production of an egg weighing 0.47 g, which is very close to the weight of the smallest egg laid by any bird - Rahn and Ar (1974) found the smallest eggs in Schönwetter's data (1960-71) weighed 0.2-0.3 g for two members of the Trochilidae (hummingbirds). Would Barn Swallows be capable of laying eggs of such a size and if so what would be the chances of survival for much smaller young? Is this why egg weight has not been reduced in this species? Finally, from what is known of shell structure and formation, is it possible to offer a plausible model which allows for changes in shell porosity independently of egg weight if conditions demand?

4. A Simple Model of the Avian Eggshell and its Formation

A simple model of the avian eggshell has been proposed by Tullett (1975) in which the basic building blocks of the shell are considered to be hexagonal columns of calcite. This model finds general support in the literature since in transverse sections of shell separate columns fitting together like a jig-saw puzzle can be observed (Becking, 1975) and, in polarised light, columns appear to run radially through the shell (Schmidt, 1957). It is surmised that the columns become larger (in both diameter and length) in eggs of increasing weight. The number of columns per unit area would therefore decrease in proportion to the weight of the egg and because the number of pores in a given area is directly related to the number of columns in that area (Tullett, 1975) the number of pores per unit area would also decrease. These changes have been observed and quantified (Tullett, 1976). The increase in shell surface area vis à vis the decrease in pores per unit area is such that the total number of pores per egg increases in proportion to egg weight [Eq. (3)].

The initial distribution of the columns is probably determined by the cohesion of secretion from groups of cells as the egg passes down the oviduct (Wyburn et al., 1973). These secretions, which are laid down on the outer surface of the outer shell membrane, probably correspond to the organic concretions observed by Fujii and Tamura (1970) and the mammillary "cores" investigated by Simkiss and Tyler (1957). The distribution of these "seeding sites" on eggshell membranes collected at different stages of egg formation has been studied by Stemberger et al. (1977) and it seems reasonable to assume a priori that one seeding site provides the nucleus for crystallisation of one column of calcite. This simple model of the shell and its formation is illustrated in Figure 2.

Pores arise as spaces between four or more columns (Tullett, 1975) which, it has been suggested (Tyler and Simkiss, 1959), are kept open by an uptake of liquid by the albumen across the developing shell. In the domestic fowl it is known that only half the volume of albumen of the egg as laid is added to the ovum in the magnum of the oviduct; the remainder is made up by the uptake of liquid during the first few hours in the shell gland. Shell formation is only proceeding slowly during this "plumping" process and the liquid uptake suggested to be responsible for keeping the pore canals open during the later more rapid phase of shell formation is probably a slow continuation of plumping. However, if shell formation is made successively to correspond more exactly in time with the process of plumping and hence lead to successively greater rates of liquid uptake during the phase of more rapid shell formation it may be possible that both the number of pore canals and their cross-sectional area increase independently of changes in egg weight. An additional mechanism would be for less albumen, with a greater liquid uptake potential, to be added in the magnum so the amount of liquid required to cross the developing shell to achieve a constant egg weight is increased. Conversely, to effect a decrease in the conductance of the eggshell independent of changes in egg weight a possible mechanism would be that more albumen, with a lower liquid uptake potential, was added in the magnum so that less liquid crosses the developing shell. It would be interesting to know if these types of mechanisms occur in birds thus releasing them from the seemingly stringent allometric relationships should conditions demand.

Acknowledgments. I would like to express my gratitude to Dr. R.G. Board (University of Bath, U.K.) who introduced me to the avian eggshell and under whose guidance the thoughts expressed herein evolved.

Fig. 2. A simple model of the avian eggshell and its formation. *Top*, seeding sites on the outer of two shell membranes. *Right*, photographic evidence of this stage during formation of a domestic fowl egg. (From Fujii and Tamura, 1970). *Centre*, initial shell deposition with beginning of formation of one hexagonal column at each seeding site. *Right*, photographic evidence. (From Fujii and Tamura, 1970). *Bottom*, shell nearing completion. Pores (*black areas*) arise as spaces between four or more columns. The scanning electron micrographs (both ×200) were kindly supplied by Dr. T. Tamura (Hiroshima University, Japan) ▶

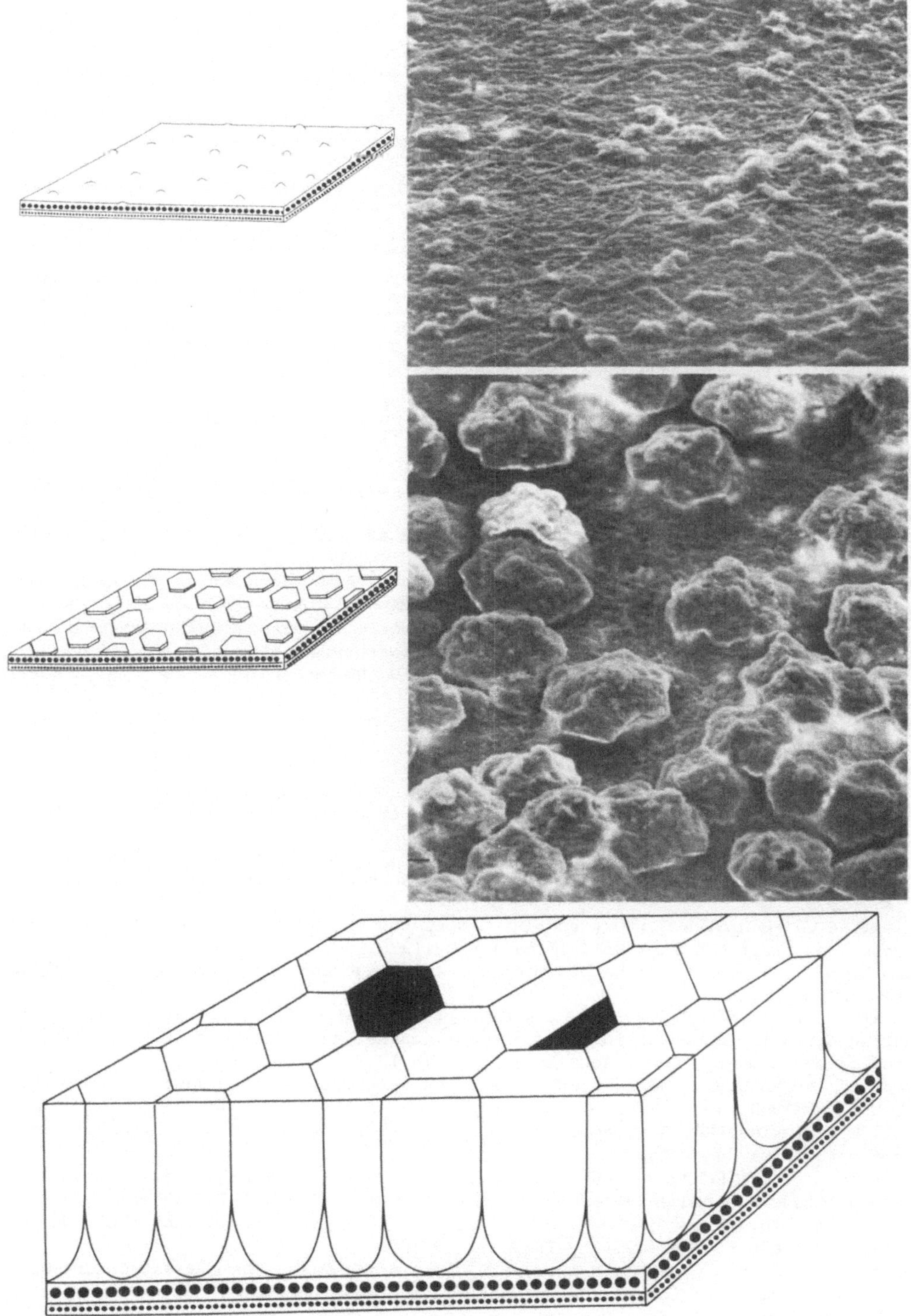

References

Ar, A., Paganelli, C.V., Reeves, R.B., Greene, D.G., Rahn, H.: The avian egg: water vapour conductance, shell thickness and functional pore area. Condor 76, 153-158 (1974)

Becking, J.H.: The ultrastructure of the avian eggshell. Ibis. 117, 143-151 (1975)

Besch, E.L., Sluka, S.J., Smith, A.H.: Determination of surface area using profile recordings. Poultry Sci. 47, 82-85 (1968)

Board, R.G., Tullett, S.G., Perrott, H.R.: An arbitrary classification of pore systems in avian eggshells. J. Zool. 182, 251-265 (1977)

Campbell, B., Ferguson-Lees, J.: A Field Guide to Birds' Nests. London: Constable, 1972

Fujii, S., Tamura, T.: Scanning electron microscopy of shell formation in the hen's egg. J. Fac. Fish. Anim. Husb. Hiroshima Univ. 9, 65-81 (1970)

Kutchai, H., Steen, J.B.: Permeability of the shell and shell membranes of hen's eggs during development. Respir. Physiol. 11, 265-278 (1971)

Lomholt, J.P.: The development of the oxygen permeability of the avian egg shell and its membranes during incubation. J. Exp. Zool. 198, 177-184 (1976)

Packard, G.C., Sotherland, P.R., Packard, M.J.: Adaptive reduction in permeability of avian eggshells to water vapour at high altitude. Nature (London) 266, 255-256 (1977)

Paganelli, C.V., Olszowka, A., Ar, A.: The avian egg: surface area, volume and density. Condor 76, 319-325 (1974)

Rahn, H., Ar, A.: The avian egg: incubation time and water loss. Condor 76, 147-152 (1974)

Rahn, H., Paganelli, C.V., Nisbet, I.C.T., Whittow, G.C.: Regulation of incubation water loss in eggs of seven species of tern. Physiol. Zool. 49, 245-259 (1976)

Schmidt, W.J.: Über den Aufbau der Schale des Vogeleies nebst Bemerkungen über kalkige Eischalen anderer Tiere. Ber. oberhess. (Naturwiss. Abt.) 28, 82-108 (1957)

Schönwetter, M.: Handbuch der Oologie, Lief. 1-19, Meise, W. (ed.). Berlin: Akademie-Verlag, 1960-1971

Simkiss, K., Tyler, C.: A histochemical study of the organic matrix of the hen eggshell. Quart. J. Microsc. Sci. 98, 19-28 (1957)

Simons, P.C.M.: Ultrastructure of the hen eggshell and its physiological interpretation. Comm. No. 175. Centr. Inst. Poultry Res. Beekbergen, the Netherlands (1971)

Stemberger, B.H., Mueller, W.J., Leach, R.M.: Microscopic study of the initial stages of egg shell calcification. Poultry Sci. 56, 537-543 (1977)

Tullett, S.G.: Regulation of avian eggshell porosity. J. Zool. 177, 339-348 (1975)

Tullett, S.G.: The Avian Eggshell: a Mediating Boundary. Unpubl. thesis, Univ. Bath, U.K., 1976

Tullett, S.G., Board, R.G.: Oxygen flux across the integument of the avian egg during incubation. Brit. Poultry Sci. 17, 441-450 (1976)

Tullett, S.G., Board, R.G.: Determinants of avian eggshell porosity. J. Zool. 183, 203-211 (1977)

Tullett, S.G., Board, R.G., Love, G., Perrott, H.R., Scott, V.D.: Vaterite deposition during eggshell formation in the cormorant, gannet and shag, and in "shell-less" eggs of the domestic fowl. Acta Zool. (Stockh.) 57, 79-87 (1976)

Tyler, C.: Avian eggs: their structure and characteristics. Int. Rev. Gen. Exp. Zool. 4, 82-127 (1969)

Tyler, C., Simkiss, K.: A study of the egg shells of ratite birds. Proc. Zool. Soc. London 133, 201-243 (1959)

Wangensteen, O.D., Rahn, H., Burton, R.R., Smith, A.H.: Respiratory gas exchange of high altitude adapted chick embryos. Respir. Physiol. 21, 61-70 (1974)

Wangensteen, O.D., Wilson, D., Rahn, H.: Diffusion of gases across the shell of the hen's egg. Respir. Physiol. 11, 16-30 (1970-1971)

Wyburn, G.M., Johnston, H.S., Draper, M.H., Davidson, M.F.: The ultrastructure of the shell-forming region of the oviduct and the development of the shell of *Gallus domesticus*. Quart. J. Exp. Physiol. 58, 143-151 (1973)

Interdependence of Gas Conductance, Incubation Length, and Weight of the Avian Egg

A. Ar and H. Rahn

Summary

Allometric relationships of incubation length (I), water vapor conductance (G_{H_2O}), and total functional pore area (Ap) as a function of egg weight (W), were established in 90 species of birds. A constant was derived which expresses the interrelationship between three variables. Thus (G_{H_2O} · I)/W = 5.13 ± 0.86 S.D. [mg/(g·torr)]. Similar constants were derived for O_2 and CO_2 conductances and discussed in relation to the metabolic requirements and water loss of eggs during incubation. It is proposed that the total quantity of water lost and the total quantity of O_2 taken up per gram of egg is independent of egg size and incubation period.

1. Introduction

Towards the end of the incubation, the O_2 and CO_2 tensions in the air cell of bird eggs, appear to be very similar among different species. Thus, inspite of differences in egg size and incubation lenght, gas tensions of about 100 and 40 torr for O_2 and CO_2, respectively, exist in the air cell (Rahn et al., 1974). These gas tensions are determined by the ratio of metabolic rate to diffusive gas conductance across the gas-filled pores of the egg shell and the gas spaces of the outer membrane (Wangensteen and Rahn, 1970/71), and play an important role in the final events leading to hatching (Visschedijk, 1968) and in the establishment of normal acid-base values of the blood.

These observations suggested an interdependence between egg size, incubation length, and shell conductance. While the *independent* relationships between the three factors water vapor conductance (G_{H_2O}), incubation time (I), and egg weight (W) had been previously established (Rahn and Ar, 1974), it was of interest to explore the *interdependence* between these three variables for a large number of species. Our findings demonstrate a simple relationship between these three parameters which can be expressed as follows: GI/W = constant. This constant, which has a coefficient of variation of ±17%, allows one to predict any one variable if the other two are known.

Tel-Aviv University, Department of Zoology, Tel-Aviv, Ramat-Aviv, Israel. State University of New York at Buffalo, School of Medicine, Department of Physiology, Buffalo, N.Y. 14214, USA. Max-Planck-Institut für experimentelle Medizin, Abteilung Physiologie, Hermann-Rein-Str. 3, 3400 Göttingen, FRG.

2. Methods

The proportionality constants between water vapor, CO_2, and O_2 conductances across the gas-filled pores of the shell and the gas spaces of the outer membrane have been recently established in vivo (Paganelli et al., 1978). The G_{H_2O} may be easily obtained using the gravimetric method (Ar et al., 1974). Mean water vapor conductances for 90 species were used to analyze their interrelationship with incubation length and egg weight. They include those previously published by Ar et al. (1974), Lomholt (1976), and Hoyt et al. (personal communication). Fresh egg weights were recorded and incubation lengths obtained from various literature sources and personal communications. Eggshell thicknesses (L) were averaged from measurements around the egg's equator and poles or taken from Schönwetter's tables (1962-1974). The total functional pore are a (A_p) for each species was calculated from G_{H_2O} and L according to Ar et al. (1974) and Rahn et al. (1976). Conventional statistical methods were applied for data processing. The symbol for the antilogarithim of the standard error of estimate is SEE which is a factor, by which the dependent variable has to be multiplied or divided, ($\div\!\!\!\times$). r^2 is square of the correlation coefficient, r. The units used throughout are: G_{H_2O} in mg(day·torr); I in days; W in g; L, in mm; and A_p in mm.

3. Results and Discussion

a) Water Vapor Conductance

Table 1 lists egg weights, incubation lengths, water conductance values, and shell thicknesses for 90 bird species, representing 15 orders, as well as the calculated ratios W:I, GI:W, and Ap. The geographic area where the fresh eggs were collected is also given.

The regression of water vapor conductance as a function of fresh egg weight for all 90 species is:

$$G_{H_2O} = 0.384\ W^{0.814} \quad (SEE = \div\!\!\!\times\ 1.242;\ r^2 = 0.971) \qquad (1)$$

The regression for incubation time *vs.* egg weight yields:

$$I = 11.64\ W^{0.221} \quad (SEE = \div\!\!\!\times\ 1.229;\ r^2 = 0.731) \qquad (2)$$

These results are very similar to those previously reported (Ar et al., 1974; Rahn and Ar, 1974) and basically confirm the egg weight relationship for G_{H_2O}, which was based on a much smaller number of observations.

The product of Equations (1) and (2) suggests an essentially linear relationship between water vapor conductance (and therefore O_2 and CO_2 conductances) multiplied by the incubation length and the egg weight. This allometrically derived relationship predicts that for each species

$$GI = cW \qquad (3)$$

where c is a constant common to all species.

Statistical analyses of this relationship indicate that the standard error of estimate is reduced when the variables are treated as *inter-*

Table 1. The number of eggs, n, and their weights, W, which were used to obtain the water vapor conductance, G. The shell thickness, L, was measured or obtained from Schönwetter (1962-74). The values for the incubation period, I, were obtained from various sources in the literature and personal communications. The ratio of egg weight to incubation period, W/I, the water vapor conductance constant, (GI)/W, and the total functional pore area of the shell, A_p, was calculated. A key to the localities where the fresh eggs were collected are as follows:
1, Buffalo, N.Y.; 2, Albany, N.Y.; 3, Buffalo Zoological Garden; 4, Maine; 5, Massachussetts; 6, San Diego Zoological Garden, California; 7, Gulf of California, Mexico; 8, Kachemak Bay, Alaska; 9, Manana Island, Hawaii; 10, Faeroe Islands, Denmark; 11, Marshall Islands, South Pacific Ocean; 12, McMurdo Sound, Antarctica; 13, Tel Aviv University Research Zoological Garden, Israel; 14, Israel; 15, Denmark.
Conductance values: for *Struthio camelus* (Hoyt et al., personal communication), for *Larus canus* and *Larus ridibundus* (Lomholt, 1976), and *Catharacta maccormicki* (Hammel and Rahn, 1978)

Order	Species	n	W g	I days	$\frac{W}{I}$	G $\frac{mg}{day \cdot torr}$	$\frac{G \cdot I}{W}$	L mm	Ap mm^2	Locality
Struthioniformes										
	Struthio camelus, Ostrich	5	1500	45	33.3	187	5.61	2.01	168	6
Rheiformes										
	Rhea americana, Rhea	2	609	40	15.2	77.7	5.10	0.91	31.6	6
Casuariiformes										
	Dromiceius novae-hollandiae, Emu	1	578	59	9.80	51.8	5.29	0.94	21.8	6
Procellariiformes										
	Puffinus pacificus, Shearwater	12	56,3	50	1.13	6.42	5.70	0.27	0.77	9
Pelecaniformes										
	Phaethon rubricauda, Tropicbird	2	67.7	44	1.54	9.56	6.21	0.36	1.54	11
	Sula leucogaster, Brown Booby	4	67.8	45	1.28	7.64	5.95	0.39	1.33	11
	Phalacrocorax auritus, Double-Crested Cormorant	8	57.7	28	2.06	8.36	4.06	0.39	1.46	4
	Phalacrocorax pelagicus, Pelagic Cormorant	18	39.3	31	1.27	6.78	5.34	0.35	1.06	8
Ciconiiformes										
	Nycticorax nycticorax, Night Heron	2	20.8	25	0.83	5.30	6.39	0.20	0.47	13
	Bubulcus ibis, Cattle Egret	2	23.2	22.5	1.03	5.57	5.40	0.20	0.50	14
	Egretta garzetta, Little Egret	2	28.5	22	1.30	7.22	5.57	0.22	0.71	13
Anseriformes										
	Cairina moschata, Muscovy Duck	4	80.2	35	2.29	12.3	5.37	0.40	2.20	1
	Anas boscas, Pekin Duck	11	82.3	28	2.94	14.5	4.93	0.31	2.01	1
	Aix sponsa, Wood Duck	5	43.3	29	1.49	8.40	5.62	0.32	1.20	3
	Somateria mollissima, Eider	5	100	25	4.00	21.4	5.35	0.38	3.64	8
	Anser domesticus, Embden Goose	11	170	28	6.07	27.7	4.56	0.67	8.30	1
	Anser anser, Grey Lag Goose	3	195	28	6.96	35.1	5.04	0.67	10.5	3

Table 1. Continued

Order	Species	n	W g	I days	$\frac{W}{I}$	G $\frac{mg}{day \cdot torr}$	$\frac{G \cdot I}{W}$	L mm	Ap mm^2	Locality
Falconiformes										
	Falco naumanni, Lesser Kestrel	2	10.8	28	0.39	2.80	7.24	0.19	0.24	13
	Falco tinnunculus, Kestrel	2	18.1	29	0.62	3.98	6.38	0.24	0.43	13
	Aquila rapax, Tawny Eagle	2	92.8	45	2.06	13.0	6.31	0.52	3.02	13
	Gyps fulvus, Griffon Vulture	1	243	58	4.20	18.5	4.40	0.68	5.62	13
Galliformes										
	Gallus gallus, Chicken	8	53.9	21	2.57	14.4	5.61	0.35	2.25	1
	Pavo muticus, Green Peafowl	14	102	29	3.52	20.9	5.94	0.55	5.14	3
	Pavo cristatus, Indian Peafowl	2	95.1	28	3.40	14.0	4.12	0.56	3.50	3
	Syrmaticus soemmerringii, Copper Pheasant	10	31.5	24	1.31	7.97	6.07	0.28	1.00	3
	Chrysolophys pictus, Golden Pheasant	10	32.1	24	1.34	6.03	4.51	0.26	0.70	3
	Lophophorus impejanus, Impeyan Pheasant	6	63.7	28	2.28	6.69	4.26	0.37	1.60	3
	Chrysolophys amherstiae, Lady Amherst Pheasant	17	29.7	24	1.24	5.99	4.84	0.27	0.72	3
	Phasianus colchicus, Ring-Necked Pheasant	12	33.8	24	1.41	6.60	4.69	0.31	0.91	1
		4	29.2	24	1.22	7.24	5.95	0.31	1.00	13
	Lophura nycthemera, Silver Pheasant	3	39.9	26	1.53	9.24	6.02	0.38	1.57	1
	Lophura swinhoei, Swinhoe's Pheasant	3	41.4	25	1.66	7.34	4.43	0.34	1.12	3
	Coturnix coturnix, Japanese Quail	10	9.61	17	0.57	3.09	5.47	0.17	0.23	1
	Meleagris gallopavo, Turkey	11	87.8	28	3.14	13.5	4.31	0.43	2.59	1
	Alectoris graeca, Rock Patridge	5	18.2	24	0.76	3.38	4.45	0.28	0.42	14
	Gallus bankiva, Bankiva	1	40.3	21	1.92	9.7	5.05	0.32	1.39	13
Gruiformes										
	Burhinus oedicnemus, Stone Curlew	2	33.5	26	1.29	4.55	3.53	0.27	0.55	13
Charadriiformes										
	Haematopus ostralegus, European Osystercatcher	2	41.5	26	1.60	6.80	4.26	0.27	0.82	10
	Charadrius vociferus, Killdeer	3	14.5	27	0.54	2.32	4.32	0.15	0.16	1
	Pluvialis apricaria, Golden Plover	3	32.6	31	1.05	5.02	4.77	0.17	0.38	10
	Numenius phaeopus, Whimbrel	4	53.5	28	1.91	9.74	5.10	0.20	0.87	10
	Catharacta skua, Skua	6	95.5	28	3.41	18.4	5.39	0.33	2.71	10
	Catharacta maccormicki, MacCormick's Skua	11	99.3	28	3.55	15.9	4.48	0.40	2.84	12
	Larus glaucescens, Glaucous-Winged Gull	15	96	27	3.56	23.1	6.50	0.35	3.61	8
	Larus marinus, Great Black-backed Gull	6	112	29	3.86	16.7	4.32	0.37	2.76	10
	Larus heermanni, Heermann's Gull	11	53.4			11.0		0.27	1.32	7
	Larus argentatus, Herring Gull	15	90.3	27	3.34	16.4	4.90	0.32	2.35	5

Larus fuscus, Lesser Black-backed Gull	6	84.9	26	3.27	16.0	4.90	0.30	2.15	10
Larus canus, Mew Gull	10	56.6	24	2.36	14.0	5.92	0.24	1.50	15
	11	46.4	24	1.93	10.9	5.64	0.24	1.17	8
Larus occidentalis, Western Gull	12	97.7	27	3.62	23.7	6.55	0.33	3.50	7
Larus ridibundus, Black-headed Gull	20	37.7	23	1.64	9.45	5.77	0.21	0.89	15
	5	28.8	22	1.31	6.36	4.85	0.23	0.65	13
Larus tridactyla, Black-legged Kittiwake	11	50.5	27	1.87	9.67	5.17	0.26	1.12	8
Sterna paradisaea, Arctic Tern	9	17.5	22	0.80	5.17	6.50	0.16	0.37	8
Sterna hirundo, Common Tern	8	21.0	21	1.00	4.30	4.30	0.19	0.37	5
Sterna albifrons, Least Tern	14	8.37	21	0.40	1.76	4.42	0.13	0.10	5
Sterna fuscata, Sooty Tern	12	33.9	29	1.17	6.76	5.78	0.23	0.69	9
Thalasseus elegans, Elegant Tern	12	40.9			9.93		0.29	1.29	7
Thalasseus maximus, Royal Tern	7	70.4	31	2.31	15.1	6.65	0.29	1.96	7
Anous stolidus, Common Noddy	12	37.3	36	1.04	6.97	6.73	0.23	0.72	11
	12	37.4	36	1.04	5.92	5.70	0.24	0.64	9
Anous tenuirostris, Black-capped Noddy	11	23.7	35	0.68	4.51	6.66	0.19	0.38	11
Gygis alba, White Tern	14	21.4	36	0.59	3.47	5.84	0.17	0.26	11
Uria aalge, Common Murre	21	115	33	3.48	20.0	5.74	0.66	5.90	8
Fratercula arctica, Common Puffin	6	59.7	38	1.57	7.99	5.09	0.31	1.11	10
Lunda cirrhata, Tufted Puffin	10	90.8	38	2.39	13.0	5.44	0.36	2.09	8
Glareola pratincola, Pratincole	16	8.45	17.5	0.48	2.15	4.45	0.15	0.14	13
Columbiformes									
Streptopelia senegalensis, Laughing Dove	6	6.63	14	0.47	2.61	5.51	0.12	0.14	14
Streptopelia decaocto, Collared Turtle Dove	3	7.45	14	0.53	2.54	4.77	0.13	0.15	14
Streptopelia turtur, Turtle Dove	2	8.30	13.5	0.61	2.21	3.59	0.14	0.14	14
Streptopelia risoria, Ringed Dove	71	8.07	13.5	0.60	2.24	3.75	0.12	0.12	13
Columba livia, Rock-Dove Pigeon	7	21.2	18	1.18	5.00	4.25	0.19	0.42	1
Psittaciformes									
Melopsittacus undulatus, Budgerigar	7	2.25	18	0.13	0.57	7.09	0.12	0.031	14
Strigiformes									
Strix aluco, Tawny Owl	5	36.1	29	1.25	6.15	4.94	0.27	0.74	14
Bubo bubo, Eagle Owl	2	69.3	35	1.98	11.3	5.70	0.35	1.77	13
Passeriformes									
Cinnyris osea, Palestine Sunbird	2	0.86	13.5	0.06	0.31	4.87	0.05	0.007	14
Prinia gracilis, Graceful Warbler	5	1.12	11.5	0.10	0.37	3.80	0.07	0.012	14
Erythropygia galactotes, Rufous Warbler	3	2.30	13.5	0.17	0.71	4.17	0.08	0.025	14
Passer moabiticus, Dead Sea Sparrow	15	1.50	13	0.11	0.54	4.68	0.09	0.022	14
Passer domesticus, House Sparrow	3	2.76	13.5	0.20	0.74	3.62	0.10	0.033	14
	23	2.67	12	0.22	0.97	4.36			1

Table 1. Continued

Order	Species	n	W g	I days	$\frac{W}{I}$	G $\frac{mg}{day \cdot torr}$	$\frac{G \cdot I}{W}$	L mm	Ap mm^2	Locality
Passeriformes										
	Muscicapa striata, Spotted Flycatcher	3	1.86	14	0.13	0.62	4.66	0.08	0.022	14
	Galerida cristata, Crested Lark	13	2.93	13	0.23	0.92	4.08	0.10	0.041	14
	Pycnonotus capensis, Common Bulbul	3	3.05	14	0.22	0.99	4.54	0.08	0.035	14
	Turdus merula, Blackbird	5	6.36	14.5	0.44	1.75	3.99	0.12	0.094	14
	Iridoprocne bicolor, Tree Swallow	5	1.72	15	0.11	0.50	4.36	0.064	0.015	1
	Troglodytes aedon, House Wren	12	1.32	15	0.09	0.40	4.55	0.078	0.014	1
	Agelaius phoeniceus, Red-winged Blackbird	19	4.11	12	0.34	1.50	4.38	0.093	0.062	2
	Molothrus ater, Brown-headed Cowbird	7	3.33	12	0.28	1.14	4.11			1
	Quiscalus quiscula, Common Grackle	3	6.31	12	0.53	2.98	5.67	0.123	0.164	1
	Spinus tristis, American Goldfish	3	1.52	13	0.12	0.56	4.79			1
	Peophila castanotis, Zebra Finch	8	0.92	12	0.077	0.39	5.09			1

dependent rather than *independent* values. The mean and standard deviation of constant c, calculated for each of the 90 species, is:

$$\frac{G_{H_2O} I}{W} = 5.13 \pm 0.86 \text{ S.D. } [mg(g \cdot torr)] \qquad (4)$$

The coefficient of variation is 17%. In more general terms, it follows that whenever two of the three variables in Equation (4) are known the third can be predicted. Thus reasonable conductance values for other species can be easily calculated from egg weights and incubation period reported in the large ornithological literature. This is illustrated in Figure 1 where the simultaneous values for G_{H_2O} as a function of W:I are plotted for convenience on logarithmic scales. A good agreement exists between the data points and the predicted line ($r^2 = 0.950$) with a slope of 1.0. In most cases the standard deviation for each species crosses the line.

The constant c of Equation (4) applies to G_{H_2O} at 25°C expressed in mg H_2O(g·torr). Similar constants for G_{O_2} and G_{CO_2} can also be calculated since within the same egg the in vivo measured G_{H_2O}, G_{O_2}, and G_{CO_2} conductances of the eggshell are proportional to their respective diffusion coefficients in air (Paganelli et al., 1978). Thus the c constant for G_{O_2} at 38°C is 5.39 cm^3 $O_{2\,STPD}$(g·torr) and for G_{CO_2} 4.12 cm^3 $CO_{2\,STPD}$(g·torr). On the basis of these considerations one may predict not only the G_{H_2O} but also the G_{O_2} and G_{CO_2} of the shell for any W:I ratio.

It is important, however, to point out certain egg species which deviate from this norm because of unusually high or low conductances. For example, eggs of the grebe, *Podiceps cristatus*, the coot, *Fulica atra* (Lomholt, 1976), the Mound-building Brush-turkey, *Alectura lathami* (Seymour and Rahn, 1978), and

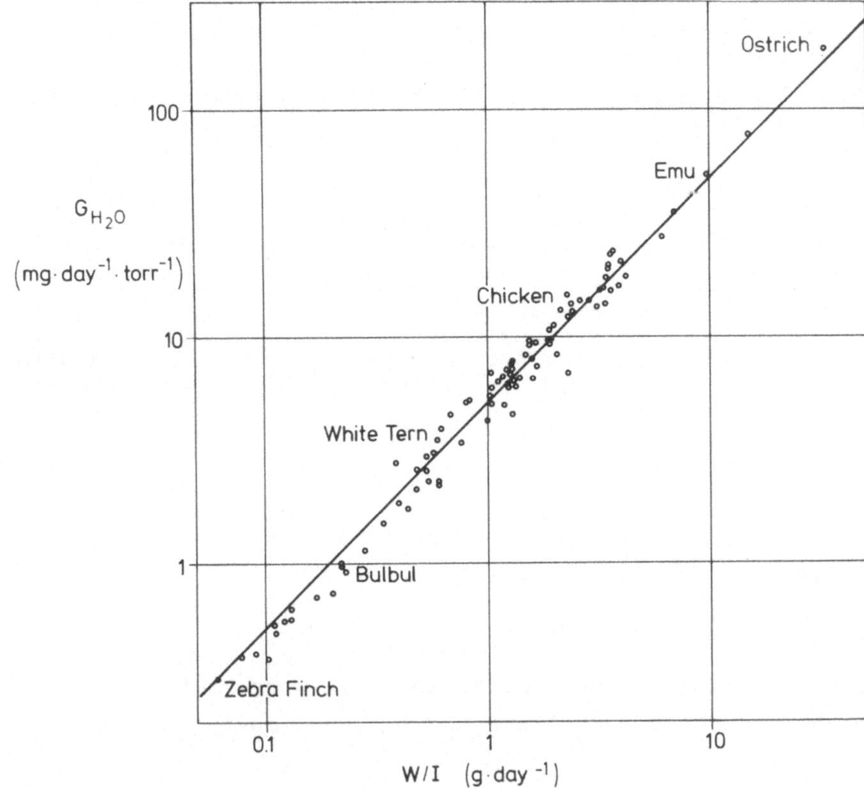

Fig. 1. Water vapor conductance of eggs, G_{H_2O}, plotted as a function of egg weight divided by incubation time, *W/I*, on log-log scales for 90 species of birds. (Data from Table 1)

the Common Loon, *Gavia immer* have unusually high conductances. These may be special adaptations to a high nest humidity, since the nest material of these species are frequently moist or soggy and in case of *Alectura* to the unusual O_2 and CO_2 concentrations of the gas surrounding the buried egg. We have refered to these birds as "wet nesters" in contrast to most species where the relative humidity of the nest's microclimate is maintained at about 50% (Rahn et al., 1977). On the other hand, eggs of the Adelie Penguin, *Pygoscelis adeliae* (Hammel and Rahn, 1978), and the Red-breasted Merganser, *Mergus serrator*, have unusually low conductance for their W:I ratio. None of these species have been listed in Table 1. They may represent adaptations to unusual ecological environments and need to be studied more intensively.

b) Incubation Water Loss

Equation (4) can also be applied to the considerations of water loss, $\dot{M}_{H_2O}$, of the egg during incubation. The daily water loss can be described as follows:

$$\dot{M}_{H_2O} = G_{H_2O} \Delta P_{H_2O} \quad \text{(mg/day)} \qquad (5)$$

where ΔP_{H_2O} is the water vapor pressure difference between the egg and the microenvironment of the nest (Rahn and Ar, 1974; Rahn et al., 1976). Assuming a typical mean water loss during incubation ($\dot{M}_{H_2O} \cdot I$) of ca. 15% of the initial weight of the egg, a value which appears to be characteristic for many species (Drent, 1975; Rahn et al., 1976), and substituting Equation (4) into (5) one solves for an average ΔP_{H_2O} ca. 30 torr during incubation.

Assuming, furthermore, a typical egg temperature of 36.2°C (Drent, 1975) such an egg will exert at full saturation a vapor pressure of 45 torr. Subtracting from this value a ΔP_{H_2O} of 30 torr yields an absolute humidity value for the average microclimate of the nest air of 15 torr, equivalent to a relative humidity of 38% at the typical nest air temperature of 34°C (Drent, 1975). This value may be compared with an average value of 19 torr ±4 S.D. (relative humidity of 48%) recently reported for the nest air of 23 species (Rahn et al., 1977).

c) Metabolism

The metabolism of the egg, $\dot{M}_{O_2}$, can be expressed in a manner similar to that of the incubation water loss [Eq. (5)], namely

$$\dot{M}_{O_2} = G_{O_2} \, \Delta P_{O_2} \quad (\text{cm}^3 {}_{\text{STPD}}/\text{day}) \tag{6}$$

where ΔP_{O_2} is the partial pressure difference, in torr, between the inside of the shell (air cell P_{O_2}) and the external environment (Rahn et al., 1974).

Recent evidence suggests that during comparable stages of embryonic development comparable differences in O_2 tension (ΔP_{O_2}) exist across the eggshell. This is based upon the recent observations of Ackerman, Whittow, and Paganelli (personal communication) who measured the changes in air-cell O_2 tension during development in the eggs of the Wedged-tailed Shearwater, *Puffinus pacificus*. These eggs weigh the same as the eggs of chickens but have an incubation period of about 50 days instead of 21 days. At comparable stages of development in these two species the ΔP_{O_2} across the eggshell were similar.

If one may apply this relationship to bird eggs in general, it follows that the time integral of ΔP_{O_2} for the period I, normalized to various incubation lengths is constant. Thus,

$$\int_{t=0}^{t=I} \Delta P_{O_2(t)} \, d\left(\frac{t}{I}\right) = \overline{\Delta P}_{O_2} \tag{7}$$

where t is any time between 0 and I, t/I is a fraction of incubation time which varies between 0 and 1 and $\overline{\Delta P}_{O_2}$ is a simplified symbolic expression of the proposed constant. This constant has the value of 16.6 torr, and represents the weighted mean of ΔP_{O_2} during the incubation. It has been calculated by integration of the $\Delta P_{O_2(t)}$ values during development of the chicken egg, the only species for which published data are available, and our analysis has been based upon the data of Wangensteen and Rahn (1970/71).

One can now calculate the product of the O_2 conductance constant, derived from Equation (4) and ΔP_{O_2} of Equation (7) to obtain the total amount of O_2 taken per g of egg.

$$\frac{G_{O_2} \cdot I}{W} \cdot \overline{\Delta P}_{O_2} = 5.39 \times 16.6 = 89 \text{ cm}^3 \ O_2 (\text{STPD})/\text{g} \tag{8}$$

The first factor, derived from Equation (4), describes the quantity of O_2 per g of egg which diffuses across the eggshell per unit ΔP_{O_2}. The second factor is the weighted mean of ΔP_{O_2} during the incubation time. Thus a typical egg regardless of size or incubation length will take up 90 cm^3 O_2 for every gram ($\approx$ 470 cal/g). A 100-g egg will consume 100 times this amount. The average quantity of O_2 required to produce 1 g of embryonic tissue will be greater by a factor of 1.54, since the typical hatching weight is 65% of the initial egg weight.

d) Functional Pore Area

The simultaneous values of G, W, and L in Table 1 allow one to calculate the increase in the functional pore are a, A_p, as a function of egg weight:

$$A_p \ (\text{cm}^2) = 9.72 \cdot 10^{-5} \cdot W^{1.249} \quad (\text{SEE} = \dot{\underset{\bullet}{x}} \ 1.305;\ r^2 = 0.979) \tag{9}$$

This relationship is similar to the equation derived previously (Ar et al., 1974) from the general comparisons of G and L as a function of egg weight.

The total functional pore area of an egg, A_p, may be calculated from the values of G and L according to Fick's first law of diffusion (Ar et al., 1974). Thus

$$A_p = (G_{H_2O} \cdot L)/23.42 \tag{10}$$

where A_p = cm^2 and L = cm.

By solving for G_{H_2O} in Equation (4) and substituting this value into Equation (10) one obtains the following expression

$$A_p = \frac{1}{I} \ (0.22 \ WL) \tag{11}$$

which tells us in general that for a given egg weight and shell thickness the total functional pore area is inversely related to the incubation length. For example, from Table 1 it will be seen that the eggs of the White Tern, *Gygis alba*, and the Rock-Dove-Pigeon, *Columba livia*, have similar egg weights (21.4 and 21.2 g) as well as shell thicknesses (0.17 and 0.19 mm). However, the White Tern has an incubation period which is twice as long as that of the pigeon, namely 36 days instead of 18 which is reflected in a pore area of 0.26 mm^2, nearly half that of 0.42 mm^2 for the pigeon.

Acknowledgment. This study was in part supported by NSF Grant No. PCM 76-20947.

References

Ar, A., Paganelli, C.V., Reeves, R.B., Greene, D.G., Rahn, H.: The avian egg: water vapor conductance, shell thickness and functional pore area. Condor 76, 153-158 (1974)

Drent, R.: Incubation. In: Avian Biology, Vol. V. Farner, D.S., King, J.R. (eds.). New York: Academic Press, 1975, p. 351

Hammel, H.T., Rahn, H.: Incubation water loss in eggs of the Adelie Penguin and the McCormick Skua. Am. J. Physiol. in press (1978)

Lomholt, J.P.: Relationship of weight loss to ambient humidity of birds eggs during incubation. J. Comp. Physiol. 105, 189-196 (1976)

Rahn, H., Ackerman, R.A., Paganelli, C.V.: Humidity in the avian nest and egg water loss during incubation. Physiol. Zool. 50, 269-283 (1977)

Rahn, H., Ar, A.: The avian egg: incubation time and water loss. Condor 76, 147-152 (1974)

Rahn, H., Paganelli, C.V., Ar, A.: The avian egg: aircell gas tension, metabolism and incubation time. Respir. Physiol. 22, 297-309 (1974)

Rahn, H., Paganelli, C.V., Nisbet, I.C.T., Whittow, G.C.: Regulation of incubation water loss in eggs of seven species of terns. Physiol. Zool. 49, 245-259 (1976)

Schönwetter, M.: Handbuch der Oologie, Lief. 1-24. Meise, M. (ed.). Berlin: Akademie Verlag, 1962-74

Visschedijk, A.H.J.: The air space and embryonic respiration. The pattern of gaseous exchange in the fertile egg during the closing stages of incubation. Brit. Poultry Sci. 9, 173-184 (1968)

Wangensteen, O.D., Rahn, H.: Respiratory gas exchange by the avian embryo. Respir. Physiol. 11, 31-45 (1970/71)

Metabolism of Avian Embryos: Comparative Ontogeny

D. F. Hoyt, C. M. Vleck, and D. Vleck

Summary

The ontogeny of O_2 consumption of avian embryos differs with size and condition at hatching. Altricial species do not exhibit the plateau phase observed in precocial species. Large precocial species exhibit a more pronounced decline in metabolic rate during the latter phase of the incubation period then do small precocial species. It is suggested that these differences may be due to differences in the pattern of growth. In spite of these differences, the rate of O_2 consumption immediately prior to pipping is roughly proportional to the 3/4 power of fresh egg mass.

Oxygen consumption of individual eggs was measured throughout incubation in the Ostrich (*Struthio camelus*), Rhea (*Rhea americana*), Emu (*Dromiceius novae-hollandiae*), Domestic Goose (*Anser anser*), Pigeon (*Columba livia*), Coturnix Quail (*Coturnix coturnix*), Village Weaverbird (*Ploceus cucullatus*), and Zebra Finch (*Taeniopygia guttata*). All measurements were made at 37.5°C, except those on the ostrich (35°C) and emu (36°C).

The pattern of O_2 consumption in the Domestic Goose and Coturnix Quail (C. Vleck et al., in prep.) are similar to that in the domestic fowl (Romijn and Lokhorst, 1951); a period of exponential increase in O_2 consumption is followed by a plateau.

The pattern of O_2 consumption in the Ostrich (Hoyt et al., submitted to The Condor), Rhea, and Emu (D. Vleck et al., in prep.) differs from the pattern in the goose and quail. Following a period of exponential increase, there is a period of decline in the rate of O_2 consumption which terminates when the process of pipping commences. The maximum rate preceding the decline is termed the "hump", and the minimum rate at the end of the decline is the "pre-pipping" rate. The rate of O_2 consumption during the hump is 20-25% higher than the pre-pipping rate.

The O_2 consumption of the eggs of the Pigeon, Village Weaverbird, and Zebra Finch (C. Vleck et al., in prep.) exhibit no plateau: the period of exponential increase in the rate of O_2 consumption terminates when the egg is pipped.

It is assumed that the O_2 consumption of avian embryos is the sum of three components: basal metabolism, growth metabolism, and muscular activity. It is suggested that the pattern of O_2 consumption in the precocial species may be the result of a decrease in growth rate and/or muscular activity.

Department of Biology, University of California at Los Angeles, Los Angeles, California 90024, U.S.A.

The relation between the pre-pipping rate of O_2 consumption and fresh egg mass is:

$$\dot{V}_{O_2} = 26.4\ (\text{Mass})^{0.726}$$

(number of species = 34; correlation coefficient $r^2 = 0.972$; standard error of the estimate for the log-transformed values = 0.083). This regression is based upon the data of: Rahn et al. (1974), Hoyt et al. (in prep.), and this report.

References

Hoyt, D.F., Board, R., Paganelli, C.V., Rahn, H.: Eggs of the Anatidae: Conductance ore geometry and metabolism. (In prep.)

Hoyt, D.F., Vleck, D., Vleck, C.E.M.: Metabolism of avian embryos: Ontogeny and temperature effects in the Ostrich. (Submitted to The Condor)

Rahn, H., Paganelli, C.V., Ar, A.: The avian egg, air-cell gas tension, metabolism and incubation time. Respir. Physiol. 22, 297-309 (1974)

Romijn, C., Lokhorst, W.: Foetal respiration in the hen. Phys. Comp. Oecol. 2, 187-197 (1951)

Vleck, C.M., Hoyt, D.F., Vleck, D.: Metabolism of avian embryos: Patterns in altricial and precocial birds. (In prep.)

Vleck, D., Vleck, C.M., Hoyt, D.F.: Metabolism of avian embryos: Ontogeny of oxygen consumption in the Emu and Rhea. (In prep.)

The Kiwi: A Case of Compensating Divergences From Allometric Predictions

W. A. Calder III

Summary

In terms of incubation time and relative egg size, kiwis are exceptions to general avian allometric relationships. These extremes thus serve as a test of the eggshell model of Rahn et al.

Reduced permeability to water vapor exactly compensates for the prolonged incubation to preclude abnormal evaporative loss. Plateau O_2 consumption ($\dot{V}_{O_2}$) is similarly reduced by the apparent reduced porosity. This does not appear to cause hypoxic stress, because the resting $\dot{V}_{O_2}$ of chicks and adults is also reduced by the same factor, compared to predictions based on body-mass.

Thus the lower metabolic demands of kiwi tissue make it possible to have a prolonged embryonic development without excessive water loss from the egg.

1. Introduction

Several characteristics of bird eggs and incubation have been correlated with the size of the adult or of the egg (Rahn and Ar, 1974; Rahn et al., 1974, 1975; Ar et al., 1974). The high correlation coefficients of these relationships suggest that natural selection has had a predominating allometric component. However, if the odd species appears to be exempted from the "rules" of scaling, we must ask how good these "rules" are and how good is our understanding of them?

The kiwis (*Apteryx* spp.) are the smallest of the flightless ratite birds, and are thought to be secondarily small, having evolved from a common ancestor with the moas in New Zealand (Cracraft, 1974). The kiwis' extreme in egg-body size proportions has caused them to be rejected in allometric analyses, though their nonconformity is sometimes noted (Huxley, 1927; Brody, 1945; Rahn et al., 1975). This egg has one of the longest incubation periods. It is adequately provisioned for this energetically, with one of the proportionately largest yolks (61%; Reid, 1971), as large as the yolk of the emu egg (Romanoff and Romanoff, 1949). However, without structural or behavioural compensation, evaporation could be excessive during this long incubation. These divergences from allometric predictions seemed worthy of further examination.

Department of Ecology and Evolutionary Biology, University of Arizona, Tucson, Arizona 85721, USA.

2. Methods

The Brown Kiwi (*A. australis*) was studied at the breeding facilities of the Otorohanga Zoological Society. Water loss and the water-vapor permeability of the shell were determined by the method of Rahn and Ar (1974). The O_2 uptake of an artificially-incubated egg, as well as the metabolic rates of juvenile and adult birds were determined in an open system with a paramagnetic analyser (Servomex OA272), with range-expansion and readout of a digital voltmeter and potentiometric recorder. Air burrow temperatures were monitored with calibrated copper-constantan thermocouples. Incubation temperatures were determined with thermocouples and telemetry (Mini-mitter model V) in water-filled dummy eggs (reinforced real and artificial shells, both with cooling constants as predicted by equations of Ricklefs, 1974). Eggshell thickness was measured with a metric micrometer after removing the membranes. Burrow air was collected in a football bladder, via tubing previously installed, and a hand pump, and passed through the paramagnetic analyser within 5 min at calibrated flow rate. Humidity in burrows and in the egg-permeability chamber was determined with appropriate narrow-range sensors of an electronic humidity meter (Aminco-Hygrodynamics), checked at various ambient humidities with a motor-driven psychrometer. Vapor pressures were calculated from relative humidity and temperature data.

3. Results and Discussion

The kiwi eggshell is significantly thinner than predicted by Equation (7) of Ar et al. (1974). The samples included eggs from three different aviaries using different diets, and three wild eggs. Variations in thickness could not be correlated with source. Comparison with the prediction (Table 1) was made from the middle of the egg. The small

Table 1. Brown kiwi eggs, incubation and metabolism

	Predicted	Measured	Individual determinations	Predicted/ measured
t_{inc}, days	44	71 (74-84)	1 a	1.61 1.67
G_{H_2O}, g/mm/day	0.043	0.026±0.004	5	0.60±0.09
$\dot{V}_{O_2}$, ml/day	2277	1423 ±48	1(7)	0.63
T_{egg}, °C	37.7	35.4 ±0.7	2(41)	-
L, cm	0.075	0.047±0.003	17	0.63±0.04
Shell mass, g	38.8	27.5	6	0.71
ΔP_{H_2O}, mm Hg	35	29.4	2(18)	0.84
Juven. $\dot{H}_m$, W	1.2-2.9	0.61-2.15	3(4)	0.65±0.14
Adult $\dot{H}_m$, W	6.73	4.08 ±0.41	5	0.61±0.10
A_p, mm^2	13.4	5.2	$(G \cdot L)_{H_2O}$	0.38

[a] Reid and Williams (1975).

ends were of similar thickness, but the shells were yet thinner at the large (air space) end (80.8±7.7% of mid-egg thickness). The especially thin blunt end (0.038 cm) may be related to method of hatching, the kiwi chick lacking an egg-tooth. One would expect the overall thickness, if proportional to mean diffusion distance, to increase the rate of evaporation from the egg. It would also reduce the resistance to buckling load which the egg could withstand, as well as the tolerance of kiwi eggshell to thinning by pesticides and polychlorinated biphenyls (PCB) (see Ar and Rahn, 1974).

Does the burrow compensate for this by providing physical protection and a higher humidity? Measurements from both an artificial burrow used for successful incubation, and a burrow, dug by captive birds in a large outdoor pen and used for incubation, indicate only slight reduction in vapor pressure difference below the typical 35 torr H_2O of Rahn and Ar (1974). While the relative humidity in the burrows was high, the vapor pressure was low due to the low temperatures: 13-15°C.

The long incubation (t_{inc}) would appear excessive, if permeability were as predicted for a typical egg of this size. Note however, that the product of the reduction in water-vapor permeability (60%) and increase in incubation time (161 to 167%) is 97 to 100% and compensates perfectly.

The water-vapor permeability could be reduced by having fewer and (or) smaller pores and/or blockage of some of the pores. These have, in fact, been noted as characteristics of kiwi eggshells without functional explanation (Tyler and Simkiss, 1959, and their citation of Nathusias). Functional pore area, A_p is proportional to $G_{H_2O} \cdot L$ (where G_{H_2O} is water vapor conductance and L shell thickness), so it must therefore be reduced to 0.60 × 0.63 = 38% that of the typical egg, predicted with Equation (8) of Ar et al. (1974).

The reduction in permeability, measured in five eggs, leads one to expect a parallel 40% reduction in O_2 permeability and therefore in plateau O_2 uptake as well. It was thus gratifying to be able to monitor O_2 uptake of the first artificially-incubated kiwi egg. This egg consumed 63±2.5% of the prediction during a plateau period from 6 days prior to the first puncture of the shell (eccentric to the air space) and did not change significantly until a second puncture on through the airspace.

This reduced O_2 uptake probably imposed no hypoxic stress on the tissues, because the metabolic rates were similarly reduced to fractions of predicted standard metabolism in that chick at age 9 1/2 days (225 g), two other chicks at 641-765 g body mass, five adult brown kiwis, and two individuals of each of the other two species of kiwis (*A. haasti*, *A. oweni*), at resting.

The temperature of a water-filled kiwi-eggshell measured with a temperature-transmitter was significantly lower (35.4±0.7°C) when incubated by male kiwis than by a bantam domestic hen (37.6°-37.8°C overnight, air temperature declining to 7 1/2°C vs. a 12° minimum for kiwi burrows). Assuming Q_{10} = 2, this temperature difference would be sufficient to increase the development time 16%. The lower egg temperature for kiwis is consistent with observation of body temperature (Farner et al., 1956) and the low metabolic rates of adults in this study.

In conclusion, it appears that the lower metabolic demands of kiwis have permitted the reduction in eggshell permeability compatible with normal evaporation over an unusually long embryonic development. While the kiwi does not fall on the allometric lines, the combined product of divergences from "typical" or predicted incubation time, and water vapor conductance, and vapor pressure gradient divided by fresh egg mass yields the "typical" fractional mass decrease value of 14 to 15%.

Acknowledgments. This study was possible because procedures for captive breeding and artificial incubation had been developed by Barry Rowe. His time, effort, and valuable advice are acknowledged with gratitude.

Otorohanga Zoological Society Scientific Publication No. 5. Supported by Grant PCM 76 - 09411, National Science Foundation (U.S.A.).

References

Ar, A., Rahn, H.: Avian egg-shell strength: size dependent predictions. Proc. 23rd I Int. Ornithol. Congr. 122 (1974)

Ar, A., Paganelli, C.V., Reeves, R.B., Greene, D.G., Rahn, H.: The avian egg: water vapor conductance, shell thickness, and functional pore area. Condor 76, 153-158 (1974)

Brody, S.: Bioenergetics and Growth. New York: Reinhold, 1945. Reprinted: New York: Hafner, 1964

Cracraft, J.: Phylogeny and evolution of the ratite birds. Ibis 116, 494-521 (1974)

Farner, D.S., Chivers, N., Rinez, T.: The body temperatures of the North Island kiwis. Emu 56, 199-206 (1956)

Huxley, J.S.: On the relation between egg-weight and body weight in birds. J. Linn. Soc. Zool. 36, 457-466 (1927)

Rahn, H., Ar, A.: The avian egg: incubation time and water loss. Condor 76, 147-152 (1974)

Rahn, H., Paganelli, C.V., Ar, A.: The avian egg: air-cell gas tension, metabolism and incubation time. Respir. Physiol. 22, 297-309 (1974)

Rahn, H., Paganelli, C.V., Ar, A.: Relation of avian egg weight to body weight. Auk 92, 750-765 (1975)

Reid, B.: Composition of a kiwi egg. Notornis 18, 250-252 (1971)

Reid, B., Williams, G.: The kiwi. In: Biogeography and Ecology in New Zealand. Kuschel, G. (ed.). 1975 The Hague: Dr. W. Junk b.v.

Ricklefs, R.E.: Energetics of reproduction in birds. In: Avian Energetics. Paynther, R.A. (ed.). 1974

Romanoff, A.L., Romanoff, A.J.: The Avian Egg. New York: Wiley, 1949

Tyler, C., Simkiss, K.: A study of the egg shells of ratite birds. Proc. Zool. Soc. London 133, 201-243 (1959)

Gas Conductance in the Eggshell of the Mound-Building Brush Turkey

R. S. Seymour and H. Rahn

Summary

Eggs of the Brush-turkey, *Alectura lathami*, are deposited inside mounds of earth and leaves where organic decomposition furnishes the necessary heat for development. In this study the 211-g eggs were exposed to a temperature of 37.3°C, a water vapor pressure of 48 torr and a partial O_2 and CO_2 pressure of 100 and 62 torr, respectively. These eggs had a water vapor conductance of 46 mg/day·torr which is 2.6 times greater than that predicted for typical avian eggs of similar size and incubation period (63 days). The larger than normal gas conductance is presumably an adaptation to the unusual gaseous nest environment and would greatly reduce the typical O_2 and CO_2 differences that normally develop across the egg shell during embryonic development.

1. Introduction

For incubation of their eggs, the birds of the megapode family rely on a variety of heat sources such as solar radiation, volcanic heat, and organic decomposition (Frith, 1962). The male Australian Brush-turkey, *Alectura lathami*, constructs an incubation mound which may be about one meter high, and the female lays as many as 35 eggs in it during an extended period (Fleay, 1937; Baltin, 1969). Heat is generated in the mound almost exclusively by organic means, and the mound temperature and humidity are regulated by the enthusiastic activities of the male bird. Compared to other birds' eggs, laid above ground, brush-turkey eggs are thus exposed to an extreme set of environmental conditions: the humidity is high and the respiratory gas pressures depart widely from atmospheric levels (Baltin, 1969). Shell conductance provides one aspect of the evolutionary adaptations of an egg to the problems of water loss and metabolic gas exchange. Therefore, it was of interest to examine the gaseous environment within the mound, to measure shell conductance, and to make comparisons between the eggs of brush-turkeys and other birds (Table 1).

2. The Mound

Temperature in the mound increased with depth below the surface such that at egg level, it averaged about 37°C. This falls within 1 S.D.

Department of Zoology, University of Adelaide, G.P.O. Box 498, Adelaide, S.A. 5001, Australia. Department of Physiology, State University of New York at Buffalo, N.Y. 14214, USA.

of the mean egg incubation temperature of 35.6°C obtained by averaging values for 27 species reviewed by Drent (1975).

The humidity was saturated because of a high water content in the mound material. In this case, water loss from the eggs could not be expected unless the eggs were warmer than the mound. Heat produced by the embryo would tend to create a temperature gradient across the shell (Drent, 1970; Kashkin, 1961), but the excellent insulation offered by the earth and organic material in the mound would make the gradient negligible. Hence water loss from the eggs is probably insignificant. In contrast, the eggs of nesting birds experience substantial humidity differences across the shell and they lose about 14-18% of their weight during incubation (Rahn and Ar, 1974; Rahn et al., 1976; Chattock, 1925; Drent, 1975).

The mound gas at egg level consisted of about 14% O_2 and 9% CO_2 (Table 1). There are no comparative data for the respiratory gases in the

Table 1. Comparison of six brush-turkey eggs and "ideal" eggs

Parameter	Units	Actual ($\bar{x}$ ± 95% CI)	Predicted[a]	Actual / Predicted
Incubation temperature	°C	37.3 ± 2.2	34	1.10
Environmental P_{H_2O}	torr	47.8 ± 6.1	20	2.39
Environmental P_{O_2}	torr	100 ± 10	136	0.74
Environmental P_{CO_2}	torr	62 ± 12	8	7.75
Egg weight	g	211 ± 3	60	3.52
Shell thickness	mm	0.30 ± 0.03	0.59	0.51
Water vapor conductance	mg/day·torr	45.7 ± 26.5	17.6	2.60
Functional pore area	mm^2	5.65 ± 3.28	6.86	0.82
Incubation time	days	63[a]	38	1.64

[a] Sources of all predicted values and actual incubation time are presented in the text. Additional measurements of eggs from a single hen in the Tel-Aviv Zoo, Israel, were kindly provided by A. Ar (personal communication). Egg weight (g) 186, S.D. ±1.6, n = 4; shell thickness (mm) 0.415, S.D. ±0.025, n = 4; water vapor conductance (mg/day·torr) 50.6, S.D. 3.6, n = 3.

nests of other birds but these can be calculated from the rates of egg metabolism and nest ventilation in several species (Rahn et al., 1976, 1977). These calculations indicate that the nest environment is not greatly different from the normal atmosphere (Table 1). Thus it appears that the megapodes (and perhaps some hole-nesting birds) expose their eggs to a markedly atypical environment. In fact, O_2 and CO_2 pressures *outside* brush-turkey eggs diverge from atmospheric levels more than those pressures normally found *inside* other bird eggs (cf. Rahn et al., 1974).

3. The Eggs

Brush-turkey eggs are about 3.5 times the expected weight for a galliform bird (Lack, 1968) and the incubation period which averages about

63 days (Fleay, 1937; Baltin, 1969) is about 60% longer than predicted by allometric analysis of incubation and egg weight (Rahn and Ar, 1974). Despite the large size, the shell is very thin, being about the same as that of a 60-g hen's egg. The allometric prediction is for a shell about twice as thick (Ar et al., 1974). Other megapodes also have large eggs with thin shells (Frith, 1962).

An index of the barrier to gas diffusion is the water vapor conductance which was measured with the standard technique involving weight loss of eggs in desiccators at constant temperature (Ar et al., 1974). Water vapor conductance in brush-turkey eggs averages about 2.6 times higher than that predicted from the relationship that shell conductance is proportional to egg weight and inversely proportional to the incubation time (Rahn and Ar, 1974; Ar and Rahn, 1978). Since shell conductance is also proportional to the ratio of total pore area to pore length or shell thickness (Paganelli et al., 1975) most of the increase in conductance can be explained by the thinning of the eggshell. The derived value of functional pore area is approximately as predicted by egg weight (Ar et al., 1974).

The eggs lack the air space normally found between the shell membranes in other birds' eggs (Baltin, 1969). However, gas spaces do form within the albumen if the eggs are removed from the mound and artificially dehydrated.

4. Implications and Conclusions

Although the mound eliminates the potential problem of egg dehydration, it also prevents the formation of an air space in the egg. Most birds require an air space to initiate pulmonary respiration prior to pipping (Romanoff, 1967). This presumably facilitates the transition from chorioallantoic to pulmonary gas exchange, a process which requires about 24 hours in domestic chickens (Visschedijk, 1968). The lack of an air space in brush-turkey eggs, and perhaps in megapode eggs in general, implies a markedly different pattern of hatching whereby the first breath is taken only after the shell is broken. It would seen advantageous for the chick to break open the shell quickly in order to breathe and this may be faciliated by the thin shell.

When a chick hatches, it may be as deep as 50 cm below the surface of mound and it must dig its way out. Emergence requires up to two days (Baltin, 1969), during which time the chick relies solely on interstitial gas for respiration. Few air-breathing vertebrates are exposed to CO_2 pressures as high as 60 torr and O_2 pressures as low as 100 torr. Those that are, such as some burrowing mammals, show marked physiological adaptations in their ventilation and blood properties. Unfortunately, there is no information about the respiratory physiology in brush-turkey chicks.

Of course the O_2 and CO_2 pressures inside the shell are even more severe than in the mound. Although high shell permeability appears to be an adaptation for reducing the gradient across the shell, it is impossible for internal gas pressures to be similar to those in typical eggs. With estimated metabolic rates [Eq. (4), Rahn et al. (1974)] and the conductance value of the shell, it can be predicted that both air cell P_{O_2} and P_{CO_2} are about 80 torr in brush-turkey eggs at time of hatching. It remains to be seen just how the unhatched chick accomplishes O_2 loading and acid-base balance under these conditions.

Finally, it is tempting to compare brush-turkeys with certain reptiles which also bury their eggs. The gaseous environments in marine green turtle nests and brush-turkey mounds are similar and both species show adaptive increases in shell porosity (Ackerman and Prange, 1972; Ackerman, 1977). However it is dangerous to view the megapode brooding behavior as the primitive condition in birds. There is good evidence for the evolution of moundbuilders from birds with normal nesting behavior (Frith, 1962). Regardless of whether the burying of eggs is a primitive or advanced condition, the high shell conductances in both instances probably represent similar adaptation to similar egg environments.

Acknowledgments. We thank Trevor Gibson and Robert Baker who kindly loaned the eggs from the Adelaide Zoo. The presentation of this paper was partly funded by a travel grant from the U.S.A. National Committee for the International Union of Physiological Sciences.

References

Ackerman, R.A.: The respiratory gas exchange of sea turtle nests (Chelonia, Caretta). Respir. Physiol. 31, 19-38 (1977)

Ackerman, R.A., Prange, H.D.: Oxygen diffusion across a sea turtle (*Chelonia mydas*) egg shell. Comp. Biochem. Physiol. 43A, 905-909 (1972)

Ar, A., Paganelli, C.V., Reeves, R.B., Greene, D.G., Rahn, H.: The avian egg: water vapor conductance, shell thickness, and functional pore area. Condor 76, 153-158 (1974)

Baltin, S.: Zur Biologie und Ethologie des Talegalla-Huhns (*Alectura lathami* Gray) unter besonderer Berücksichtigung des Verhaltens während der Brutperiode. Z. Tierpsychol. 26, 524-572 (1969)

Chattock, A.P.: On the physics of incubation. Phil. Trans. Roy. Soc. London, Ser. B 213, 397-450 (1925)

Drent, R.H.: Functional aspects of incubation in the Herring Gull. Behaviour, Suppl. 17, 1-132 (1970)

Drent, R.H.: Incubation. In:Avian Biology, Vol. 5. Farner, D.S., King, J.R. (eds). New York: Academic Press, 1975, pp. 334-420

Fleay, D.H.: Nesting habits of the Brush-Turkey. Emu 36, 153-163 (1937)

Frith, H.J.: The Mallee-Fowl. Sydney: Angus and Robertson, 1962, pp. 136

Kashkin, V.V.: Heat exchange in bird's eggs on incubation. Biophysics 6, 97-107 (1961)

Lack, D.: Ecological Adaptation for Breeding in Birds. London: Methuen, 1968, pp. 408

Paganelli, C.V., Ar, A., Rahn, H., Wangensteen, O.D.: Diffusion in the gas phase: the effects of ambient pressure and gas composition. Respir. Physiol. 25, 247-258 (1975)

Rahn, H., Ackerman, R.A., Paganelli, C.V.: Humidity in the avian nest and egg water loss during incubation. Physiol. Zool. 50, 269-283 (1977)

Rahn, H., Ar, A.: The avian egg: incubation time and water loss. Condor 76, 147-152 (1974)

Rahn, H., Paganelli, C.V., Ar, A.: The avian egg: air-cell gas tension, metabolism and incubation time. Respir. Physiol. 22, 297-309 (1974)

Rahn, H., Paganelli, C.V., Nisbet, I.C.T., Whittow, G.C.: Regulation of incubation water loss in eggs of seven species of terns. Physiol. Zool. 49, 245-259 (1976)

Romanoff, A.L.: Biochemistry of the Avian Embryo. New York: Wiley, 1967, pp. 3987

Visschedijk, A.H.J.: The air space and embryonic respiration. II. The times of pipping and hatching as influenced by an artificially changed permeability of the shell over the air space. Brit. Poultry Sci. 9, 185-196 (1968)

Acid-Base Balance During Eggshell Formation

P. Mongin

Summary

One of the two main metabolic factors involved in the eggshell formation is the acid-base balance. Under normal physiological conditions the shell deposition induces metabolic acidosis regulated at the pulmonary level by hyperventilation to reduce the blood P_{CO_2} and at the kidney level by an increase of the titratable acidity in the urine. From the data obtained by different experimental disturbances of the acid-base status it is concluded that a chronic decrease of the blood bicarbonate level reduces the calcium carbonate deposition on the eggshell and vice versa.

Besides the lungs and the kidneys, other organs are implicated in the acid-base balance such as shell gland, skeleton, and the glandular stomach. Following different acid-base status the composition of uterine fluid and of the egg albumen is given. A working model is also presented for the shell gland mucosa based upon new results on the intracellular composition. Finally a general survey tempts to make some connections between shell gland physiology and that of glandular stomach.

1. Introduction

The eggshell of a standard hen's egg of 55 g weight about 5 g. It is made of calcite cristals deposited on a protein matrix of polymucosaccharides.

The main role of the eggshell is to protect the egg against any physical and chemical aggressions from the outside and its main quality appeared with the breeding of strains for egg laxing. Laying tests between 1900 and 1911 reported that the best birds layed 110 eggs per year while the average laying rate of a commercial flock of 25,000 birds is by now 250 eggs per year. The calcium carbonate exported by the egg want up from 500 g to 1250 g/year. Now the amount of calcium deposited within a year represents 20 times the total calcium reserves in the skeleton.

In fact the eggshell is made of 40% calcium and 60% carbonate ions. Until the early sixties most the work was devoted to the calcium metabolism, and information concerning the acid-base balance in birds was quite limited.

Station de Recherches Avicoles, Institut National de la Recherche Agronomique, C.R. Tours-Nouzilly, 37380 Monnaie (France).

Buckhardt (1933) supplemented a diet for layers with either lactic acid or hydrochloric acid in equivalent amounts and found that lactic acid did not produce any sign of acidosis, while HCl decreases nitrogen and phosphorus retention, increases urinary ammoniac excretion, and leads to a negative calcium balance. Solum and Schuster (1934) studied the effect of different calcium salts and concluded that the best results obtained with calcium carbonate or lactate are due to the fact that they prevent an acidosis. The bad results with calcium chloride are attributed to an alteration of the acid-base balance.

Heller and Cursell (1937) studied blood composition as a function of the age and claimed that the ratio between chloride concentrations in red blood cells and plasma is lower in the layers compared to immature birds; so that the acid-base should move towards an alcalosis during the sexual maturation.

Common (1941) showed that the alcalin reserves increase at the sexual maturity. The level tends to decrease at each egg layed but remains higher than that of the immature bird or cock.

In 1945, Gutoska and Mitchell working on uterine carbonic anhydrase found a higher activity in the laying fowl compared to the immature female. Furthermore, diamox injected intravenously inhibited eggshell formation. At that time they formulated a first working hypothesis: the eggshell carbonate ions could come from the plasma bicarbonate ions.

This summarizes the information about the acid-base balance in birds at the end of fifties. This paper concerns the role of the acid-base balance in the formation of the particular calcified tissue which is the eggshell. A working model of the shell gland is given together with an overall synthetic view including most of its main functions.

2. General Acid-Base Balance and Eggshell Deposition

a) Normal Physiological Status

Under normal physiological conditions eggshell deposition induces metabolic acidosis partially compensated by ventilatory alcalosis (Mongin and Lacassagne, 1964, 1966). The plasma pH drops from 7.52 just before the calcification down to 7.41 12 h later (Table 1). At the same time, the bicarbemia goes from 27.5 down to 20.7 mEq/l. It appears that eggshell formation may affect the general acid-base balance of the laying bird possibly through the secretion of the carbonate ions by the uterus.

This acidosis is regulated both by the pulmonary and renal functions. The decrease of the blood pH is partly compensated by a decrease of the plasma P_{CO_2} due to a hyperventilation (Table 2). The respiratory rhythm of the bird during the night, i.e., during the shell deposition, is 11.8/min, while in the same bird outside the times of egg formation it is 7.2 c/min, the same as that in the cock (Mongin and Lacassagne, 1966).

The kidneys also correct the systemic metabolic acidosis (Fig. 1) by forming an acid urine without bicarbonates and having a relatively high net acidity (titratable plus NH_4^+) (Anderson, 1967; Simkiss, 1969; Mongin, 1969).

Table 1. Variations of venous blood pH, bicarbonate, and P_{CO_2} during the egg formation[a] in the oviduct (Mongin and Lacassagne, 1964)

Hours after oviposition	Number of hens	Mean pH	HCO_3^- (mEq/l)	P_{CO_2} (mm Hg)
0	6	7.522 ± 0.013	29.6 ± 0.8	37.7 ± 1.0
3.30	7	7.530 ± 0.015	31.2 ± 1.4	38.1 ± 1.4
5	7	7.533 ± 0.015	31.5 ± 1.1	38.2 ± 1.7
10	7	7.523 ± 0.013	27.5 ± 1.3[b]	34.7 ± 1.4[b]
15	7	7.478 ± 0.012[b]	24.6 ± 0.7[d]	34.4 ± 1.0[c]
20	8	7.454 ± 0.020[d]	22.4 ± 0.7[d]	33.4 ± 1.3[d]
22	7	7.411 ± 0.016[d]	20.7 ± 0.3[d]	32.1 ± 1.3[d]
25	6	7.435 ± 0.018	23.7 ± 1.3	35.5 ± 1.4
30	3	7.483 ± 0.020	29.1 ± 0.9	38.7 ± 1.7

[a] Five hours after the previous oviposition, the forming egg enters the shell gland. Between 5 and 12 h, the egg albumen is pumpled. Shell deposition occurs between 10 and 22 h.
[b] Significatively different from the value at 5 h : $P < 0.05$.
[c] Significatively different from the value at 5 h : $P < 0.01$.
[d] Significatively different from the value at 5 h : $P < 0.001$.

Table 2. Respiratory rythm of laying hens during or in absence of shell deposition compared to that of the cock in the same conditions (Mongin and Lacassagne, 1966)

Animal number	With shell deposition	Without shell deposition	Cock
1	10.5 ± 0.4	7.7 ± 0.1	7.7 ± 0.1
2	10.7 ± 0.2	7.2 ± 0.2	6.3 ± 0.2
3	11.3 ± 0.3	6.4 ± 0.1	7.0 ± 0.1
4	13.2 ± 0.4	7.7 ± 0.1	7.4 ± 0.3
5	13.5 ± 0.2	7.3 ± 0.1	7.3 ± 0.1
Mean	11.8 ± 0.6	7.2 ± 0.2	7.2 ± 0.2

All these regulatory phenomena are consistant with the hypothesis of a metabolic acidosis induced by a calcium carbonate deposition in utero.

b) Disturbed Physiological Status

Metabolic modification. An experimental metabolic acidosis, induced by ingestion of either NH_4Cl or HCl, considerably reduces the eggshell secretion (Hall and Helbacka, 1959; Hunt and Aitken, 1962; Sauveur, 1968). However, the potential acid load brought into the diet is not the only determinant factor. The anions that accompany the protons play a decisive role in the development of the acidosis by hindering or promoting the renal elimination of the H^+ ions. As shown in Table 3, with the same acid load equivalent to 3% of NH_4Cl in the diet, the reduction of shell deposition is less important with SO_4^{--} than with Cl^- anions.

On the other hand, a metabolic alcalosis facilitates the $CaCO_3$ deposition in utero. The easiest experimental method involves the induction of a hypochloremic alcalosis by limiting the ingestion of Cl^- ions without modification of the Na intake (Mongin et al., 1972). The anion-

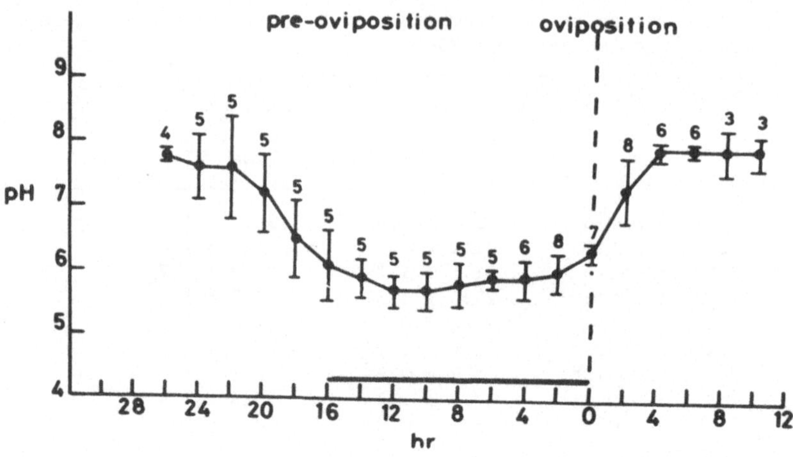

Fig. 1. Urine pH during egg formation (Anderson, 1967)

Table 3. Effect of the anions during induced metabolic acidosis on acid-base balance and eggshell quality (Sauveur, 1971)

	pH	P_{CO_2} mm Hg	HCO_3^- mEq/l	Shell index g/100 cm^2
Control	7.4	38.4	21.7	7.70
NH_4CL	7.16	39.0	12.7	5.59
$(NH_4)SO_4$	7.29	40.4	18.0	7.34
HCL	7.18	38.8	13.1	5.61
H_2SO_4	7.35	35.4	17.9	7.11

Table 4. Hypochloremic alkalosis induced by a reduction of the dietary chloride and its effect on shell quality

Chloride in the diet %	Plasma HCO_3^- mEq/l	Eggshell index g/100 cm^2
0.24	24.8	6.93
0.12	29.3	7.51

cation unbalance stimulates renal bicarbonate regeneration and consequently increases the plasma bicarbonate level (Table 4).

These two characteristic methods use the kidney as a regulatory organ to disturb the acid-base balance so that the consequences at the level of the eggshell can be observed. Other methods particularly those employing the pulmonary system can also bring interesting results.

Ventilatory modifications. The bird has no sweat glands. To overcome hypothermia, it must evaporate water through the lungs. The resulting hyperventilation leads to a decrease of the plasma P_{CO_2} which induces a ventilatory alcalosis followed by a renal compensation. The acid-base balance is readjusted to a lower level than normal and available bicarbonates for the eggshell are reduced (Mongin and Lacassagne, 1964;

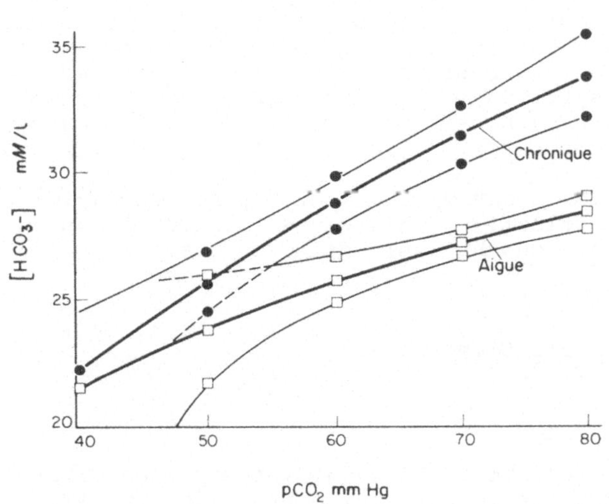

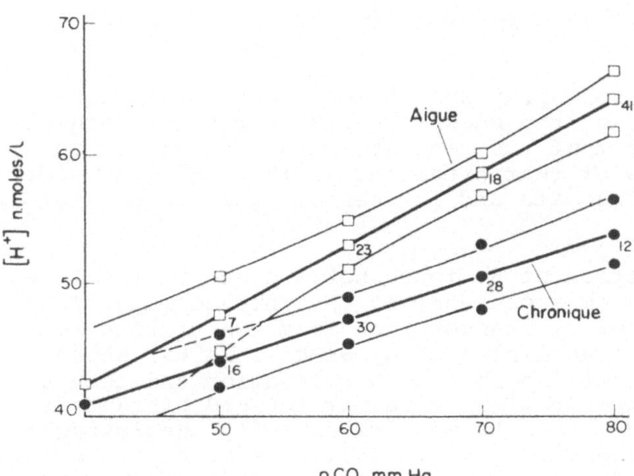

Fig. 2. Plasma HCO_3^- and H^+ ions concentrations as a function of the blood P_{CO_2} during acute and chronic hypercapnia in layers (Sauveur and Mongin, 1972)

Calder and Schmidt-Nielsen, 1966). The mechanism should explain, at least in part, the fragility of eggshell during the hot season.

Conversely, hypercapania should increase $CaCO_3$ deposition. As expected acute hypercapnia reduces the eggshell strength since such a treatment lowers the buffer capacity of the blood (Helbacka et al., 1963). However under chronic hypercapnia, i.e., after a full renal compensation, the eggshell strength is improved (Franck and Burger, 1965; Mongin, 1968, 1969). A complete study of the acute and chronic status has been done by Sauveur and Mongin (1972) and shows how, under the same P_{CO_2}, bicarbemia is increased after 21 days of exposition to an environment enriched in CO_2 (Fig. 2).

c) Conclusion

As far as the acid-base balance is a limiting factor in the deposition of $CaCO_3$ on the eggshell, all the causes which increase the plasma bicarbonate level should improve the eggshell quality. This is the case with metabolic alcalosis and ventilatory acidosis. Moreover, pH is not a good criterion and bicarbemia remains the best one. Furthermore the acute or chronic aspect of the treatment is important. An acute ventilatory acidosis has an inauspicious effect on eggshell quality; the small elevation of the plasma bicarbonate produces results of a purely chemical nature and no renal compensation. A steady state is not reach.

Even it is possible to establish a causal relationship between acid-base balance and eggshell calcification, this gives no information on the nature of the exchanges through the shell gland mucosa.

3. Mechanisms of Transfer in Shell Gland

a) Acid-Base Balance in Shell Gland

From the hypothesis of Gutoska and Mitchell (1945), the plasma bicarbonate ions should pass through the mucosal cells and react to give CO_3^{2-} and H_2CO_3. The later one being eliminated from the uterine fluid by mucosal reabsorption and dehydrated into CO_2 and H_2O under the action of the carbonic anhydrase.

As an alternative, Simkiss (1961) proposes that the carbonate ions derive from the metabolic CO_2 of the mucosal cells. Perfusion with Ca and labeled bicarbonate shows that the ratioactive ^{14}C deposited on the shell represents only 20% of that predicted if the CO_3^{2-} ions came directly from the plasma bicarbonate and it seems largely to consist of CO_2 of metabolic origin.

Under such conditions, production of H^+ ions should occur at the level of the CO_2 hydration and lead to a decrease of the intracellular pH either by in situ generation or by reabsorption from the uterine fluid. A first measurement of the intracellular pH by Simkiss (1969) shows a rapid fall at the beginning of the shell mineralisation and then a gradual restoration towards the initial values at the end of the calcification. An attempt by the author to measure directly the intracellular pH with micro pH electrodes completely failed because it was not possible to stick the electrodes into the mucosal cells from the luminal side. With the same DMO technique as Simkiss but with an other extracellular space marker (Mongin and Carter, 1976) no variations of the intracellular pH could be detected (Mongin and Carter, 1977). So this question remains open.

b) Acid-Base Balance in the Uterine Fluid

The amount of bicarbonate in the shell gland fluid is quite important and varied between 90 mEq/l (El Jack and Lake, 1967) and 68 mEq/l (Mueller et al., 1969). The acid-base balance of both the uterine fluid (Mongin and Sauveur, 1970) and the albumen (Sauveur and Mongin, 1971) is affected by the egg formation and becomes acidotic during shell secretion. Most treatments known to reduce shell thickness decreases the pH of the shell gland fluid usually with a concommitent decrease of the bicarbonate concentration (Mueller et al., 1969). But these two parameters are not sufficient to predict the effect on the shell deposition. For example under spironolactone, an aldosterone antagonist, the pH decreases and the HCO_3^- concentration increases; while both increase with 0.05% of aminophylline in the diet. In the two cases a reduction of the shell thickness is observed (Mueller et al., 1969).

All these observations are always expressed in terms of concentration and not in terms of bicarbonate (HCO_3) secretion rate which would be a better criterion. However, it should be pointed out that the acid-base status of the shell gland fluid can be observed without measurable variations of the systemic acid-base balance, indicating an alteration of the cellular mechanism involved in calcium carbonate secretion.

c) Mechanism of Transfer. A Working Model

The intracellular composition of the shell gland mucosa varies during egg formation in the oviduct (Mongin and Carter, 1977). Just prior to the calcification, the mucosal cell is hydrated. Its total water content increases from 4.8 to 6.7 g of H_2O/g of dry matter. This water accumulation apparently results from a Na transfer into the cell as shown by a constant linear relationship between intracellular Na and water content (Fig. 3).

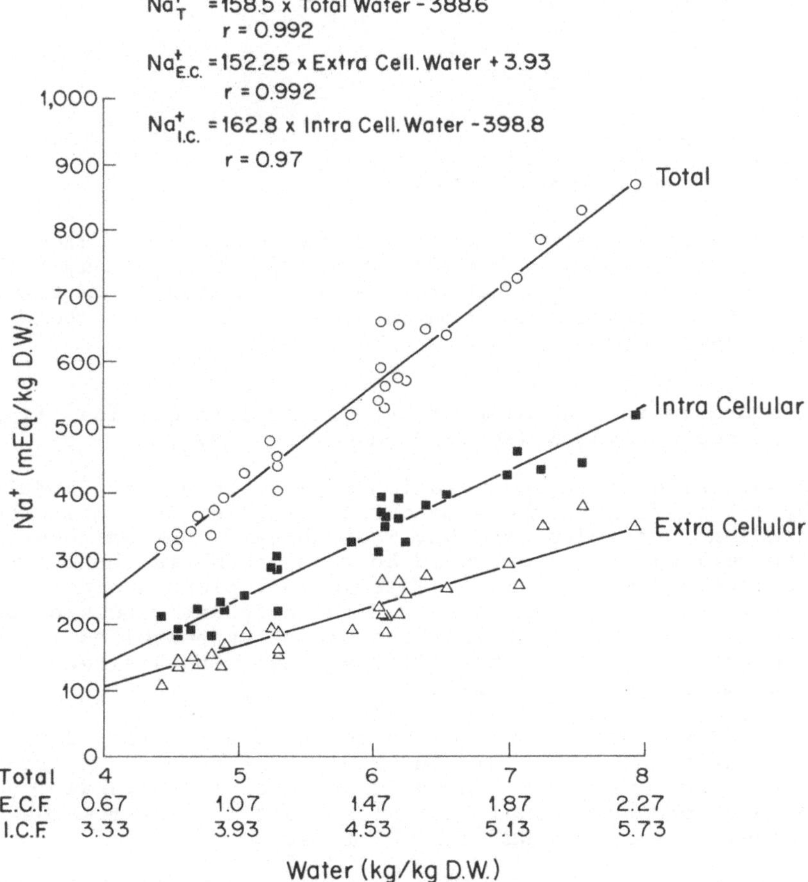

Fig. 3. Relationship between Na and H_2O contents of the shell gland mucosa. Whatever is the physiological stage of the egg formation there is always a straight-line relationship (Mongin and Carter, 1971)

On the other hand, the Cl content of the cell is always proportional to its Na content at a given physiological stage (Fig. 4); but from one physiological stage to the next one it varies. Everything happens as if some of the Cl^- which enters the cell was reabsorbed through a Na-independant pathway.

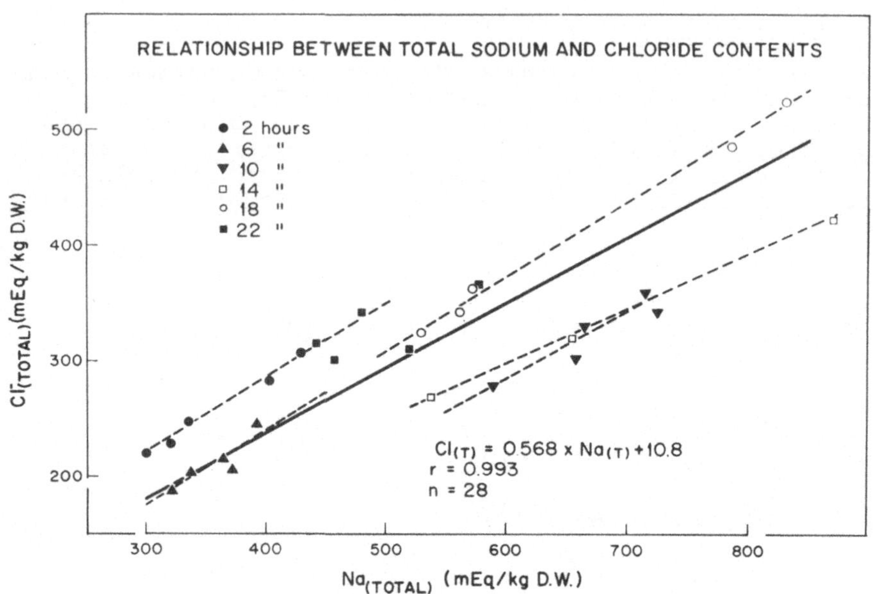

Fig. 4. Relationship between total Na and Cl contents of the uterine mucosa. Each symbol represents a physiological stage of the egg formation expressed in hours after the ovulation. For a given physiological stage, a Na accumulation in the tissue leads to a proportional Cl accumulation; but from one stage to next one, the mucosa loses Cl without Na movements (Mongin and Carter, 1977)

From these results and taking the Simkiss hypothesis into consideration a working model has been proposed (Mongin and Carter, 1977; Fig. 5).

The main process suggested is an accumulation of Na within the glandular cells followed by a bulk flow in the same direction under the influence of the osmotic gradient created by the Na. Then the Cl^- ions which have penetrate into the cell by bulk flow would be reabsorbed via the intercellular channels together with H^+ ions. This would explain the metabolic acidosis observed either at the level of the efferent venous blood of the shell gland (Hodges, 1966) or in the systemic venous blood (Mongin and Lacassagne, 1964, 1966) and all the consequences already described at the pulmonary and renal levels.

Under such an hypothesis, the H^+ ions reabsorbed as counter ions for Cl^- would be of an intracellular origin and would come from the hydration of the CO_2 under the action of the cellular carbonic anhydrase. But this reaction also produces HCO_3^- ions which would diffuse towards the lumen of the shell gland with Na^+ and Cl^- ions.

None of these transfer processes is electrogenetic and all of them occur in the glandular cells.

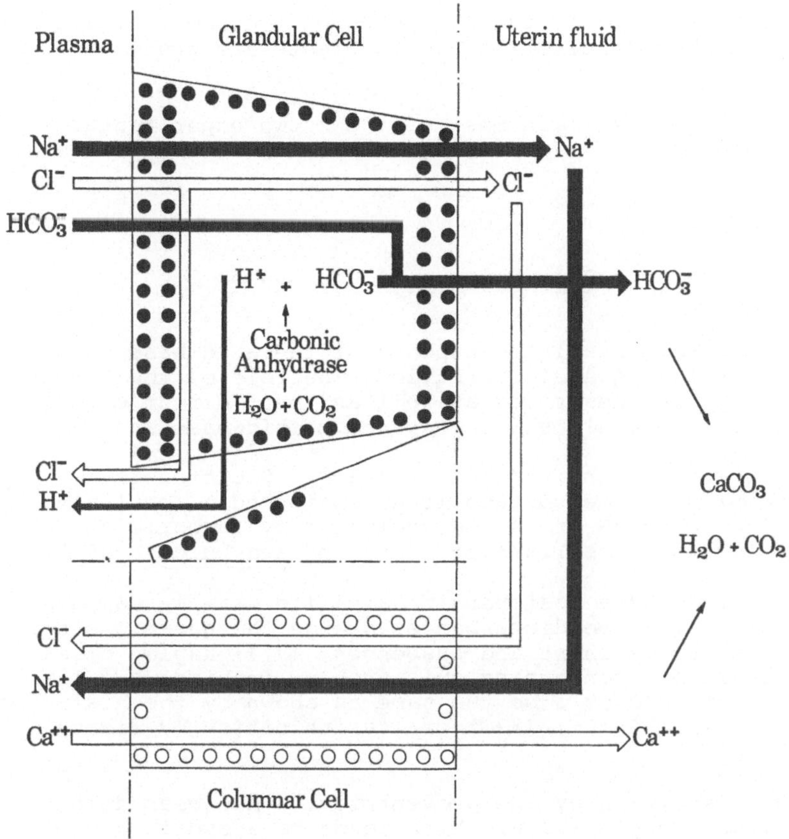

Fig. 5. A working model of the shell gland mucosa. For details see the text (Mongin and Carter, 1977)

Up to that point, the shell gland fluid would present an excess of Na^+ and Cl^- on one side, and an important Ca^{2+} deficit on the other side. A process of exchange diffusion between the luminal sum of (Na^+ + Cl^-) and the extracellular Ca^{2+} ions through the columnar cells would provide the uterine fluid with calcium.

Calcium and bicarbonates ions would coprecipitate as $CaCO_3$ and excess CO_2 would diffuse back to the cells.

This model takes into consideration most of the results already known. It gives a specific role to each of the two cellular types: the glandular cells packed around a micro duct which empty the cellular secretion into the shell gland lumen; and the columnar cells which line the luminal side of the mucosa. The carbonic anhydrase is only located in the glandular type and present in the soluble fraction of the tissue homogenate (Bernstein et al., 1968). Autoradiographies of the mucosa show that ^{45}Ca accumulates into the columnar cells during calcification although it did not have any differential affinity for a cellular type before (Gay and Schraer, 1967). The transmucosal potential is always low and does not vary during egg formation (Leonard, 1969). A Na pump is described in the shell gland tissue and the process of exchange diffusion does not discriminate between Ca and Sr (Droti et al., 1964). Finally, the inhibition of the carbonic anhydrase by diamox, enhances the intracellular Cl content and leads to an increase of the Cl concentration in the uterine fluid at the beginning of the shell calcification (Mongin, unpublished results); those two facts are predicted from Figure 5.

While this model may roughly explain most of the known experimental data, it must be considered as a working hypothesis and not as a definitive conclusion.

4. General Aspects of the Acid-Base Balance. A Conclusion

Several organs are involved in all the aspects of the acid-base balance regulation and hence in the eggshell formation. The role of the lungs and kidneys has already been described as well as their direct effect on the HCO_3 fraction of the shell via the systemic acid-base status.

However, other organs may act indirectly or locally. Skeleton is a wellknown source of buffers under metabolic acidosis and a local alcalinisation of the cellular pH of the osteoclastes by carbonic anhydrase inhibition blocks Ca mobilisation (Siegmund and Dulce, 1960).

Gastric HCl secretion induces a postprandial alcaline wave in mammals and conversely typical acid-base disturbances change the gastric secretion rate in rat and sheep (Richart and Lasbennes, 1971; Kaplan et al., 1972). No data are available concerning this problem in birds but the mechanism of acid secretion should be the same as shown by the histamine stimulation (Angelucci and Linari, 1970) or its inhibition by diamox (Ojha and Ahmed, 1967).

Recently a cyclic HCl secretion by the proventriculus has been demonstrated in laying birds (Mongin, 1976) which leads to a cyclic dissolution of the dietary $CaCO_3$ (Fig. 6). It is tempting to make a connection between this observation and the cyclic $CaCO_3$ secretion in utero. In both, HCO_3 secretion occurs either on the serosal side or on the luminal side of the mucosa and in both, HCl secretion, if not at least Cl^- ion, is the counter part. Furthermore the Ca^{2+} ion is always implicated either to coprecipitate with HCO_3^- or to stimulate the gastric HCl secretion, at least in mammals (Reeder et al., 1971).

So we may wonder whether a common stimulus would not be the trigger of the two secretary processes.

The classical Ca metabolism hormones, PTH and calcitonin, should be involved but the role of calcitonin remains mysterious in birds. Why should this hormone not be involved in gastro-intestinal physiology? What are the possible relationships between calcitonin and gastrin? An answer to these questions may be found in the near future and lead to a concept of closer relationship of Ca metabolism and acid-base balance.

References

Anderson, R.S.: Acid base changes in the excreta of the laying hens. Vet. Rec. 80, 314-315 (1967)

Angelucci, L., Linari, G.: The action of caerulein on gastric secretion of the chicken, Europ. J. Pharmacol. 11, 204-216 (1970)

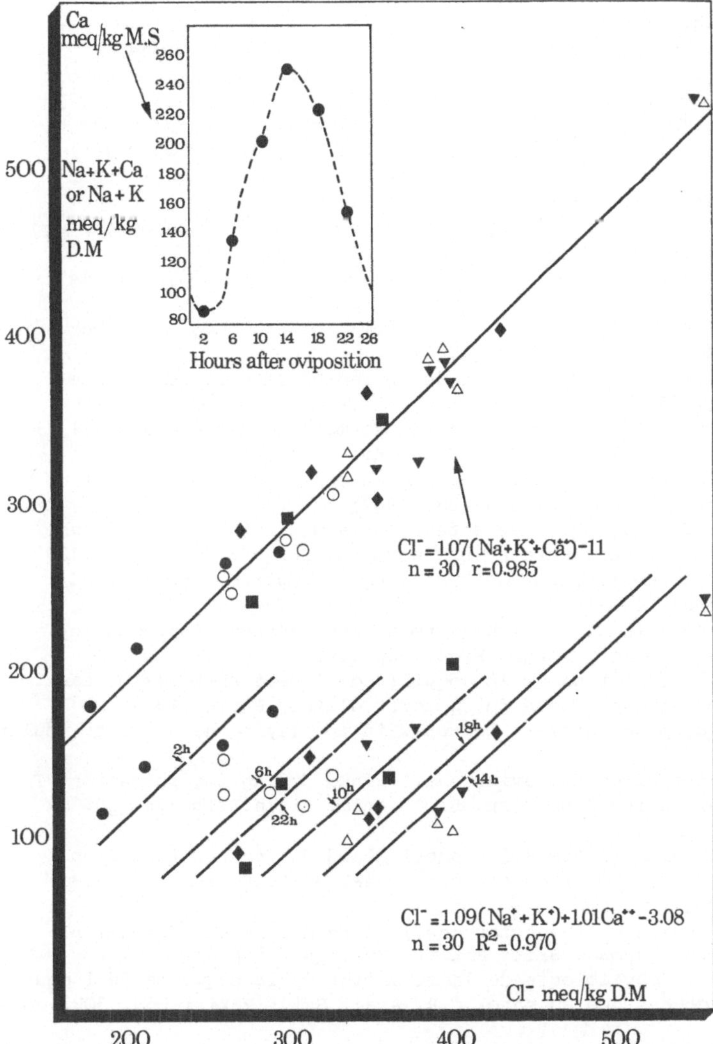

Fig. 6. Mineral composition of the proventriculus and gizzard (liquid phase) content at different physiological stages of the egg formation (Mongin, 1976).
At each physiological stage (Na + K) is well correlated with Cl but when Cl increases from 2 to 14 h the mean of (Na + K) does not change.
When (Na + K + Ca) is considered, only one single straight-line relationship is found with Cl indicating that Ca dissolution is correlated with HCl secretion. Finally this Ca dissolution varies with the physiological stage (enclosed chart)

Bernstein, R.J., Nevalcrinen, T., Schraer, R., Schraer, H.: Intracellular distribution and role of carbonic anhydrase in the avian shell gland mucosa. Biochim. Biophys. Acta 159, 367-376 (1968)

Burckhardt, H.: Der Einfluß einiger Säuren auf den Ca-P und N Stoffwechsel und das Säuren-Basegleichgewicht bei Hühnern. Rep. Versuchsanstalten Liebefed Bern. Nutr. Abstr. Rev. 3, 1094 (1933)

Calder, W.A., Schmidt-Nielsen, K.: Evaporative cooling and resporatory alcalosis in the pigeon. Proc. Nat. Acad. Sci. US 55, 730-756 (1966)

Common, R.H.: Observations of the mineral metabolism of pullets. V. Acide base equilibrium and reproductive activity. J. Agr. Sci. 31, 281-294 (1941)

Drori, D., Volcani, R., Feige, Y., Shalmon, E., Passy, M.: Factors affecting the Ca^{45} and Sr^{89} transfer in the laying hen. Poultry Sci. 43, 486-491 (1964)

El Jack, M.H., Lake, P.L.: The content of the principal inorganic ions and carbon dioxide in uterin fluids of domestic fowl. J. Reprod. Fert. 13, 127-132 (1967)

Franck, F.R., Burger, R.E.: The effect of carbon dioxide inhalation and sodium bicarbonate ingestion on egg shell deposition. Poult. Sci. 44, 1604-1606 (1965)

Gay, C.V., Schraer, H.: Autoradiographic localization of calcium in the mucosal cells of the avian oviduct. Calc. Tiss. Res. 7, 201-211 (1971)

Gutowska, M.S., Mitchell, C.A.: Carbonic anhydrase in the calcification of the egg shell. Poultry Sci. 24, 159-168 (1945)

Hall, K.N., Helbacka, N.V.: Improving albumen quality. Poultry Sci. 38, 111-114 (1959)

Helbacka, N.V., Casterline, J.L., Smith, C.J.: The effect of high CO_2 atmosphere on the laying hen. Poultry Sci. 42, 1082-1084 (1963)

Heller, V.G., Pursell, L.: Repartition of chloride in red blood cells and plasma. J. Biol. Chem. 118, 549 (1937)

Hodges, R.D.: Changes in blood pO_2 and pH during the formation of the egg shell in laying hens. Proc. 13th World's Poultry Sci. Cong. Kiev, 314-319 (1966)

Hunt, J.R., Aitken, J.R.: The effect of ammonium and chloride ions in the diet of hens on shell quality. Poultry Sci. 41, 434-438 (1962)

Kaplan, E.L., Peskin, G.W., Jaffe, B.M.: The effects of acute metabolic acid-base changes on secretion of gastrin and parathyroid hormone. Surgery 72, 53-59 (1972)

Leonard, E.: The transmural potentials in the functional oviduct of the hen. J. Physiol. (London) 203, 83-84 (1969)

Mongin, P.: Keeping laying hens in CO_2 enriched atmosphere. Action on production and egg shell quality. Eur. Poultry Congr. Proc. 565 (1968)

Mongin, P.: Physiologie de la formation de la coquille de l'oeuf chez les oiseaux. Rôle de l'équilibre acido-basique. Thèse de doctorat d'Etat, Paris, 1969

Mongin, P.: Composition of crop and gizzard contents in the laying hens. Brit. Poultry Sci. 17, 499-507 (1976)

Mongin, P., Carter, N.W.: Studies of the avian shell gland during egg formation. I. The extracellular space of laying hens. Ann. Biol. Anim. Bioch. Biophys. 16, 649-654 (1976)

Mongin, P., Carter, N.W.: Studies of the avian shell gland during egg formation. II. Aqueous and electrolytic composition of the mucosa. Brit. Poultry Sci. 18, 337-349 (1977)

Mongin, P., Carter, N.W.: Studies of the avian shell gland during egg formation. III. Intracellular pH of the mucosa. Brit. Poultry Sci. 18, in press (1977)

Mongin, P., Lacassagne, L.: 2. Physiologie de la formation de la coquille de l'oeuf de poule et équilibre acido-basique du sang. C.R. Acad. Sci. (Paris) 258, 3093-3094 (1964)

Mongin, P., Lacassagne, L.: 3. Physiologie de la formation de la coquille de l'oeuf de poule et ventilation pulmonaire. C.R. Acad. Sci. (Paris) 261, 4228-4230 (1965)

Mongin, P., Lacassagne, L.: 4. Equilibre acido-basique du sang et physiologie de la formation de la coquille de l'oeuf. Ann. Biol. Anim. Biochim. Biophys. 6, 93-199 (1966)

Mongin, P., Lacassagne, L.: 5. Rythmes respiratoire et physiologie de la formation de la coquille de l'oeuf. Ann. Biol. Anim. Biochim. Biophys. 6, 101-111 (1966)

Mongin, P., Sauveur, B.: Composition du liquide utérin et de l'albumen durant le séjour de l'oeuf dans l'utérus chez la poule domestique. C.R. Acad. Sci. (Paris) 270, 1715-1718 (1970)

Mongin, P., Sauveur, B., Amin, M.: Alcalose hypochlorémique chez la poule pondeuse. Ann. Biol. Anim. Biochim. Biophys. 12, 57-67 (1972)

Mueller, W.J., Brubaker, R.L., Caplan, M.D.: Egg shell formation and bone respiration in laying hens. Federation Proc. 28, 1851-1856 (1969)

Ojha, K.N., Ahmed, Q.: The inhibitory effect on some thiazide diuretics and acetazolamide on the histamine induced gastric secretory response in pigeons. Ind. J. Physiol. Pharmacol. 11, 53-61 (1967)

Reeder, D.D., Conlee, J.L., Thompson, J.C.: Calcium carbonate antiacid and serum gastrin concentration in duodenal ulcer. Surgical Forum 22, 308-320 (1971)

Reichcut, E., Lasbernes, F.: Influence de l'hypercapnie sur la secrêtion gastrique chez le rat. Rev. Eur. Etudes Clin. Biol. 16, 818-821 (1971)

Sauveur, B.: Acidoses métaboliques expérimentales chez la poule pondeuse. I. Action sur l'équilibre acido-basique du sang et l'excrétion rénale des électrolytes. Ann. Biol. Anim. Biochim. Biophys. 91, 379-391 (1969)

Sauveur, B., Mongin, P.: Etude comparative du fluide utérin et de l'albumen de l'oeuf in utero chez la poule. Ann. Biol. Anim., Biochim. Biophys. 11, 213-224 (1971)

Sauveur, B., Mongin, P.: Contribution à l'étude de l'hypercapnie aiguë et chronique chez la poule. Comp. Biochem. Physiol. 41A, 869-875 (1972)

Siegmund, P., Dulce, H.J.: Zur Biochemie der Knochenauflösung. I. Einfluß des Carboanhydratase-inhibitors (Diamox) auf den Calciumstoffwechsel von Legehennen. Hoppe-Seyler's Z. Physiol. Chem. 320, 149-159 (1960)

Simkiss, K.: Calcium metabolism and avian reproduction. Biol. Rev. 36, 321-367 (1961)

Simkiss, K.: Intracellular pH during calcification. Biochem. J. 111, 647-652 (1969)

Solum, A.S., Shuster, M.: Without title. Tiernährung 6, 515 (1934)

Development of the Avian Respiratory and Circulatory Systems

H.-R. Duncker

Summary

The chorioallantoic membrane, with blood capillaries incorporated into the chorionic epithelium adjoining the inner shell membrane, is the first respiratory system of the avian embryo. The internal structures of the lung and airsacs develop to their final number and topographical positions two or three days before hatching, at which time they are aerated. Although gas exchange is further carried out by the chorio-allantois, the lung increases its potential gas exchanging ability by the sprouting of the initial layer of air capillaries around the para-bronchial blood capillaries. Then the embryo hatches.

Embryos in sauropsidian eggs live in a closed system with respect to most substances. Many reptilian eggs take up water from the surroundings, but the calcareous shelled eggs of some reptiles and of all birds are independent in relation to water uptake; their water loss is regulated by the eggshell porosity. However, all sauropsidian eggs have to take up O_2 and give off CO_2. The small embryos in the first days of incubation are sufficiently supplied by diffusion from the shell pores through the albumen to the embryo and the circulatory system of the yolk sac. But the increasing O_2 demand forces then the rapid development of a vascular system for the embryonic gas exchange, which is taken over by the chorioallantoic membrane (Rahn et al., 1974).

The chorionic epithelium, derived from the extraembryonic ectoderm and developed under the inner shell membrane, rests directly on the extraembryonic mesenchyme. The allantoic outgrowth from the caudal intestine, also covered by extraembryonic mesenchyme, contacts the chorion in chick embryos at the fifth day of incubation. Both layers of mesenchyme fuse with each other forming the chorioallantoic membrane, which contains the allantoic vascular system supplied by the two allantoic arteries and the allantoic vein. The capillary net grows very quickly simultaneously with the fusion of the chorion and the allantois from the 5th to the 11th day. At day 11 the chorioallantois lies under most of the inner eggshell membrane (Romanoff, 1960) (Fig. 1). Further increase in the expansion of the chorioallantoic membrane proceeds slowly up to the 20th day (Fig. 2). However, the location of the capillaries is of major importance in the chorioallantoic membrane. Up to the 10th day, the capillaries lie under the two-layered chorionic epithelium, and then they begin to be incorporated into the epithelium. The capillaries are always covered by a basal lamina, also when they are totally surrounded by chorionic epithelial cells. This shifting of the capillaries into the epithelium is complete at day 12 in chicken embryos, decreasing the thickness of the tissue barrier between the inner shell

Zentrum für Anatomie und Cytobiologie der Justus-Liebig-Universität Giessen, Aulweg 123, D-6300 Lahn-Giessen, Federal Republic of Germany.

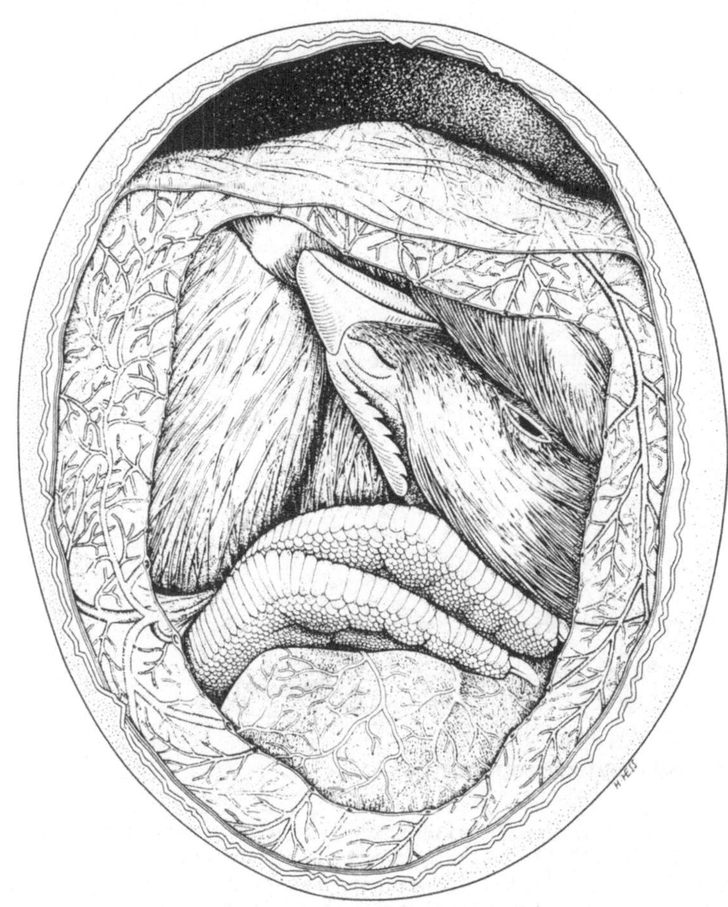

Fig. 1. Domestic fowl. Chick in the egg just before ingestion of ammiotic fluid and aeration of the amnion. The gas exchange is carried out by the chorioallantoic membrane, which underlies the entire internal shell membrane. Below in the egg, the yolk sac

membrane and the capillary lumen from 4.3 μm at day 8 to less than 0.8 μm after day 12 (Fig. 2). The capillaries at this stage are only separated from the inner shell membrane by very thin processes of the chorionic epithelial cells. However, the volume of the intraepithelial capillaries continues to increase up to the 19th day, then decreases slightly until hatching. The development of the chorioallantoic membrane in chick embryos was reviewed and qualitatively and quantitatively reinvestigated by Fitze-Gschwind (1973). The decrease in capillary volume proceeds parallel to the increasing reactivity of the chorioallantoic vascular system. From day 19 in chicken, the time of aeration of the amnion, the vessels react to irritation by constriction, as can be observed during the opening of the egg.

The chorioallantoic vascular system is totally responsible for the gas exchange function of the embryo until day 19 in the chicken embryo (Visschedijk, 1968; Rahn et al., 1974). Beginning at day 4 in chick embryos and at day 5 in duck embryos, the lungs develop out of the lung bud stage. These lung buds grow into the craniodorsal pleuro-peritoneal

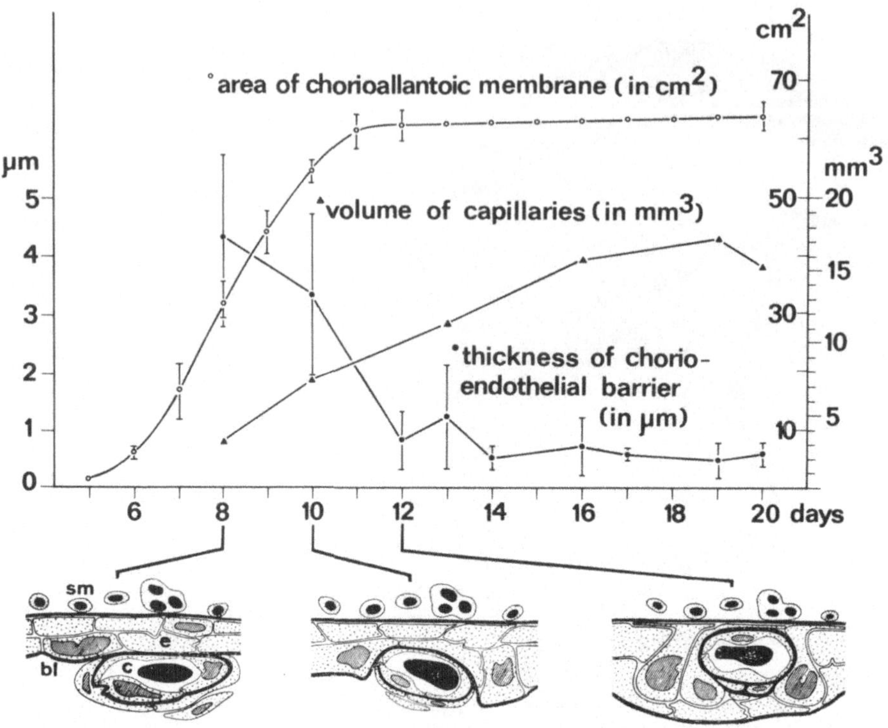

position of the capillaries in the chorionic epithelium

Fig. 2. Domestic fowl. Development of the chorioallantoic membrane, compiled after figures and data from Fitze-Gschwind (1973). Allantois and chorionic epithelium begin to fuse at day 5 of incubation. Capillaries (*c*) are incorporated into the chorionic epithelium (*e*) after day 8, their endothelial cells and pericytes surrounded by the basal lamina (*bl*). They reach their final position nearest the inner shell membrane (*sm*) at day 12

cavity (Fig. 3) and are only connected dorsomedially to the mediastinum and more ventrally to the dorsal pericardium. At this stage the liver and the kidney are relatively well developed, in accordance with the functional needs of the embryo, whereas the lungs develop slowly. Firstly, three series of secondary bronchi sprout from the primary bronchus, soon reaching their final number and position in the lung (Fig. 4). During further growth and ramification of the secondary bronchi the first parabronchi originate as epithelial buds, elongating and obtaining a tubular lumen (Fig. 5). Both the paleopulmonic and the neopulmonic parabronchi originate at the same time in this way. The parabronchi from the medioventrobronchi and those from the mediodorso- and the lateroventrobronchi grow toward one another and anastomose in the medial plane, always one with two or more from the other side, making up the paleopulmonic parabronchial system (Fig. 6). At this time the bronchial connections to the airsacs begin to sprout from the primary bronchus, and the medioventro- and the lateroventrobronchi, as well as from the lateral lung margin. The sprouts at the lung margin form more or less defined saccobronchi of varying size for the paleopulmo lateral to the lung hilus and for the neopulmo at the posterior airsac ostia. This development of the bronchial system and the outgrowth of the airsac

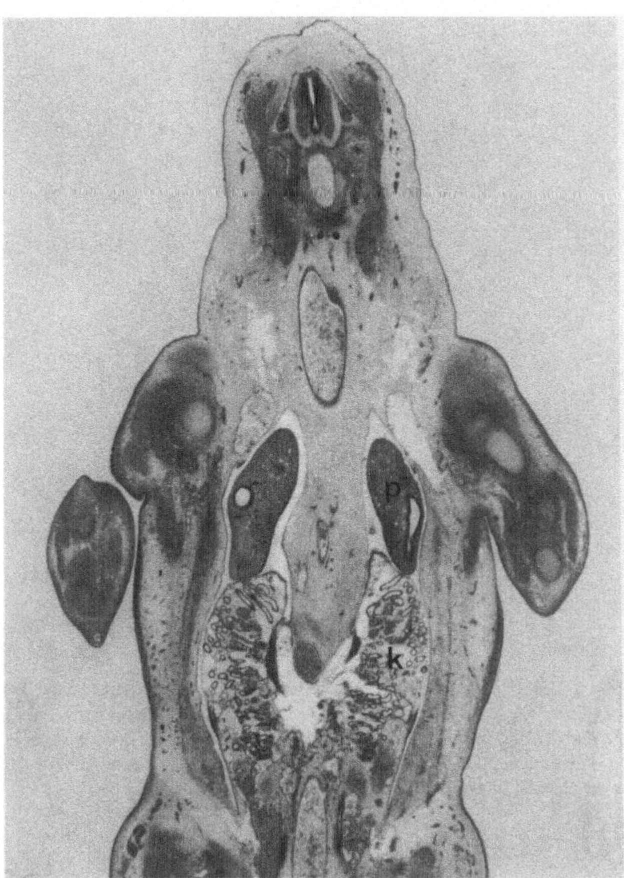

Fig. 3. Khaki-Campbell duck. Embryo 6 days old. Horizontal section through neck and dorsal thoraco-abdominal region with wing buds. Lung buds (*p*) in longitudinal section in the dorsal pleuroperitoneal cavity above the kidneys (*k*). Susa, paraffin, 8 µm, hematoxylin-eosin, ×12.5

connections has been described by Juillet (1912) and by Locy and Larsell (1916a, b); the airsac development in domestic fowl was investigated by Selenka (1866), and in domestic duck by Delphia (1958).

The development of the postpulmonary septum has already taken place, before the airsacs begin to sprout, at day 8 in the chick embryo and day 10 in the duck embryo. The postpulmonary septum begins as an extension of the transverse septum, which originates out of the caudal wall of the pericardium and the yolk sac stalk, and into which the early liver anlage invades. This transverse septum extends mediodorsally as the mediastinal wall or dorsal mesentery and is closed dorsolaterally by the cranial nephric fold on each side. With the development and closure of the postpulmonary septum the lung grows together with the septum, beginning at the adhesion to the dorsal pericardium. At the time at which the airsacs begin to emerge the lung is attached along its total ventral surface to the postpulmonary septum; however, a pleural space still exists at the medial and dorsolateral lung surface. All airsacs except the abdominal invade the postpulmonary septum, more or less simultaneously, splitting the postpulmonary septum into the horizontal septum, which lies directly beneath the lung, and the oblique septum between the airsacs and the intestines (Poole, 1909;

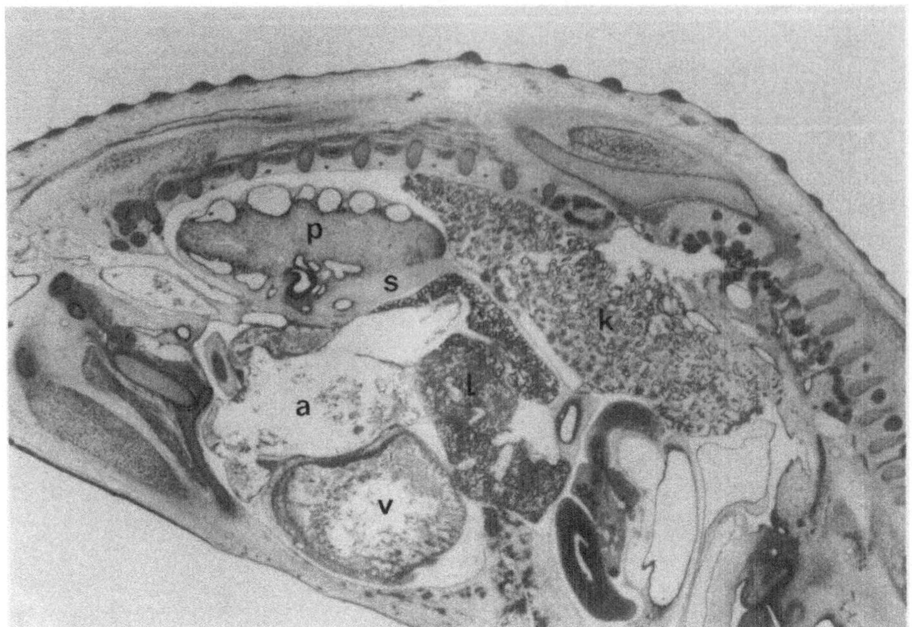

Fig. 4. Khaki-Campbell duck. Embryo 10 days old. Paramedian sagittal section through the thoraco-abdominal cavity. Lung (*p*) with its secondary bronchi at the beginning of parabronchial sprouting. It is located in the pleural cavity, and adheres ventrally to the postpulmonary septum (*s*). Beneath it, the heart with atrium (*a*) and ventricle (*v*), behind it, liver (*l*), and dorsal kidney (*k*). Susa, paraffin, 8 µm, hematoxylin-eosin, ×10

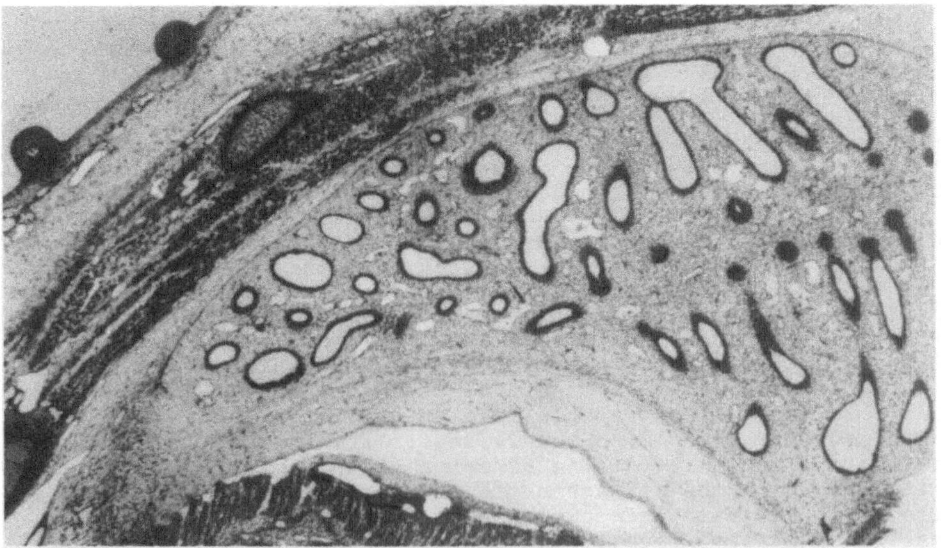

Fig. 5. Domestic fowl. Embryo 10 days old. Cross section through the lung, adhering to the thoracic wall above and to the horizontal septum below. Parabronchi are spouting out from the secondary bronchi at the dorsal and ventral lung surface. Glutaraldehyde, epon, 1 µm, toluidine blue-pyronine, ×36

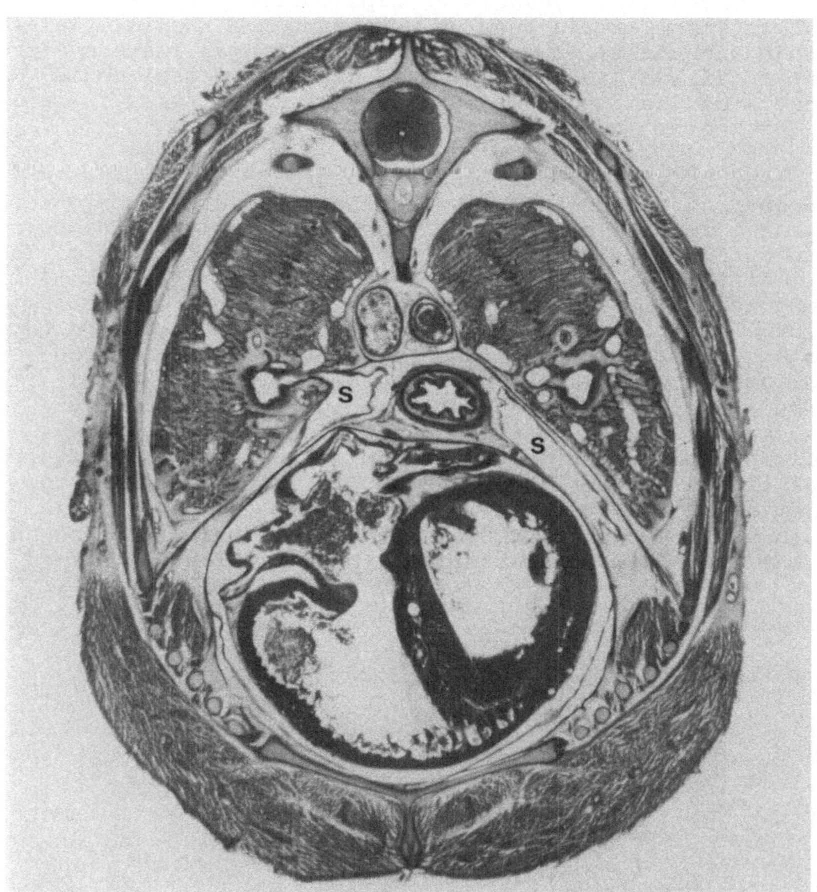

Fig. 6. Khaki-Campbell duck. Embryo 14 days old. Total cross section through the anterior thoracic region. The lungs lie in the pleural cavity, ventrally attached to the postpulmonary septum, into which the anterior thoracic airsacs (*s*) are invading, splitting it into the horizontal septum beneath the lung and the oblique septum, which adheres in this region to the pericardium. In the lung numerous parabronchi develop from the secondary bronchi and join in the medial anastomosing plane. Susa, paraffin, 100 µm, Goldner trichrome, ×10.5

Fig. 6). Only the abdominal airsac passes through the postpulmonary septum. It expands in the abdominal peritoneal cavity, ventral to the kidneys and is attached to the dorsal and dorsolateral abdominal body wall, also covering the kidneys. Thus the typical series of airsacs arrives at its final topographical relationship. Pneumatization of the skeleton occurs much later in the adult. During the outgrowth of the airsacs, the lung becomes attached to the thoracic wall, first at the medial side, and then at the laterodorsal surface. A few days before hatching, only small islets of the pleural space on the dorsolateral side remain, and may persist into adult life.

In spite of the nearly complete adhesion of the lung with the walls of the pleural cavity, the pulmonary circulation remains totally separated from the systemic circulation, with the only exception given by the bronchial circulation (Abdalla and King, 1976). Despite examination of large series of histological preparations of the process of lung adhesion, I never observed blood vessels or capillaries penetrating the former pleural covering of the lung and entering the thoracic wall. In some hundred injection casts in which the pulmonary capillary system was nearly completely filled, no one capillary or other vessel could be found that penetrated the lung surface at any location. This is in contradiction to the recent observations of Burger and Estavillo (1977).

During the development of the airsacs, the parabronchi also undergo a further development. At the beginning of the last third of embryonic development, they have reached their final number and position. Number and manner of interconnection of the parabronchi are species specific. The parabronchi are at first long tubes consisting of a cuboidal epithelium, surrounded by a mesenchymal tissue rich in cells. The parabronchial epithelium then forms cup-like excavations into the mesenchyme, the later atria. Smooth muscle rings or nets develop around the parabronchial lumen bordering the atrial openings. The beginning of the outgrowth of infundibula can be also observed at this time, especially in chick and duck embryos. Although a few days before hatching, the vascular system of the lung is developed in the following basic pattern. In the mesenchymal tissue between the parabronchi, arterioles in large number can be observed. Small venules lie directly at the base of the growing atria. Both venules and arterioles retain this position throughout development, which is their characteristic position in the para-

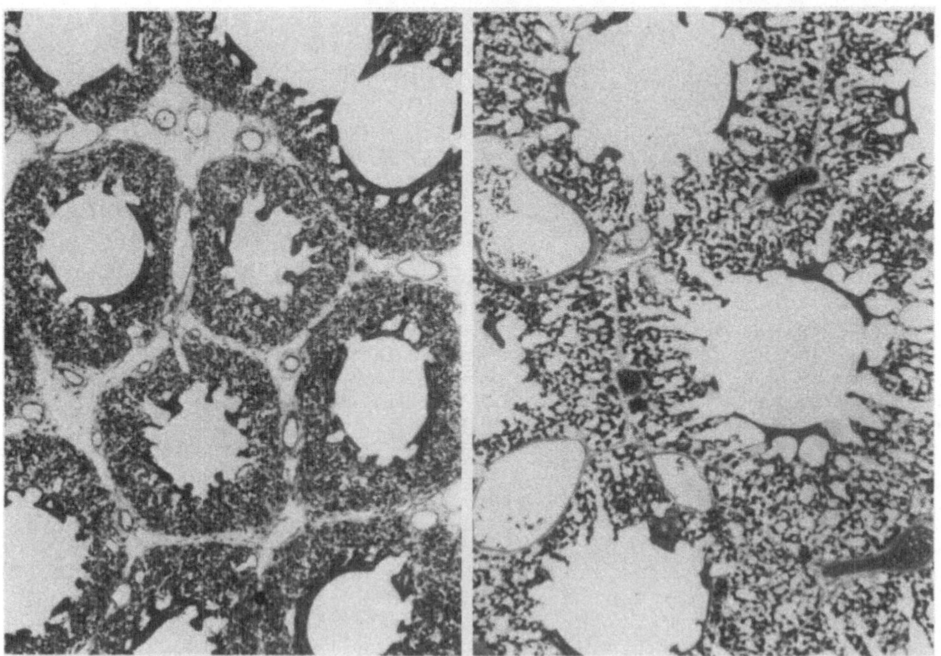

Fig. 7. Domestic fowl, lung, parabronchi in cross section. *Left*, embryo 19 days old, just before aeration of the amnion. The parabronchial mantle contains numerous capillaries. The mesenchyme between them is invaded by the growing infundibula. Between the parabronchi, the supplying vessels. *Right*, chick just after hatching. There is a well-developed network of air capillaries around blood capillaries in the parabronchial wall. This meshwork completely embeds the vessels between the parabronchi. Glutaraldehyde, epon, 1 µm, toluidine blue-pyronine, both ×78

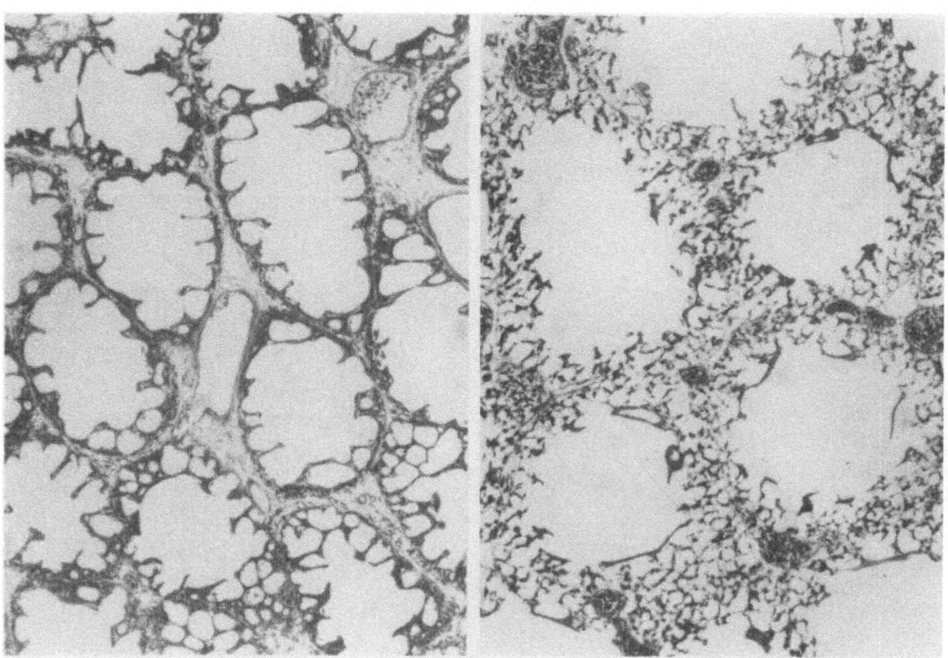

Fig. 8. Domestic pigeon, lung, parabronchi in cross section. *Left*, embryo 16 days old, just before aeration of the amnion. The atria of the parabronchi are well developed, surrounded only by a thin mesenchyme with few capillaries. *Right*, nestling just after hatching. A primary meshwork of blood capillaries and air capillaries in the parabronchial mantle is developed. Glutaraldehyde, epon, 1 μm, toluidine blue-pyronine, both ×78

bronchi of the adult. At this stage the capillary connections between arterioles and venules are rapidly increasing. Even at this time, two or three days before hatching, a major difference exists in the parabronchial structure between precocial and altricial birds (Figs. 7 and 8). In altricial birds the atria of the parabronchi are less developed and are surrounded by a relatively thin mesenchymal sheet, whereas precocial birds possess large atria and a thick mesenchymal mantle (Figs. 10 and 11).

Also at this developmental stage, two or three days before hatching, the embryo ingests the amnion fluid and the remainder of the albumen. This so-called "breakfast of the chicken" is necessary for osmotic reasons. Whether or not the air-space membrane is perforated, the amniotic cavity is aerated from this time on. Air may also penetrate through the air-filled inner shell membrane. Respiratory movements which were previously seldom are now regular and serve to aerate the bronchial system, the airsacs, and the parabronchi. This onset of the lung ventilation initiates a rapid development of the parabronchial mantle. The few capillaries between arterioles and venules increase enormously in number and in length in the last two or three days before hatching, to form a dense capillary layer. At the same time the infundibula sprout from the atria and give off numerous air capillaries, which surround the developing blood capillaries, forming a three-dimensional network of gas exchange tissue (Figs. 7, 8, 10, and 11). In this way the lung becomes functional, increasingly taking over the gas

Fig. 9. Domestic fowl, lungs, fixed in situ with Bouin's fluid. All lungs are shown at natural size to demonstrate their absolute growth, ×1.0

exchange, while the gas exchange by the chorioallantois simultaneously decreases (Rahn et al., 1974). At the end of this process the chick can hatch.

This overlap of gas exchange by the chorioallantois and the ventilation of the lung with air is a unique avian developmental characteristic. It results from the fact that the firmly attached rigid lung located in the pulmonary cavity cannot be expanded or inflated even at the moment of hatching. This structural peculiarity of the avian lung was the phylogenetical basis for the development of air capillaries which, because of their small diameter (3-10 μm) and their resulting high surface tension, cannot be inflated out of a collapsed state. They can only sprout as open tubes from the infundibula. The sprouting air capillaries may be air or fluid filled. This fluid can be taken up by the blood capillaries, when the surface tension is lowered by the fully developed surfactant film in final embryological stages (Petrick, 1967; Petrick and Riedel, 1968). This process takes place in the embryonic development of all avian species. For the air capillary development their highly evolved lungs require an ontogeny in a calcareous egg, the only situation in which lung ventilation with air and gas exchange by a chorioallantois can occur simultaneously. Therefore birds cannot develop viviparously (Duncker, 1971).

Fig. 10. Domestic fowl, lungs, parabronchi in cross section. All scanning electron micrographs are taken at the same magnification to demonstrate the absolute growth of the parabronchi. Glutaraldehyde instillation, critical point method, all ×38 ▶

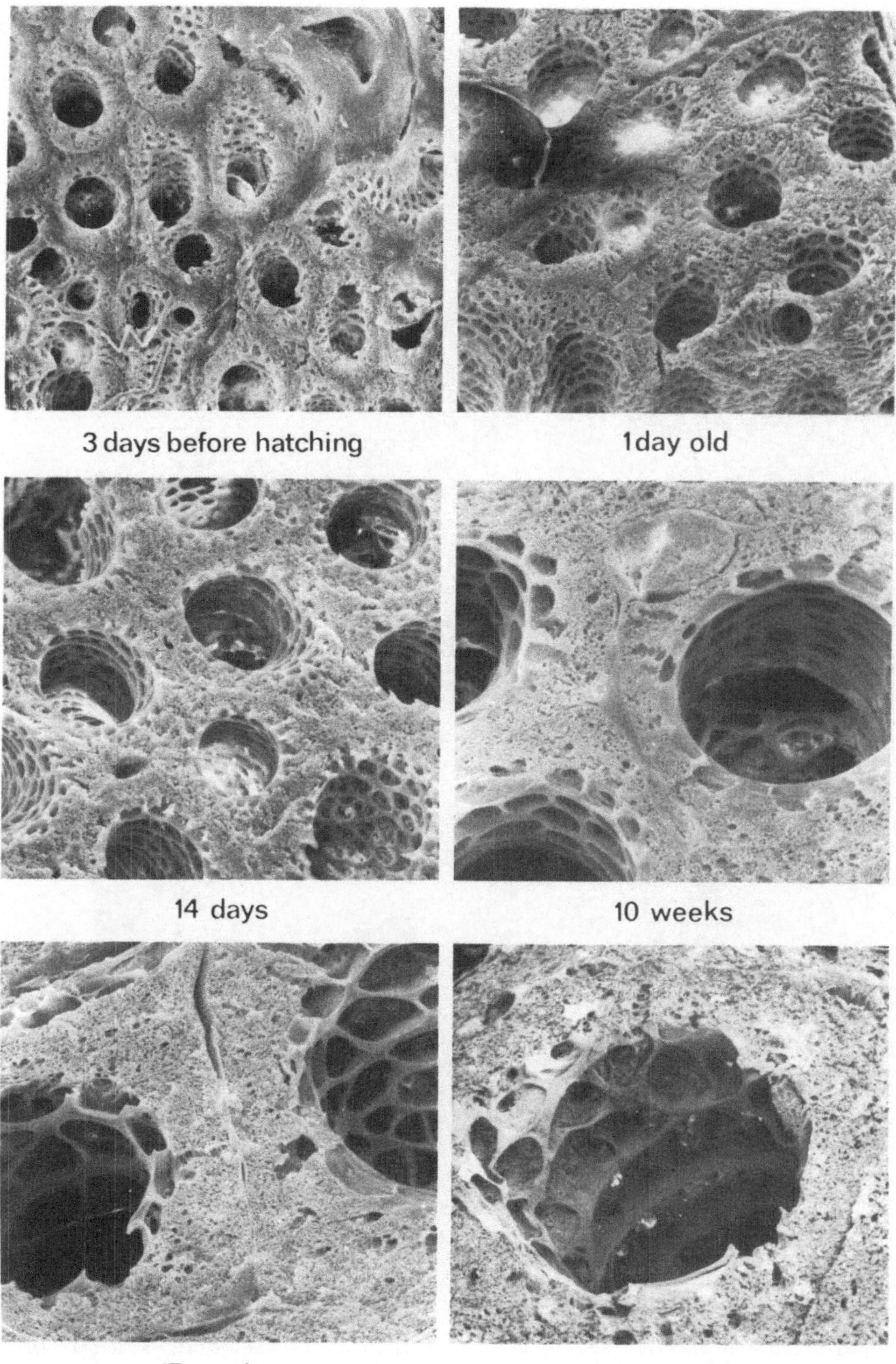
3 days before hatching
1day old
14 days
10 weeks
17 weeks
1 year old

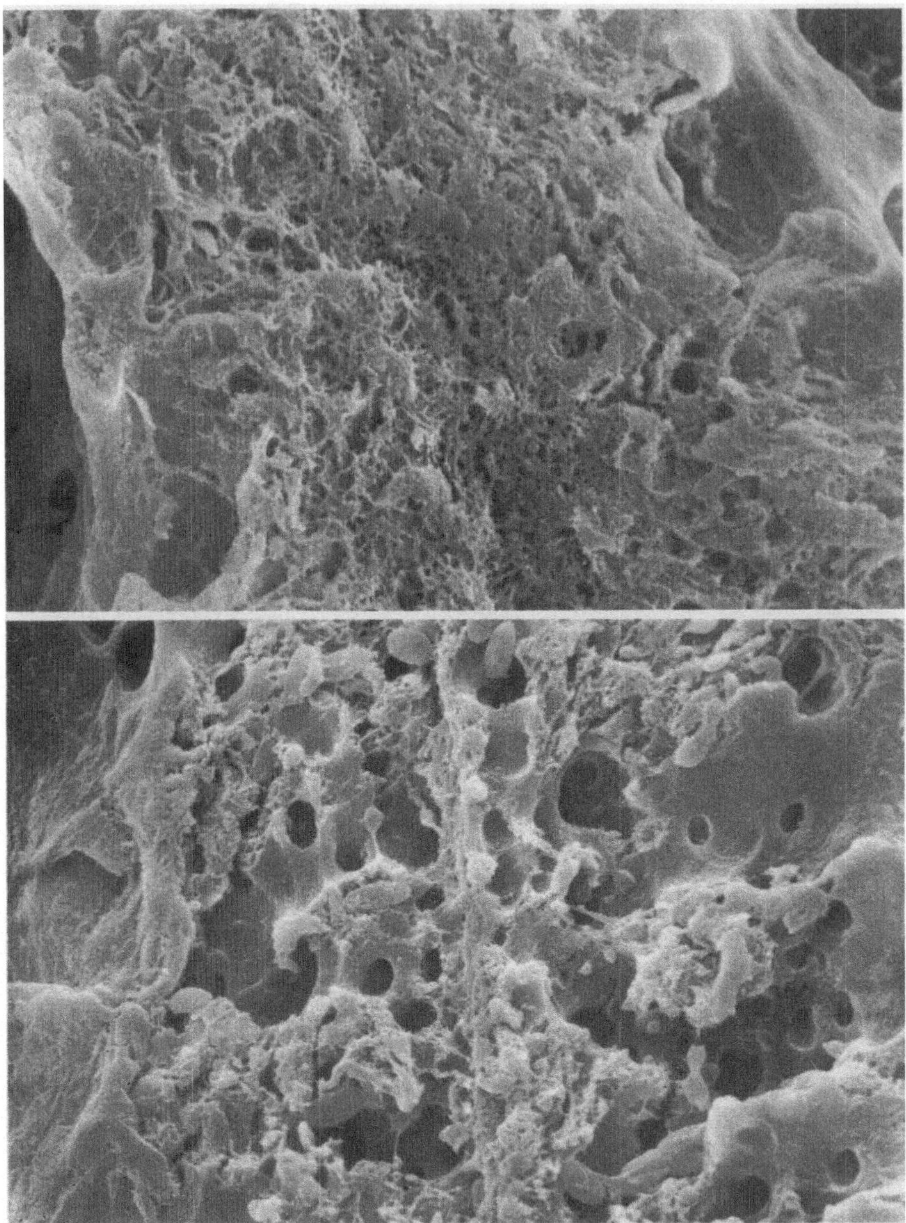

Fig. 11. Domestic fowl, lungs, higher magnification of the parabronchial wall from the two micrographs in the top row of Figure 10. *Top*, embryo lung 3 days before hatching. The parabronchial mantle consists of mesenchyme with few capillaries. The lumina and atria of two adjacent parabronchi are seen at the *lower left* and *upper right*. *Bottom*, chick lung, just after hatching. The parabronchial mantle is composed of the infundibula, out of which the air capillaries have grown around the well-developed blood capillaries, making up the respiratory tissue. Glutaraldehyde instillation, critical point method, both ×680

Besides this general principle in avian lung development, there are, however, some differences in lung structure at the time of hatching between precocial and altricial birds. The parabronchial structure with well-developed atria and a mantle of blood and air capillaries several layers thick is only found in precocial birds such as fowls and ducks (Figs. 7, 10, and 11). They have a relatively large exchange surface at the time of hatching, in relation to their motor activity and behaviour. However, the altricial birds demonstrate in a smaller lung, smaller parabronchi with flat atria, and a thin layer of blood and air capillaries, which is only one or two capillary diameters thick (Fig. 8). The growth of the parabronchi to the adult dimensions occurs in the nestling period, and is completed when the bird is able to fly. In precocial birds which have a thick blood and air capillary layer at hatching, an intensive growth of the parabronchi also occurs with general body growth after hatching. Both the luminal and atrial diameter of the parabronchi and the thickness of the blood capillary/air capillary network increase, but to a relatively lesser extent than in altricial birds.

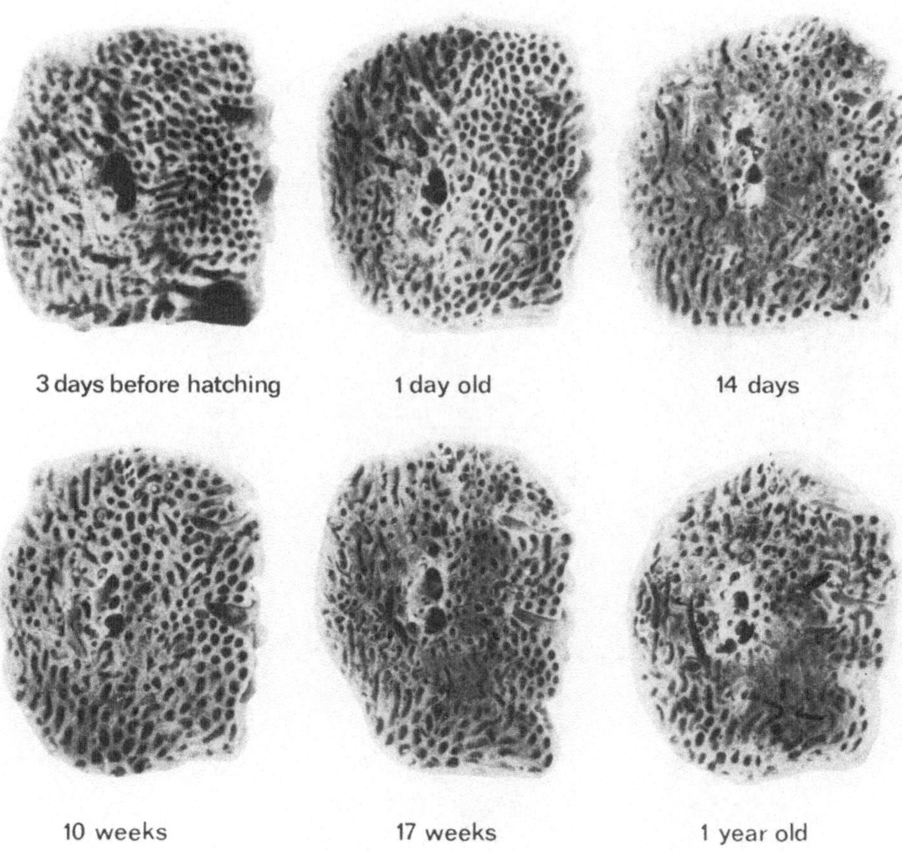

Fig. 12. Domestic fowl, the same lungs as in Figure 9, sectioned parallel to the anastomosing plane. Palaeopulmonic parabronchi in cross section showing their total number. Photographs of the smaller lungs are magnified to give them the apparent dimensions of the adult lung. The total number of parabronchi, 243 ± 8, does not further increase from 3 days before hatching the adulthood; only a proportional growth of lumen diameter and wall thickness takes place

The air capillaries developed in the final embryological phase do not differ in any respect from those of hatched or adult animals of the same species. Their diameter and the dimension of their very thin epithelial covering including its surfactant remains unchanged. Also the very close relation to the endothelium of the blood capillaries does not change. In avian species with high activity rates and O_2 consumption such as in pigeons, song birds, or hummingbirds, this barrier is the thinnest found in vertebrates, less than half of the minimum mammalian barrier thickness. This extreme thinness is only possible in rigid air capillaries in constant volume lungs. This minimal barrier thickness is present from the beginning of the air capillary formation. Only at the very periphery of growing parabronchi can some air capillaries be observed in a developing stage in which their thin wall does not contact the blood capillary endothelium, but contains a very electron-lucent mesenchymal interstitial substance. After growth has been completed, these interstitial spaces are lacking.

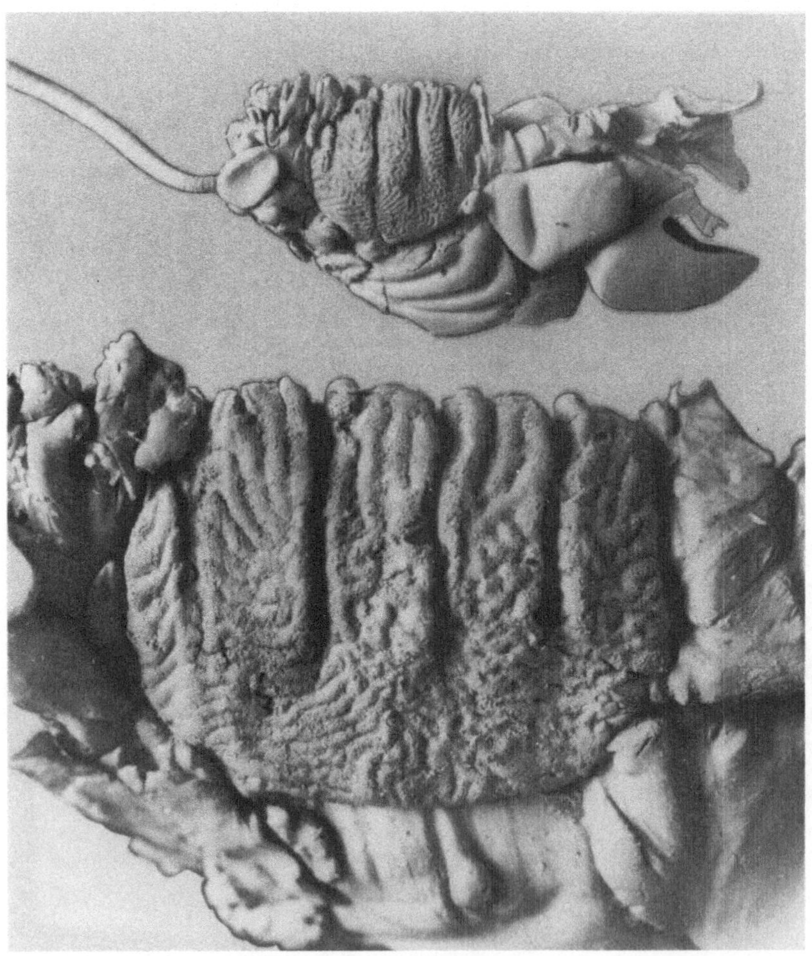

Fig. 13. Domestic fowl, silicone rubber casts of the lung airsac system. The number and position of branches of the mediodorsobronchi and of the neopulmonal parabronchi, seen at the lateral lung surface, does not change during growth. *Top*, chick 1 day old, total cast of the lungs and airsacs. *Bottom*, adult fowl, 1 year old, lung region of a total cast, both ×1.8

In all birds after hatching only a proportional growth of all existing structures of the lung occurs (Figs. 9, 10, and 12). Apart from new blood and air capillaries, no new structural elements are developed. This means that in precocial, as well as in altricial birds, the number, branching pattern, and topographical position of the secondary bronchi (Fig. 13) as well as the number, position, and interconnections of the parabronchi (Fig. 12) are fixed already at the time the embryo aerates the amnion (Duncker, 1972). This specialized embryological development of the fixed rigid avian lung is quite different from the development of the mammalian lung in which the terminal airways undergo intensive growth and important structural changes after birth (Reid, 1967; Boyden, 1975). The development of the rigid avian lung is not only an anatomical preadaptation for the evolutionary development of air capillaries, but this development also insures the special arrangement of the bronchial system with its sites and angles of origin of the secondary bronchi over all growth phases from the late embryonic period up to adulthood. This makes possible the specific aerodynamically directed air flow in the avian lung, namely the unidirectional flow in the paleopulmo.

References

Abdalla, M.A., King, A.S.: The functional anatomy of the bronchial circulation of the domestic fowl. J. Anat. 121, 537-550 (1976)

Boyden, E.A.: Development of the human lung. In: Practice of Pediatrics, Vol. IV, Ch. 64, 1-17. Hagerstown: Harper and Row, 1975

Burger, R.E., Estavillo, J.A.: Pulmonary circulation - vertebral venous interconnections in the chicken. Anat. Rec. 188, 39-44 (1977)

Delphia, J.M.: The origin of the air sacs in the White Pekin-Duck, *Anas platyrhynchos* L. Proc. North Dakota Acad. Sci. 12, 86-92 (1958)

Duncker, H.-R.: The lung air sac system of birds. Ergebn. Anat. Entwickl.-Gesch. 45, Heft 6, 1-171 (1971)

Duncker, H.-R.: Die Festlegung des Bauplanes der Vogellunge beim Embryo und das Problem ihres postnatalen Wachstums. Verh. Anat. Ges. 66, 273-277 (1972)

Fitze-Gschwind, V.: Zur Entwicklung der Chorioallantoismembran des Hühnchens. Ergebn. Anat. Entwickl.-Gesch. 47, Fasc. 1, 7-52 (1973)

Juillet, A.: Recherches anatomiques, embryologiques, histologiques et comparatives sur le poumon des oiseaux. Arch. Zool. Exp. Gen., 5. ser. IX, 207-371 (1912)

Locy, W.A., Larsell, O.: The embryology of the birds lung. Based on observations of the bronchial tree. Am. J. Anat. 19, 447-504 (1916a)

Locy, W.A., Larsell, O.: The embryology of the birds lung. Based on observations of the domestic fowl. Part II. 3. The air sacs and the reccurent bronchi. Am. J. Anat. 20, 1-44 (1916b)

Petrik, P.: The ultrastructure of the chicken lung in the final stages of embryonic development. Folia Morph. (Prag) 15, 176-186 (1967)

Petrik, P., Riedel, B.: A continuous osmiophilic noncellular membrane at the respiratory surface of the lungs of fetal chickens and of young chicks. Lab. Invest. 18, 54-62 (1968)

Poole, M.: The development of the subdivisions of the pleuroperitoneal cavity in birds. Proc. Zool. Soc. Lond. 49, 210-235 (1909)

Rahn, H., Paganelli, C.V., Ar, A.: The avian egg: air cell gas tension, metabolism and incubation time. Respir. Physiol. 22, 297-309 (1974)

Reid, L.: The embryology of the lung. In: Development of the Lung. de Reuck, A.V.S., R. Porter (eds.). London: Churchill Ltd., 1967, pp. 109-130

Romanoff, A.L.: The Avian Embryo. Structural and Functional Development. New York: Macmillan, 1960

Selenka, E.: Beitrag zur Entwicklung der Luftsäcke des Huhnes. Z. wiss. Zool. 16, 178-182 (1866)

Visschedijk, A.H.J.: The air space and embryonic respiration. I. The pattern of gaseous exchange in the fertile egg during the closing stages of incubation. Brit. Poultry Sci. 9, 173-184 (1968)

Gas Transfer in the Chorioallantois

H. Tazawa

Summary

The fine reticulate arrangement of the chorioallantoic capillary plexus was observed microscopically. The gas analyses of blood flowing through the chorioallantois were performed in embryos incubated for 10 to 18 days, and the reactions of chorioallantoic capillary blood with gases were investigated with a microphotometric reaction apparatus. Using the data of the embryonic gas exchange through the chorioallantois, the O_2 delivery to the tissues was estimated on the basis of a simple blood circulation model.

1. Introduction

Analogous to fetal mammals and univentricular amphibians, avian embryos have double circulation for gas exchange and transport. The embryonic systems are separated from the maternal body, and their respiratory organ, the chorioallantoic membrane, develops in a site which is accessible experimentally. Respiratory conditions can easily be modified by changing the environmental conditions and the chorioallantoic membrane and blood can easily be sampled. These facts are advantageous in studying the gas exchange of embryos, which contributes to the comparative study on the respiratory physiology.

The allantoic membrane envelops the whole embryo from the middle term of incubation and gas exchange occurs evenly through the membrane till the start of pulmonary respiration. This paper is concerned with the gas exchange properties of the chorioallantois and the O_2 delivery to the tissues as investigated by blood gas analyses performed between the 10th and the 18th day of incubation.

2. Respiratory Organ

a) The Chorioallantoic Membrane

In the first stage of incubation some exchange of gases occurs through the vitelline vascular plexus prior to the formation of the allantois, which is initially established in 4-day embryos (21). As incubation proceeds, the chorioallantoic membrane spreads widely just beneath the shell membranes and envelopes the entire embryo by the last half of

Department of Physiology, Yamagata University School of Medicine, Yamagata 990-23, Japan.

the incubation period. The outer surface of the allantoic membrane is well vascularized, being 6 to 10 μ thick, where the gas exchange of embryo takes place (Fig. 1). The arterioles and venules interdigitate in the mesodermal layer parallel with the air capillary plexus. The capillaries form a fine network (34, 43) (Fig. 2). Since the width of

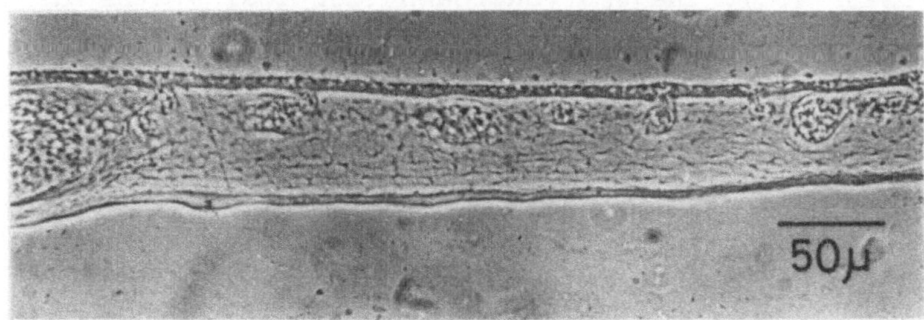

Fig. 1. Cross-section of the chorioallantoic membrane obtained by rapid freezing and freeze-drying method (34). The capillary plexus forms a monolayer on the outer surface of the membrane, and the pre- and post-capillary vessels are observed within the membrane

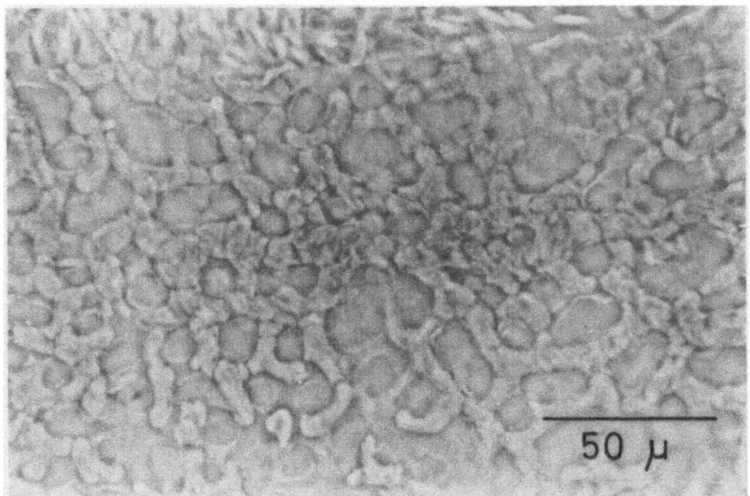

Fig. 2. Horizontal view of the chorioallantoic membrane. All the surface of the membrane is vascularized, forming the reticulate capillary plexus. The red blood cells are deformed within the capillary walls

the reticulate capillaries is approximately the size of the red blood cell (RBC), the elliptic corpuscles are variously deformed within the walls and flow in a zigzag fashion through the capillaries. Therefore, the pathway of the blood cells from the arteriole to the venule is remarkably lengthened and the stirring effect may also be enhanced. The time during which the RBCs are in contact with environmental air is prolonged and the diffusing capacity of the chorioallantoic membrane consequently increased.

b) Effect of the membrane on Gas Exchange

The rate of gas exchange through the shell is given by the product of shell permeability, shell area, and gas tension difference across the shell (44). When before incubation starts, either the broad ends or narrow ends of eggs are covered with an expoxy cement thus partially preventing gas diffusion (31), the embryos reach identical states of hypoxia and hypercapnia independently of the site covered. The reduction of respiratory area increases the hematocrit value and retards the main phase of embryonic growth, which takes place after 15 days of incubation. Table 1 shows hatchability and changes in blood gases

Table 1. Hatchability and blood gas status of incubated eggs under various experimental conditions of the eggshell

	Number of eggs	Number of chickens	Hatchability (%)	Blood gas status[a]	
Group 1	16	15	94	normoxia	normocapnia
Group 2	18	10	56	hypoxia	hypercapnia
Group 3	14	8	57	hypoxia	hypercapnia
Group 4	18	16	89	normoxia	hypocapnia
Group 5	14	7	50	moderate hypoxia	hypocapnia

Group 1, Control eggs; Group 2, Broad end is covered with cement; Group 3, Narrow end is covered with cement; Group 4, Shell over the air space is opened 3 to 4 mm wide; Group 5, Narrow end is covered with cement and shell over the air space is opened 3 to 4 mm wide.

[a] Blood gas status of group 1 is defined as normoxia and normocapnia, and the status of other groups is determined by comparing with group 1.

of the incubated eggs whose shell are partially covered with cement and/or opened for an area of 3 to 4 mm over the air space (32). The results were as follows below. 1. The gas exchange through the shell and the chorioallantoic membrane is fairly evenly distributed over the entire eggshell regardless of the position of the air space. This suggests that the chorioallantoic capillary plexus is evenly distributed over the whole egg and that the air space has no respiratory function at this period of development. 2. Whether or not blood P_{CO_2} is raised or lowered, the embryonic mortality is markedly higher when blood P_{O_2} is lowered by diminishing the respiratory area. The blood O_2 deficiency is lethal for developing embryos. 3. CO_2 accumulated in the air space and blood may not play a crucial role in pipping of the shell, because the embryos whose blood P_{CO_2} is extremely low (Group 4) pip and hatch normally.

3. Respiratory Properties in the Chorioallantois

Figure 3 shows a simplified blood circulation model for chicken embryos (40). Because the chorioallantoic membrane is situated just beneath the shell membrane, it is easy to withdraw blood samples and to make microscopic observations of the membrane. This made possible the recent studies on gas exchange based on the blood gas analyses (4, 5, 10, 12, 13, 28-33).

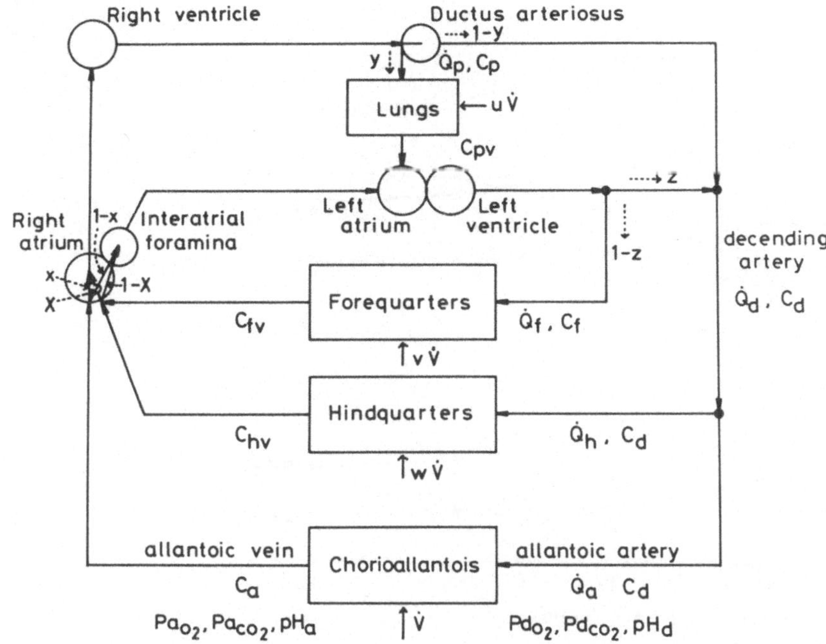

Fig. 3. A simplified blood circulation model similar to the mammalian fetus. Blood flows from the venae cavae ($\dot{Q}_f$ and $\dot{Q}_h$) proceed to the right atrium in which they become confluent with the O_2-rich blood draining the chorioallantois ($\dot{Q}_a$). Some blood is shunted through the interatrial foramina to the left ventricle (fractions $1-X$ and $1-x$) and distributed to the brain, heart, and upper extremities, i.e., forequarters. Other blood goes into the right ventricle (fractions X and x), out the pulmonary artery, and through the ductus arteriosus (fraction $1-y$) to supply the lower extremities and viscera, i.e., hindquarters, together with some output from the left ventricle (fraction z). The chorioallantois and the hindquarters are supplied with the same blood flowing in the descending artery: the double circulation to the tissues and the respiratory organ

a) Blood Gas Tensions and pH

The blood circulation in the descending artery bifurcates into the arteries of the legs and the allantois. The former supplies the O_2 to the tissues and the latter takes up O_2 when passing through the chorio-allantoic capillaries. Figure 4 shows gas tensions and pH in the allantoic artery and in the vein (33). It should be noted that the P_{O_2} in the allantoic arterial blood (Pd_{O_2}) is very low. The changes in P_{O_2}, P_{CO_2}, and pH with age indicate how the eggshell restricts diffusion for O_2 and CO_2 during development; O_2 tends to decrease and CO_2 to accumulate in blood toward the end of incubation.

b) Oxygen Capacity and Oxygen Content

Blood in the allantoic vein is highly oxygenated through the chorio-allantoic capillaries (28, 29, 40). On the other hand, the O_2 content of blood in the allantoic artery is very low and the O_2 saturation is estimated to be as small as 20% of capacity (Fig. 5). This implies

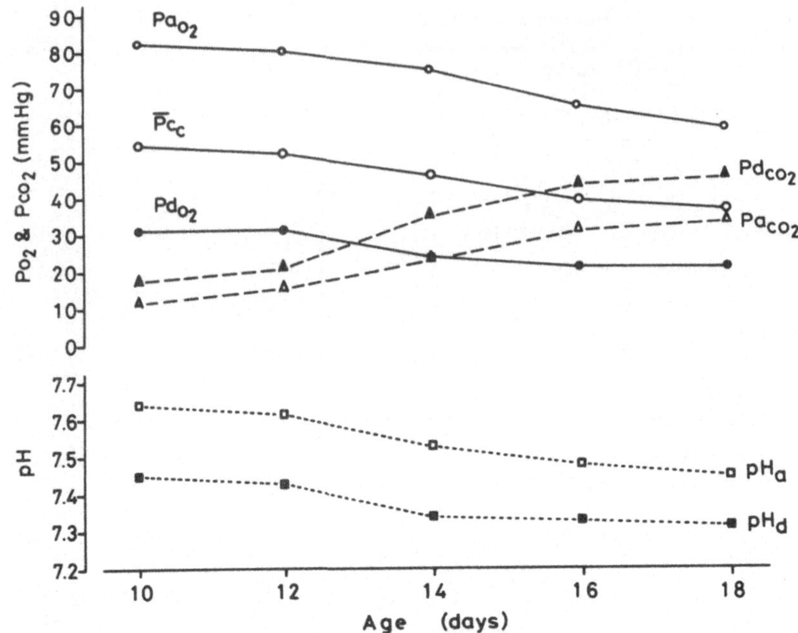

Fig. 4. Gas tensions and pH of blood in the allantoic artery and vein, and mean capillary P_{O_2} ($\overline{P}c_c$). The blood in the allantoic vein is well arterialized through the chorioallantoic capillary bed; the subscript a indicates arterialized blood in the allantoic vein. The blood in the allantoic artery, which is fiburcated from the descending artery, is shown by using the subscript d. The tissues of the hindquarters are also emptied with this blood

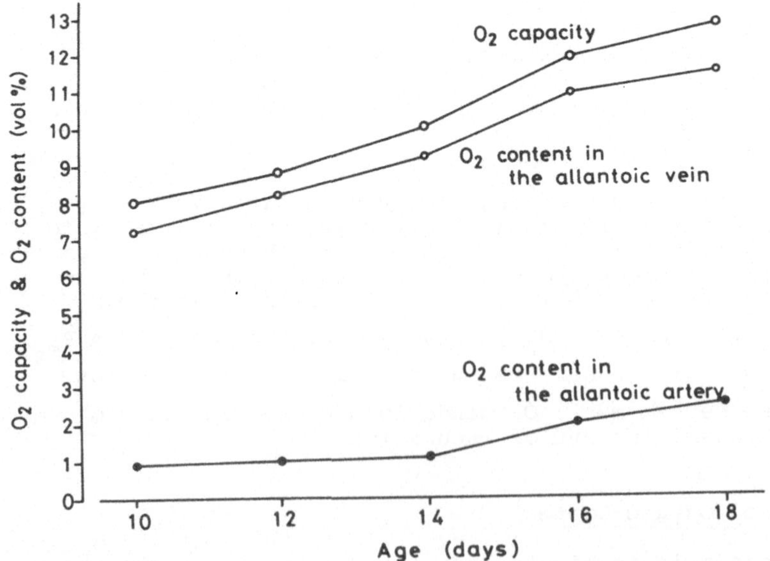

Fig. 5. Blood O_2 content in the allantoic vein and artery, and O_2 capacity during development. Both the O_2 capacity and O_2 content have a linear correlation to the embryonic weight (40)

that the tissues in the hindquarters are supplied with blood having very low O_2 content.

c) Oxygen Dissociation Curve for Capillary Blood

The dissociation curve for embryonic blood has been measured by several groups of investigators (2, 11, 14, 16, 36). The leftward displacement of the dissociation curve as compared to chick's or hen's blood is observed in the late period of incubation. In addition, the O_2 affinity is enhanced during development in accordance with the progress of hypoxia (Fig. 6).

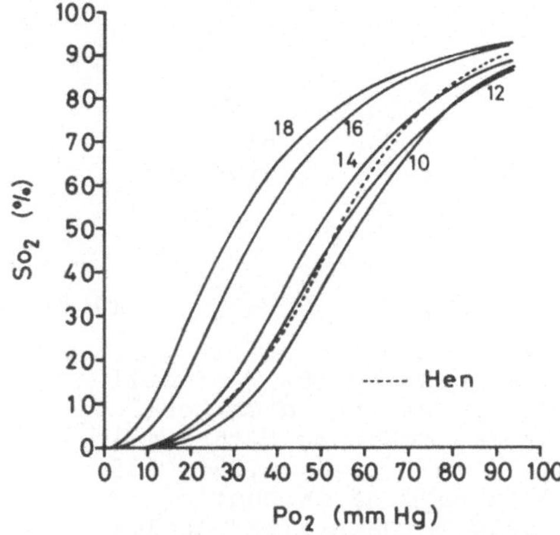

Fig. 6. O_2 dissociation curves for the chorioallantoic capillary blood of embryos during development and for hen's blood. The *curves* for embryos are constructed from the modified Hill's equations (36) at the allantoic arterial pH (pH_d) and the hen's curve (*dotted line*) is determined at pH 7.50 by Chiodi and Terman (3). The *number written beside the curve* indicates the days of incubation

d) O_2 Uptake and Blood Flow Rate

Before the start of pulmonary respiration, all O_2 used in the developing embryos is taken up through the chorioallantoic capillary plexus. O_2 uptake increases with age; however, the value per unit weight decreases to the level of adult hens. The blood-flow rate and the O_2-transport capacity in the allantoic circulation have a trend similar to that of the O_2 uptake (Table 2).

Table 2. The arteriovenous O_2 difference, O_2 uptake, blood flow rate, and O_2 transport capacity in the allantois

Age (days)	$Ca_{O_2}-Cd_{O_2}$ (vol %)	$\dot{V}_{O_2}$ (ml/min)	$\dot{Q}_a$ (ml/min)	$\dot{Q}_a \cdot Ca_{O_2}$ (ml/min)
10	6.3	0.08 (40.0)	1.3 (635)	0.09 (45.7)
12	7.2	0.15 (35.7)	2.1 (496)	0.17 (40.7)
14	8.1	0.25 (28.1)	3.1 (347)	0.28 (31.9)
16	8.9	0.36 (24.2)	4.1 (271)	0.44 (29.6)
18	9.0	0.42 (19.5)	4.7 (217)	0.54 (25.0)
Hen	5.7	24.3 (15.2)	430 (267)	43 (26.9)
Human	4.1	306 (4.1)	7540 (100)	1447 (19.2)

Number in parentheses indicates the value per kilogram body weight in ml/min/kg. The values of hen and the human beings are from Piiper et al. (22) and Bartels et al. (1), respectively.

4. Reaction of the Chorioallantoic Capillary Blood with Gases

The ease with which samples can be taken at the chorioallantoic membrane makes it possible to determine the reaction of respiratory gases with RBCs in the chorioallantoic capillary using a microphotometric reaction apparatus (18, 20, 35, 37-39).

a) Reaction with CO

The diffusing capacity for gases in the chorioallantoic capillary plexus is in general expressed by the following formula, as in the lung (17, 19),

$$D_{gas} = \dot{Q} \cdot Ht \cdot \overline{Fc}_{gas} \cdot t_c \qquad (1)$$

where $\dot{Q}$ is the blood flow rate, Ht the hematocrit value, and $\overline{Fc}_{gas}$ the time average of gas velocity factor with the red corpuscle for a period of contact time t_c. The velocity factor of CO reaction (Fc_{CO}) is calculated from the curve.

$$Fc_{CO} = \frac{O_2 \text{ capacity}}{Ht} \cdot \frac{dS_{CO}/dt}{100 \, P_{CO}} \cdot \qquad (2)$$

where dS_{CO}/dt is the rate of increasing CO saturation (S_{CO}) of cell hemoglobin. The value of O_2 capacity divided by Ht indicates "cell O_2 capacity" which represents the O_2 or CO volume combining with 1 ml RBCs (19). The values during development are 0.47, 0.44, 0.39, 0.35, and 0.33 ml/ml RBC for 10, 12, 14, 16, and 18 days of incubation. P_{CO} is CO tension of the reaction gas, which could be maintained as low as 1 to 4 mmHg. Since the CO replacement rate depends upon P_{O_2} value of the reaction gas mixute, the P_{O_2} should also be maintained at the same level of allantoic venous blood P_{O_2}. At 16 days of incubation, the Fc_{CO} value measured at 46 mmHg P_{O_2}, which is close to the mean capillary P_{O_2}, is 8.4 μl CO/ml RBC/s/mmHg (35). Referring to Equation (1), the contact time is evaluated, because the CO diffusing capacity in the chorioallantoic capillary is assessed from the data on the total CO diffusing capacity of the incubated eggs, $D_{CO, total}$= 3.3 μl/min/mmHg, (42). The D_{CO} in the chorioallantoic capillary plexus is 5.9 μl/min/mmHg and the contact time evaluated in 16-day-old embryos is 0.51 s at 46 mmH P_{O_2} (35)

b) Reaction with O_2

By measuring oxygenation reaction of RBCs in the chorioallantoic capillaries it is possible to evaluate the contact time as well as the oxygenation velocity factor (Fc_{ox}). The Fc_{ox} is calculated from the reaction curve similarly to Equation (2). Since the RBC is oxygenated by taking up O_2 during contact time through the capillaries, the change in oxygen saturation, S_{O_2}, depends on the contact time, the rate of RBC oxygenation, and the gas compositions of venous blood and air space. Thus, if the change in S_{O_2} of blood in the capillary is measured in vitro under the same conditions of P_{O_2} and P_{CO_2} as observed in vivo, the time during which the S_{O_2} increases from the venous level to the

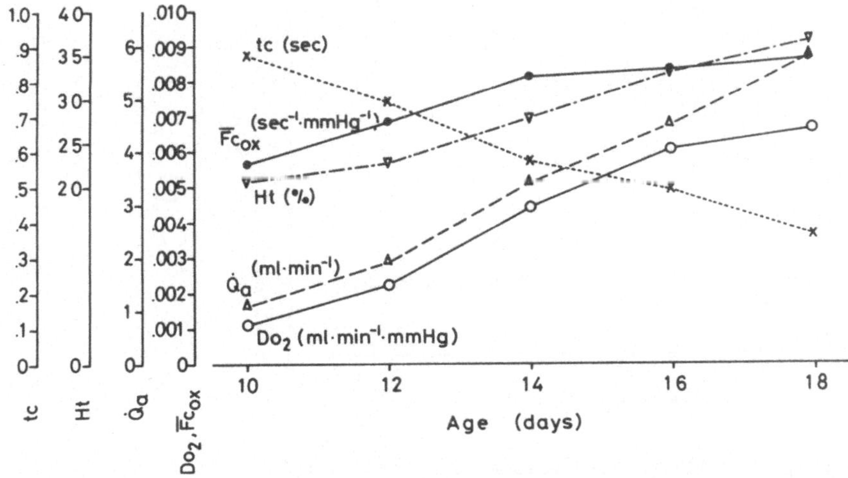

Fig. 7. Diffusing capacity for O_2 in the chorioallantoic membrane (D_{O_2}) estimated from the relationship of $D_{O_2} = \dot{Q}_a \cdot Ht \cdot \overline{Fc}_{ox} \cdot t_c$, and the individual values of blood flow rate ($\dot{Q}_a$), hematocrit value (Ht), time average of oxygenation velocity factor ($\overline{Fc}_{O_2}$) during contact time (t_c). The contact time is estimated as the oxygenation time changing from the allantoic arterial S_{O_2} to the allantoic venous S_{O_2} (39)

arterial level should reflect the contact time in vivo. Therefore, using the relationship expressed in Equation (1), the diffusing capacity for O_2 in the chorioallantoic capillary plexus could be estimated, because Fc_{ox} and t_c are determined at the same time. Figure 7 shows the estimated D_{O_2} together with the measured values for $\dot{Q}$, Ht, $\overline{Fc}_{ox}$, and t_c during development (39). The D_{O_2} increases with age. However, the value converted per kg weight of embryo decreases at the last period of incubation; 0.46, 0.42, 0.47, 0.38, and 0.29 ml/min/mmHg/kg weight for 10, 12, 14, 16, and 18 days of incubation. Although these values are lower than that of the adult hen - 1.1 to 1.5 ml/min/mmHg for 1.6-kg hens (26), they are very close to the value per kg body weight reported in man, e.g., the D_{O_2} estimated by a rebreathing procedure is about 15 ml/min/mmHg for man (23).

The chorioallantoic capillary blood volume (V_c) is evaluated from the relationship of $V_c = (\dot{Q}/60) \cdot t_c$ and the capillary bed area (A_c) is also estimated by assuming that the thickness of capillary layer is 8 μ. They increase during development, which is consistent with the increase in number of the RBCs existing in the capillary plexus as observed with a microscope (34) (Fig. 8). The values of V_c converted per kg weight for embryos near hatching, i.e., 1.5 ml/kg weight of embryo at 18 days of incubation, are similar to the values estimated for the human lung, i.e., 1.6 ml/kg weight of body (27). The capillary bed area during the last days of incubation occupies about 70-80% of the total eggshell area which is assumed about 60 cm^2.

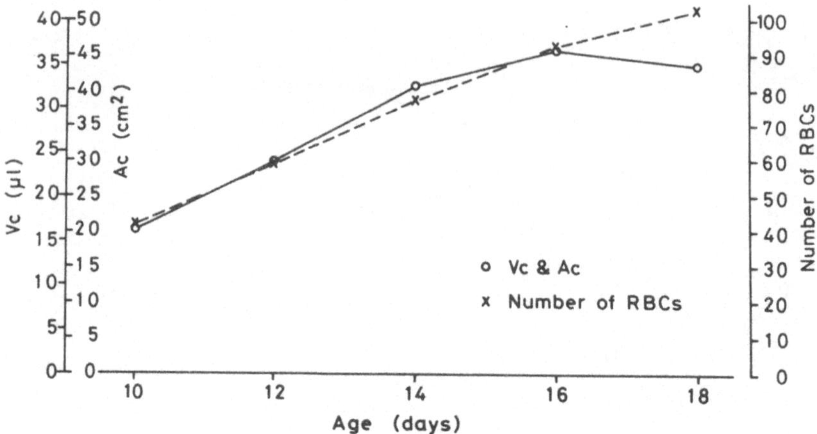

Fig. 8. Chorioallantoic capillary blood volume (V_C), capillary bed area (A_C), and number of red blood corpuscles in the field of microscopic vision of 80 μ × 80 μ

c) Comparison between the Rates of Oxygenation and CO Reaction

In 16-day-old embryo, the Fc_{ox} determined at 40 mmHg P_{CO_2} is influenced by S_{O_2} as follows (37):

$$Fc_{ox} = (9.65+0.1071\ S_{O_2} - 0.00216\ S_{O_2}{}^2)\cdot 10^{-3}\ \text{ml } O_2/\text{ml RBC/s/mmHg} \tag{3}$$

The Fc_{ox} depends on blood S_{O_2} which increases from about 17% to 88% during contact time. Using the O_2 dissociation curve, the time average of Fc_{ox} during contact time ($\overline{Fc}_{ox}$) is calculated to be 8.3 μl O_2/ml RBC/s/mmHg by Bohr's integral procedure. The Fc_{ox} at hypoxic level in a range of S_{O_2} between 17% and 60% is about 10 μl O_2/ml RBC/s/mmHg.

On the other hand, the Fc_{CO} of the capillary blood is represented empirically as a function of the reaction gas P_{O_2} (35).

$$Fc_{CO} = 2.31/(P_{O_2} + 232)\quad \text{ml CO/ml RBC/s/mmHg} \tag{4}$$

Because the blood P_{O_2} increases during contact time, the Fc_{CO} is also a function of time and varies with the blood S_{O_2}. Figure 9 shows the Fc_{O_2} and Fc_{CO} versus S_{O_2}. The change in Fc_{CO} against S_{O_2} is small, implying that the $\overline{Fc}_{CO}$ can be obtained at a first approximation by measuring the reaction at the mean capillary P_{O_2}. The Fc_{CO} determined at 46 mmHg P_{O_2} is 8.4 μl CO/ml RBC/s/mmHg (35), which is the same value as the $\overline{Fc}_{CO}$ for the S_{O_2} range of 17% to 88%. As a result, both the values of $\overline{Fc}_{ox}$ and $\overline{Fc}_{CO}$ under physiological conditions are almost identical, indicating that the D_{O_2} and D_{CO} become almost equal in the chorioallantoic capillary plexus of 16-day-old embryos. On the other hand, the ratio of Fc_{CO} at 100 mmHg P_{O_2} (7 μl CO/ml RBC/s/mmHg from Eq. 4) to Fc_{ox} at hypoxia turns out to be 0.7 in the capillary blood of embryos, which is almost the same value as that for the human red cell suspension (19).

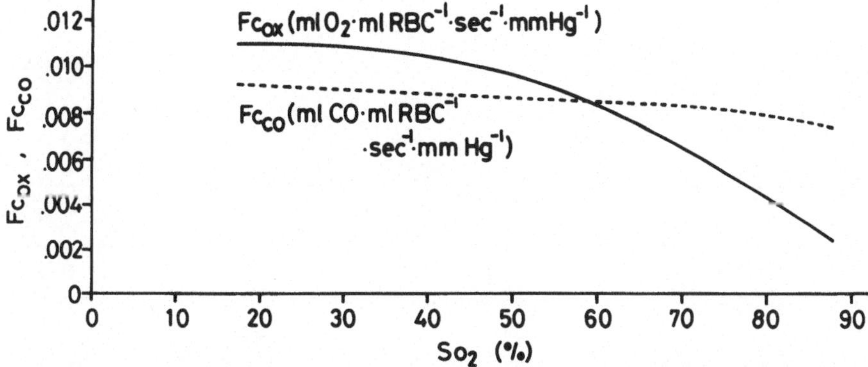

Fig. 9. Fc_{ox} and Fc_{CO} plotted against S_{O_2} in 16-day-old embryo. The Fc_{ox} maintains a rather constant value at low range of S_{O_2} and then decreases with S_{O_2}. The change in Fc_{CO}, obtained from Equation (4) by referring to the O_2 dissociation curve, is small against S_{O_2}. Both the time average values, $\overline{Fc}_{ox}$ and $\overline{Fc}_{CO}$, are identical

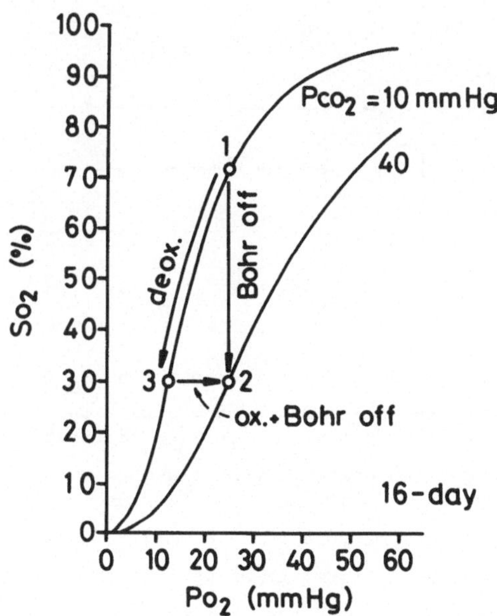

Fig. 10. O_2 dissociation curves for capillary blood of 16-day-old embryos at P_{CO_2}s of 10 to 40 mmHg. The change in reaction gas mixture from 1 to 2 induces the O_2 release, decreasing the S_{O_2}; i.e., the Bohr-off reaction. The time when the S_{O_2} reaches half level (i.e., halftime) is about 1 s, whereas the halftime of deoxygenation reaction made by changing the gas from 1 to 3 is about 0.5 s. The change in gas mixture from 3 to 2 induces the cooperative reactions of oxygenation and Bohr-off shift

d) Effect of CO_2: Bohr Shift

The effect of CO_2 on the O_2 affinity of RBCs designated as Bohr effect and the rate of the affinity change, Bohr shift, is measured in the chorioallantoic capillary blood with the microphotometric reaction apparatus (38). The abrupt increase in P_{CO_2} around RBCs at constant P_{O_2}

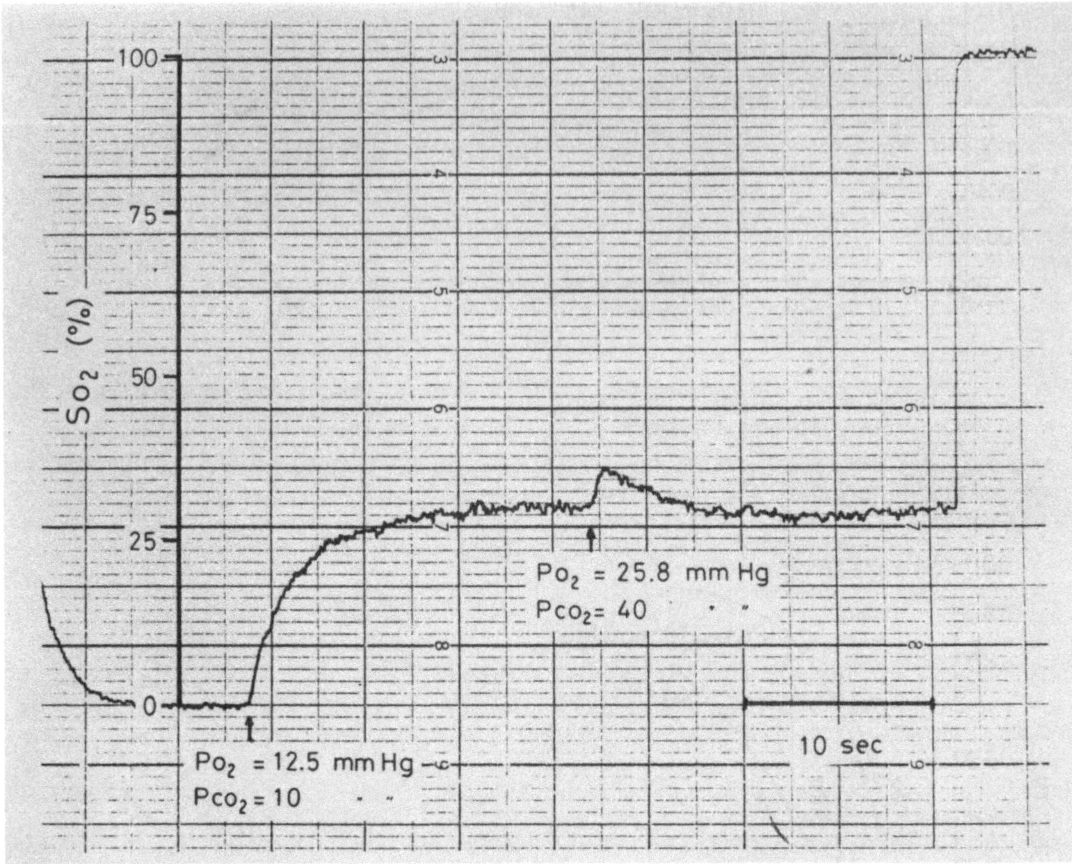

Fig. 11. Cooperative reactions of oxygenation and Bohr-off shift. Although the S_{O_2} level is identical at a steady state when changing the gas mixture from 3 to 2 in Figure 10, the kinetic record of reaction shows a transient change in S_{O_2}. At first, the RBC in the chorioallantoic membrane is deoxygenated completely by flushing the O_2 free gas. Then, the gas mixture with 12.5 mmHg P_{O_2} and 10 mmHg P_{CO_2} is flushed over the membrane. After reaching equilibrium, the gas mixture of 25.8 mmHg P_{O_2} and 40 mmHg P_{CO_2} is flushed. The *record* shows that the O_2 is first taken up into the cell by the P_{O_2} increase from 13 to 26 mmHg (oxygenation) and then released with a slower rate by the P_{CO_2} increase from 10 to 40 mmHg (Bohr-off reaction), returning the S_{O_2} level to the initial

induces the decrease in blood S_{O_2}; Bohr-off shift (Fig. 10). The half-time is about 1 s, which is almost twofold the deoxygenation reaction made by the abrupt decrease in P_{O_2}. The cooperative kinetic reactions of O_2 and Bohr shift shows that oxygenation due to the P_{O_2} increase appears first and then by the Bohr-off shift reaction S_{O_2} recovers to the initial level at a slower rate (Fig. 11). This suggests that the Bohr effect is unexpectedly small over the short time of 1 s or so during which the gas is in transit through the exchange area.

5. Oxygen Delivery to the Tissues

The blood coming out of the heart perfuses in parallel the systemic tissues and the respiratory organ. In animals having such double circulation, selective streaming occurs in the heart so that the O_2 saturation in the systemic artery may become higher than in the artery to the respiratory organ; i.e., the fetal mammals (6, 7) and the frogs with a single ventricle (8, 9, 15, 41). Because the arterial O_2 content is very low in chicken embryos, the O_2 supply to the tissues is assumed to be improved by a similar selective streaming and by a large blood flow rate.

Table 3. Definitions for parameters

Symbol	Definition
$\dot{Q}$	Blood flow rate (ml/min)
C	O_2 content (ml/ml blood)
$\dot{V}$	O_2 uptake through the chorioallantois (ml/min)
Subscript	**Definition**
p	Pulmonary circulation or pulmonary arterial blood
pv	Pulmonary venous blood
f	Circulation in the forequarters or arterial blood to the forequarters
fv	Venous blood from the forequarters
d	Descending artery to the hindquarters and the chorioallantois
h	Circulation in the hindquarters
hv	Venous blood from the hindquarters
a	Allantoic circulation or allantoic venous blood
Fraction	**Definition**
u	$\frac{O_2 \text{ consumption in the lung}}{O_2 \text{ uptake through the chorioallantois, } \dot{V}}$
v	$\frac{O_2 \text{ consumption in the forequarters}}{O_2 \text{ uptake through the chorioallantois, } \dot{V}}$
w	$\frac{O_2 \text{ consumption in the hindquarters}}{O_2 \text{ uptake through the chorioallantois, } \dot{V}}$
x	$\frac{\text{Flow component to the right ventricle deriving from the confluent flow } \dot{Q}f+\dot{Q}h}{\text{Total blood flow rate entering into the right ventricle}}$
x	$\frac{\text{Flow component to the right ventricle deriving from } \dot{Q}_a}{\text{Total blood flow rate entering into the right ventricle}}$
y	$\frac{\text{Flow component to the lungs deriving from the right ventricular output}}{\text{Output from the right ventricle}}$
z	$\frac{\text{Flow component to the descending artery deriving from the left ventricular output}}{\text{Output from the left ventricle}}$

a) Oxygen and Blood Flow Distribution

Table 3 summarizes the definitions for parameters used in the blood circulation model in Figure 3. Assuming that the selective streaming is performed as the fractional separation of blood in the interatrial foramina, the ductus arteriosus and the aorta as shown in Figure 3, the following equations are derived for the model. Blood flow rate in the descending artery ($\dot{Q}_d$) is the sum of that in the hindquarters ($\dot{Q}_h$) and that in the allantoic circulation ($\dot{Q}_a$).

$$\dot{Q}_d = \dot{Q}_h + \dot{Q}_a \tag{5}$$

These blood flows and those in the lungs ($\dot{Q}_p$) and forequarters ($\dot{Q}_f$) are related to the O_2 consumption and O_2 difference by Fick's principle.

$$\dot{Q}_p = u \cdot \dot{V}/(C_p - C_{pv}) \tag{6}$$

$$\dot{Q}_f = v \cdot \dot{V}/(C_f - C_{fv}) \tag{7}$$

$$\dot{Q}_h = w \cdot \dot{V}/(C_d - C_{hv}) \tag{8}$$

and

$$\dot{Q}_a = \dot{V}/(C_a - C_d) \tag{9}$$

The blood flow rates and the O_2 quantity in the lungs, forequarters, and hindquarters are agian expressed using the ratios of separation (X, x, y, and z).

$$y \cdot [x \cdot \dot{Q}_a + X \cdot (\dot{Q}_f + \dot{Q}_h)] = \dot{Q}_p \tag{10}$$

$$(1-z) \cdot \{[(1-x) \cdot \dot{Q}_a + (1-X) \cdot (\dot{Q}_f + \dot{Q}_h)] + y \cdot [x \cdot \dot{Q}_a + X \cdot (\dot{Q}_f + \dot{Q}_h)]\} = \dot{Q}_f \tag{11}$$

$$(1-y) \cdot [x \cdot \dot{Q}_a + X \cdot (\dot{Q}_f + \dot{Q}_h)] + z \cdot \{[(1-x) \cdot \dot{Q}_a + (1-X) \cdot (\dot{Q}_f + \dot{Q}_h)] + y \cdot [x \cdot \dot{Q}_a + X \cdot (\dot{Q}_f + \dot{Q}_h)]\} = \dot{Q}_d \tag{12}$$

$$y \cdot [x \cdot \dot{Q}_a \cdot C_a + X \cdot (\dot{Q}_f \cdot C_{fv} + \dot{Q}_h \cdot C_{hv})] = \dot{Q}_p \cdot C_p \tag{13}$$

$$(1-z) \cdot \{[(1-x) \cdot \dot{Q}_a \cdot C_a + (1-X) \cdot (\dot{Q}_f \cdot C_{fv} + \dot{Q}_h \cdot C_{hv})] + y \cdot [x \cdot \dot{Q}_a + X \cdot (\dot{Q}_f + \dot{Q}_h)] \cdot C_{pv}\} = \dot{Q}_f \cdot C_f \tag{14}$$

and

$$(1-y) \cdot [x \cdot \dot{Q}_a \cdot C_a + X \cdot (\dot{Q}_f \cdot C_{fv} + Q_h \cdot C_{hv})] + z \cdot \{[(1-x) \cdot \dot{Q}_a \cdot C_a + (1-X) \cdot (\dot{Q}_f \cdot C_{fv} + \dot{Q}_h \cdot C_{hv})] + y \cdot [x \cdot \dot{Q}_a + X \cdot (\dot{Q}_f + \dot{Q}_h)] \cdot C_{pv}\} = \dot{Q}_d \cdot C_d \tag{15}$$

Since we know only C_a, C_d, and $\dot{V}$, the following are further assumed to solve the above equations. 1. From the data compiled by Romanoff (24, 25), the O_2 uptake through the chorioallantois is consumed in the lungs, forequarters, and hindquarters by fraction of 0.05, 0.60, and 0.35, respectively (i.e., u = 0.05, v = 0.6, and w = 0.35). 2. As observed in fetal lambs (6), the S_{O_2} of blood emptying the forequarters is 10% higher than that of the hindquarters and the S_{O_2} of blood from the right ventricle is 90% of that of the descending artery (i.e., $C_f = C_d + 0.1$ O_2 capacity and $C_p = 0.9\ C_d$). 3. The venous O_2 content of the forequarters and hindquarters is identical and is 40% less than the O_2 content of pulmonary venous blood (i.e., $C_{fv} = C_{hv} = 0.6\ C_{pv}$). In 16-day-old embryos, the values measured are C_a (11.0 vol%), C_d (2.0 vol%), and $\dot{V}$ (0.36 ml/min) and the results solved from the above equations are shown in Figure 12. The blood flow rate to the tissues is twofold that through the chorioallantoic membrane, which may compensate for the low O_2 content of blood supplying the tissues. The O_2-transport capacity expressed by the product of $Q \cdot C_{O_2}$ is 0.27 and 0.18 ml/min for the forequarters and hindquarters, respectively. Assuming that the weight of the forequarters and hindquarters is equal to half of the embryonic weight, the O_2-transport capacity per unit weight of tissues is estimated to be 24 and 36 ml/min/kg weight. On the other hand, hens have a systemic O_2-transport capacity of 27 ml/min/kg weight (22) and

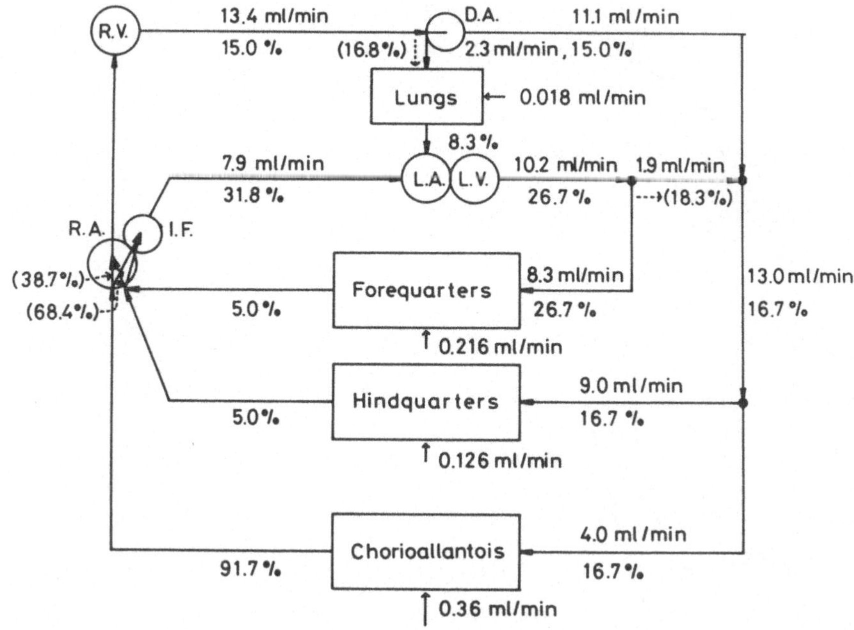

Fig. 12. S_{O_2} and blood flow distribution in 16-day-old embryos estimated from the data obtained in the allantois and assumed from the results of fetal lambs. The S_{O_2} shown in % is obtained by dividing the calculated O_2 content by O_2 capacity (11.9 vol %). The values in the parentheses show the percentage distribution ratio of blood to the interatrial foramina, the ductus arteriosus, and the aorta. The O_2 consumption in the lungs, forequarters, and hindquarters, and the O_2 uptake through the chorioallantois are also shown (in ml/min)

fetal lambs are estimated to have about 10 ml/min/kg weight of tissues (6). In chicken embryos, an ample O_2 transport to the tissues seems to be maintained through the large blood flow rate.

b) Deoxygenation in the Tissues

The O_2 flux through the eggshell depends on the P_{O_2} difference between the environmental air (P_o) and air space (P_s) and the diffusing capacity in the eggshell (D_s). Similarly, the quantity of O_2 taken up by the chorioallantoic capillary blood depends on the difference between the air space P_{O_2} and mean chorioallantoic capillary P_{O_2} ($\bar{P}c_c$) and the diffusing capacity for oxygenation in the chorioallantoic capillary bed (D_{ox}).

$$\dot{V} = D_s \cdot (P_o - P_s) = D_{ox} \cdot (P_s - \bar{P}c_c) \tag{16}$$

On the other hand, the O_2 release in the tissues is proportional to the difference between the mean systemic capillary P_{O_2} (i.e., $\bar{P}c_f$ and $\bar{P}c_h$ for the forequarters and hindquarters, respectively) and tissue P_{O_2} ($\bar{P}_t$). The proportional factor is defined as the diffusing capacity for deoxygenation in the tissues (i.e., D^f_{deox} and D^h_{deox} for the forequarters and hindquarters, respectively). That is,

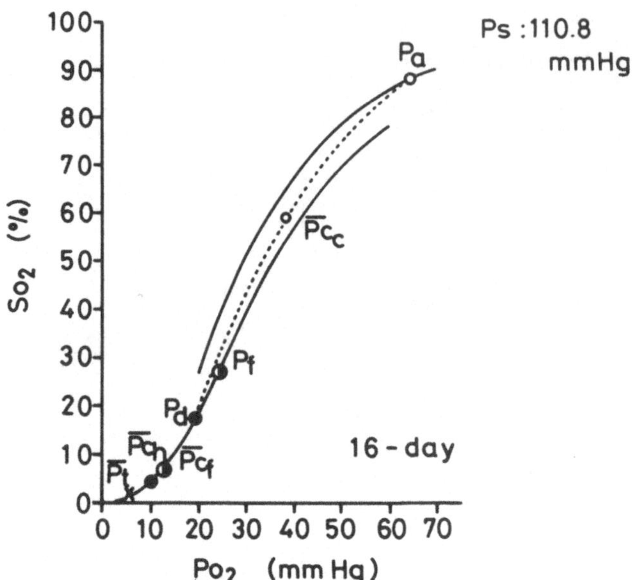

Fig. 13. Blood P_{O_2} plotted in the O_2 dissociation curves at the allantoic venous blood pH (*left side*) and the allantoic arterial pH (*right side*) of 16-day-old embryo. P_a, allantoic venous O_2 tension; P_d, allantoic arterial O_2 tension; P_f, arterial O_2 tension of the forequarters; $\overline{P}_c$, the mean capillary O_2 tension estimated by Bohr's integral procedure; i.e., $\overline{P}c_c$, $\overline{P}c$ in allantoic circulation, $\overline{P}c_f$, $\overline{P}c$ in the forequarters and $\overline{P}c_d$, $\overline{P}c$ in the hindquarters; $\overline{P}_t$, the mean O_2 tension in the tissues which is assumed to be 1 mmHg less than the systemic venous blood P_{O_2}

$$v \cdot \dot{V} = D^f_{deox} \cdot (\overline{P}c_f - \overline{P}_t) \tag{17}$$

and

$$w \cdot \dot{V} = D^h_{deox} \cdot (\overline{P}c_h - \overline{P}_t) \tag{18}$$

Figure 13 shows the blood oxygen pressures estimated from the results in Figure 12, where the mean P_{O_2} in the tissues ($\overline{P}_t$) is assumed to be 1 mmHg less than the systemic venous blood P_{O_2}. The mean capillary P_{O_2} for chorioallantoic and systemic circulations is estimated by Bohr's integral procedure. The P_s values are derived from the data of Wangensteen and Rahn (44). Using the individual value of P_{O_2} shown in Figure 13, the diffusing capacities defined by Equations (16-18) are calculated as

$$D_s = 10,\ D_{ox} = 5,\ D^f_{deox} = 30 \text{ and } D^h_{deox} = 29\ \mu l/min/mmHg.$$

The D_{deox} values both in the forequarters and hindquarters are much larger than D_{ox}. This means that the resistivity for deoxygenation in the tissues ($1/D_{deox}$) appears to be very small compared with that for oxygenation in the chorioallantois ($1/D_{ox}$) in association with the large blood flow rate through the tissues.

Analogously to the diffusing capacity for O_2 as expressed by Equation (1), the D_{deox} may also be related to the tissue blood flow rate ($\dot{Q}_t$), hematocrit, deoxygenation velocity factor in the tissues ($\overline{F}c^t_{deox}$) and transit time through the tissue capillaries (t^t_c) as

$$D_{deox} = \dot{Q}_t \cdot Ht \cdot \overline{Fc}^t_{deox} \cdot t^t_c \qquad (19)$$

Assuming that the deoxygenation velocity in the tissues is identical with that measured in the chorioallantoic capillaries, the transit time through the tissue capillaries is expressed by

$$t^t_c = \frac{D_{deox}}{D_{ox}} \cdot \frac{\dot{Q}_a}{\dot{Q}_t} \cdot \frac{\overline{Fc}_{ox}}{\overline{Fc}_{deox}} \cdot t_c \qquad (20)$$

Using the results obtained above, the transit time through the tissue capillaries of the forequarters (t^f_c) is estimated to be about 1.53 s and that of the hindquarters (t^h_c), 1.36 s. From these values, the capillary volume in the tissues becomes 0.212 and 0.204 ml for the forequarters and hindquarters, respectively. According to Yosphe-Purer et al. (45), the total blood volume of 16-day-old embryo is about 2.9 ml. As a result, the total tissue capillary volume extends to about 14% of total blood volume in 16-day-old embryo. This value is much larger than that in man, and therefore may contribute to maintenance of the O_2 supply to the tissues under condition of extremely low P_{O_2}.

References

1. Bartels, H., Beer, R., Fleischer, E., Hoffheinz, H.J., Wenner, J., Witt, I.: Bestimmung von Kurzschlußdurchblutung und Diffusionskapazität der Lunge bei Gesunden und Lungenkranken. Pfl. Arch. 261, 99-132 (1955)
2. Bartels, H., Hiller, G., Reinhardt, W.: Oxygen affinity of chicken blood before and after hatching. Respir. Physiol. 1, 345-356 (1966)
3. Chiodi, H., Terman, J.W.: Arterial blood gases of the domestic hen. Am. J. Physiol. 208, 789-800 (1965)
4. Dawes, C., Simkiss, K.: The acid-base status of the blood of the developing chick embryo. J. Exp. Biol. 50, 79-86 (1969)
5. Dawes, C., Simkiss, K.: The effects of respiratory acidosis in the chick embryo. J. Exp. Biol. 55, 77-84 (1971)
6. Dawes, G.S., Mott, J.C., Widdicombe, J.G.: The fetal circulation in the lamb. J. Physiol. 126, 563-587 (1954)
7. Dawes, G.S., Mott, J.C.: Changes in O_2 distribution and consumption in foetal lambs with variations in umbilical blood flow. J. Physiol. 170, 524-540 (1964)
8. DeLong, K.T.: Quantitative analysis of blood circulation through the frog heart. Science 138, 693-694 (1962)
9. Emilio, M.G., Shelton, G.: Factors affecting blood flow to the lungs in the amphibian, *Xenopus laevis*. J. Exp. Biol. 60, 567-579 (1974)
10. Erasmus, B.W., Howell, B.J., Rahn, H.: Ontogeny of acid-base balance in the bullfrog and chicken. Respir. Physiol. 11, 46-53 (1970/71)
11. Farooqui, A.M., Huehns, E.R.: Oxygen dissociation studies of red cells from very small human and chicken embryos. In: Int. Symp. Struktur Funktion Erythrocyten. Berlin: Akademie-Verlag, 1972, pp. 217-220
12. Freeman, B.M., Misson, B.H.: pH, pO_2 and pCO_2 of blood from the foetus and neonate of Gallus Domesticus. Comp. Biochem. Physiol. 33, 763-772 (1970)
13. Girard, H.: Respiratory acidosis with partial metabolic compensation in chick embryo blood during normal development. Respir. Physiol. 13, 343-351 (1971)
14. Hall, F.G.: Hemoglobin function in the developing chick. J. Physiol. 83, 222-228 (1934)
15. Johansen, K., Ditadi, S.S.F.: Double circulation in the giant toed, Bufo paracremis. Physiol. Zool. 39, 140-150 (1966)
16. Misson, B.H., Freeman, B.M.: Organic phosphates and oxygen affinity of chick blood before and after hatching. Respir. Physiol. 14, 343-352 (1972)
17. Mochizuki, M., Fukuoka, J.: The diffusion of oxygen inside of the red cell. Jap. J. Physiol. 8, 206-224 (1958)

18. Mochizuki, M., Tazawa, H., Ono, T.: Microphotometry for determining the reaction rate of O_2 and CO with red blood cells in the chorioallantoic capillary. In: Oxygen Tansport to Tissue. Vol. II, Bruley, D.F., Bicher, H.I. (eds.). New York: Plenum, 1973, pp. 997-1006
19. Mochizuki, M.: Graphical Analysis of Oxygenation and CO-Combination Rates of the Red Cells in the Lung. Tokyo: Hirokawa, 1975, pp. 1-72
20. Ono, T., Tazawa, H.: Microphotometric method for measuring the oxygenation and deoxygenation rate in a single red blood cell. Jap. J. Physiol. 25, 93-107 (1975)
21. Patten, B.M.: Foundations of Embryology. New York: McGraw Hill, 1964, pp. 244-259
22. Piiper, J., Drees, F., Scheid, P.: Gas exchange in the domestic fowl during spontaneous breathing and artifitial ventilation. Respir. Physiol. 9, 234-245 (1970)
23. Piiper, J., Cerretelli, P., Rennie, D.W., DiPrampero, P.E.: Estimation of the pulmonary diffusing capacity for O_2 by a rebreathing procedure. Respir. Physiol. 12, 157-162 (1971)
24. Romanoff, A.L.: The Avian Embryo. New York: Macmillan, 1960, pp. 1111-1140
25. Romanoff, A.L.: Biochemistry of the Avian Embryo. New York: John Wiley & Sons, 1967, p. 291
26. Scheid, P., Piiper, J.: Analysis of gas exchange in the avian lung: Theory and experiments in the domestic fowl. Respir. Physiol. 9, 246-262 (1970)
27. Smith, T.C., Rankin, J.: Pulmonary diffusing capacity and the capillary bed during Valsalva and Müller maneuvers. J. Appl. Physiol. 27, 826-833 (1969)
28. Tazawa, H.: Measurement of O_2 content in microliter blood samples. J. Appl. Physiol. 27, 826-833 (1969)
29. Tazawa, H.: Measurement of respiratory parameters in blood of chicken embryo. J. Appl. Physiol. 30, 17-21 (1971)
30. Tazawa, H., Mikami, T., Yoshimoto, C.: Respiratory properties of chicken embryonic blood during development. Respir. Physiol. 13, 160-170 (1971a)
31. Tazawa, H., Mikami, T., Yoshimoto, C.: Effect of reducing the shell area on the respiratory properties of chicken embryonic blood. Respir. Physiol. 13, 352-360 (1971b)
32. Tazawa, H.: Gas exchange in chicken embryo. In: Advances in Respiratory Physiology and Controls, Monograph Series, No. 20. Mochizuki, M. (ed.). Res. Inst. Appl. Electri., Hokkaido Univ. 1972, pp. 1-15
33. Tazawa, H.: Hypothermal effect on the gas exchange in chicken embryo. Respir. Physiol. 17, 21-31 (1973)
34. Tazawa, H., Ono, T.: Microscopic observation of the chorioallantoic capillary bed of chicken embryo. Respir. Physiol. 20, 81-89 (1974)
35. Tazawa, H., Ono, T., Mochizuki, M.: Reaction velocity of carbon monoxide with blood cells in the chorioallantoic vascular plexus of chicken embryos. Resp. Physiol. 20, 161-170 (1974)
36. Tazawa, H., Ono, T., Mochizuki, M.: Oxygen dissociation curve for chorioallantoic capillary blood of chicken embryo. J. Appl. Physiol. 40, 393-398 (1976a)
37. Tazawa, H., Ono, T., Mochizuki, M.: Oxygenation and deoxygenation velocity factors of chorioallantoic capillary blood. J. Appl. Physiol. 40, 399-403 (1976b)
38. Tazawa, H., Mochizuki, M.: Rates of oxygenation and Bohr shift of capillary blood in chick embryos. Nature (London) 261, 509-511 (1976a)
39. Tazawa, H., Mochizuki, M.: Estimation of contact time and diffusing capacity for oxygen in the chorioallantoic vascular plexus. Respir. Physiol. 28, 119-128 (1976b)
40. Tazawa, H., Mochizuki, M.: Oxygen analyses of chicken embryo blood. Respir. Physiol. 203-215 (1977)
41. Tazawa, H., Mochizuki, M.: Analysis of oxygen transport in bullfrogs. In: Oxygen Transport to Tissue. Vol. III, Silver, I.A. (ed.). New York: Plenum Press, in press 1977
42. Temple, G.F., Metcalfe, J.: The effects of increased incubator oxygen tension on capillary development in the chick chorioallantois. Respir. Physiol. 9, 216-233 (1970)
43. Wangensteen, O.D., Wilson, D., Rahn, H.: Diffusion of gases across the shell of the hen's egg. Respir. Physiol. 11, 16-30 (1970/71)

44. Wangensteen, O.D., Rahn, H.: Respiratory gas exchange by the avian embryo. Respir. Physiol. 11, 31-45 (1970/71)
45. Yosphe-Purer, Y., Fendrich, J., Davies, A.M.: Estimation of the blood volumes of embryonated hen eggs at different ages. Am. J. Physiol. 175, 178-180 (1953)

Respiratory Function of Embryonic Chicken Hemoglobin

R. Baumann and F. H. Baumann

Summary

The O_2 affinity of embryonic chicken blood increases continuously between 8 to 16 days of development. During this period the O_2 halfsaturation pressure (P_{50}) of whole blood (at 37°C) drops from 75 mmHg at pH 7.4 at day 8 to 35 mmHg at day 16. At the same time there is an almost complete replacement of the three embryonic hemoglobins by the adult chicken hemoglobins.

ATP is present in a high concentration in the erythrocytes of embryonic chicken blood and its concentration decreases from 11 mmol/l LBC at day 8 to 7 mmol/l LBC at day 14. Both embryonic and adult hemoglobins show strong interaction with ATP, however the O_2 affinity of embryonic hemoglobin is always higher than that of adult hemoglobin under comparable conditions. The low O_2 affinity encountered at day 8 is primarily due to the high concentration of ATP and the low intracellular pH (6.72). The subsequent increase in O_2 affinity up to day 16 seems to be mainly caused by a concomitant increase in red cell pH.

1. Introduction

The first investigation of hemoglobin function in embryonic and fetal blood of chicken dates back to 1934 when Hall investigated the O_2 affinity of chicken hemoglobin solutions prepared from blood obtained from fetuses of 10 days of incubation onward. He found that hemoglobin-O_2 affinity decreased with increasing age of the chicken embryo. This is exactly the opposite of what has been reported for whole blood. Lomholt (1975) as well as Tazawa et al. (1976) have reported that the lowest O_2 affinity was found at the earliest developmental stages covered by these investigations, that is at 8 to 10 days of incubation. The values reported for the O_2 halfsaturation pressure at this stage exceed those found in adult chicken blood, though these are still among the highest reported for warm-blooded animals. The erythrocytes of 8- to 10-day-old embryos contain large amounts of ATP (Bartlett and Borgese, 1976), but recent investigations of Cirotto and Geraci (1975) indicate a low reactivity of embryonic hemoglobin towards organic phosphates as compared to adult hemoglobin. Organic phosphates alone can not, therefore, be responsible for the low O_2 affinity at early developmental stages. The present study tried to analyse further the factors involved in the regulation of hemoglobin function of chicken embryo, and it is limited to the period between 5 to 14 days of development, when most of the transition from embryonic to adult hemoglobin types occurs (Bruns and Ingram, 1973).

Physiologisches Institut, Medizinische Hochschule Hannover, 3000 Hannover, Karl-Wiechert-Allee 9, Federal Republic of Germany.

2. Materials and Methods

Fertile eggs from White Leghorn chickens were used and incubated at 37.5^{o}. Timing of the eggs began at the moment of incubation.

a) Blood Sampling

Whole blood was obtained by venipuncture of embryos of 7 days and older. Blood from 5- and 6-day-old embryos was obtained by nicking the blood vessels and gently sucking off the blood with a Pasteur pipette. Blood samples used for the determination of ATP and 2,3 DPG were immediately transferred into liquid nitrogen. ATP was determined using the Boehringer Test-kit and 2,3 DPG with the method of Ericson and de Verdier (1972).

b) Hemoglobin Electrophoresis

Hemoglobin electrophoresis was carried out on polyacrylamide gel slabs (Drysdale et al., 1971) using LKB Ampholine pH 7-9.

c) Preparation of Hemoglobin Solutions

Red cells washed repeatedly in isotonic saline solution were lysed by sonification. The hemoglobin solutions were prepared and stripped of organic phosphates as described elsewhere (Bauer et al., 1975).

d) Determination of O_2 Affinity in Whole Blood and Hemoglobin Solutions

The O_2 halfsaturation pressure of whole blood (P_{50}) at 37°C was determined using the Lex-O_2-con (Lexington Instruments, Waltham, Mass.) as described previously (Baumann et al., 1975). O_2 equilibrium curves of hemoglobin solutions were measured spectrophotometrically in a diffusion chamber system (Sick and Gersonde, 1969) in 3×10^{-4} M Hb_4 solutions. pH was measured with the Radiometer microglasselectrode G 297/G2 and pH meter 64. Cell pH was measured as described previously (Baumann et al., 1975).

3. Results

a) Hemoglobin Electrophoresis

The change in hemoglobin pattern with increasing age of the embryo is shown in Table 1, which contains the results obtained by quantitative densitometry of the gel slabs, and the results are derived from at least four different runs. Blood from 5- and 6-day-old embryos contains three fractions with an electrophoretic mobility different from adult hemoglobin, and no adult hemoglobin. The embryonic fractions correspond to hemoglobins E, P, and P' in the classification of Brown and Ingram (1974), whereas A and D represent the major and minor fraction of adult hemoglobin respectively. At day 12 adult hemoglobin already accounts for 60% of the total hemoglobin, it should be noted however that the ratio between minor (D) and major (A) fraction of adult hemoglobin is different from the one found in adult chicken.

Table 1. Change in hemoglobin pattern with age of chicken embryo

Chicken embryo age in days	P $_{(P + P')}$ %	E (H) %	A	D
5	73.15	24.83		
6	70.8	23.7	trace	trace
8	37.12	21.65	20.83	20.43
9	32.97	16.38	24.79	22.41
10	29.19	15.13	28.54	27.83
12	19.36	15.17	35.72	29.74
adult	-	-	69.3	30.7

P, P', and E are the embryonic hemoglobins; E has the same electrophoretic mobility as Hb H, the hatching hemoglobin. A is the major fraction, D the minor fraction of adult chicken hemoglobin.

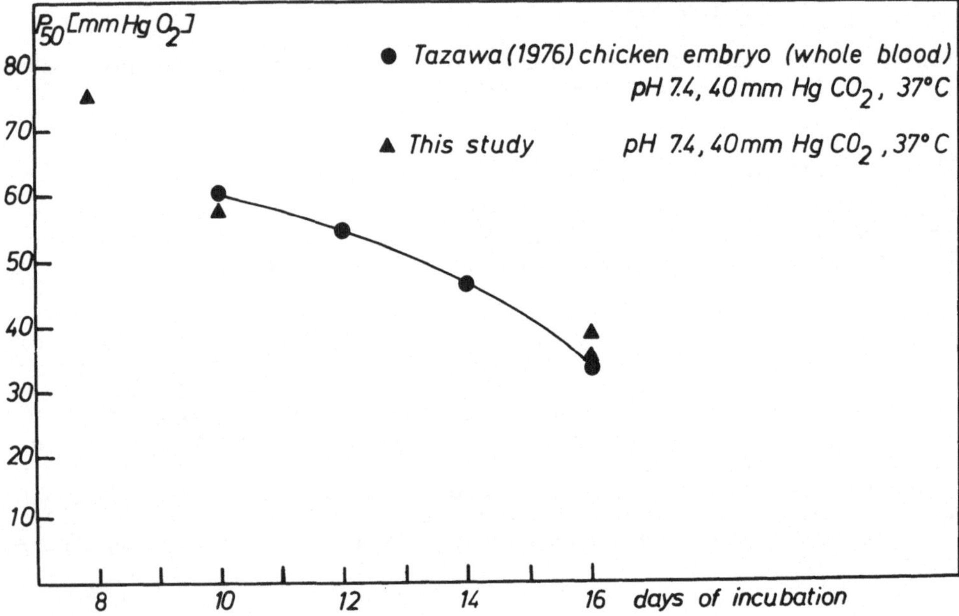

Fig. 1. Oxygen affinity expressed as O_2 halfsaturation pressure (P_{50}) of embryonic chicken blood

b) Whole Blood O_2 Affinity

The lowest O_2 affinity (P_{50} = 75 mmHg at a plasma pH of 7.4 and 37°C) was found in the blood of 8-day-old chicken embryos. This value decreased to 35 mmHg at day 16 in close agreement with the results of Tazawa et al. (1976) (see Fig. 1). At a constant extracellular pH of 7.4 the corresponding values for cell pH were 6.74 for day 8 and 7.12 pH for day 16.

c) ATP and 2,3 DPG Concentrations

Figure 2 shows concentrations for erythrocytes from 7- to 14-day-old embryos. As can be seen ATP is the dominant organic phosphate with an intracellular concentration of 10-11 mM/l RBC, whereas the concentra-

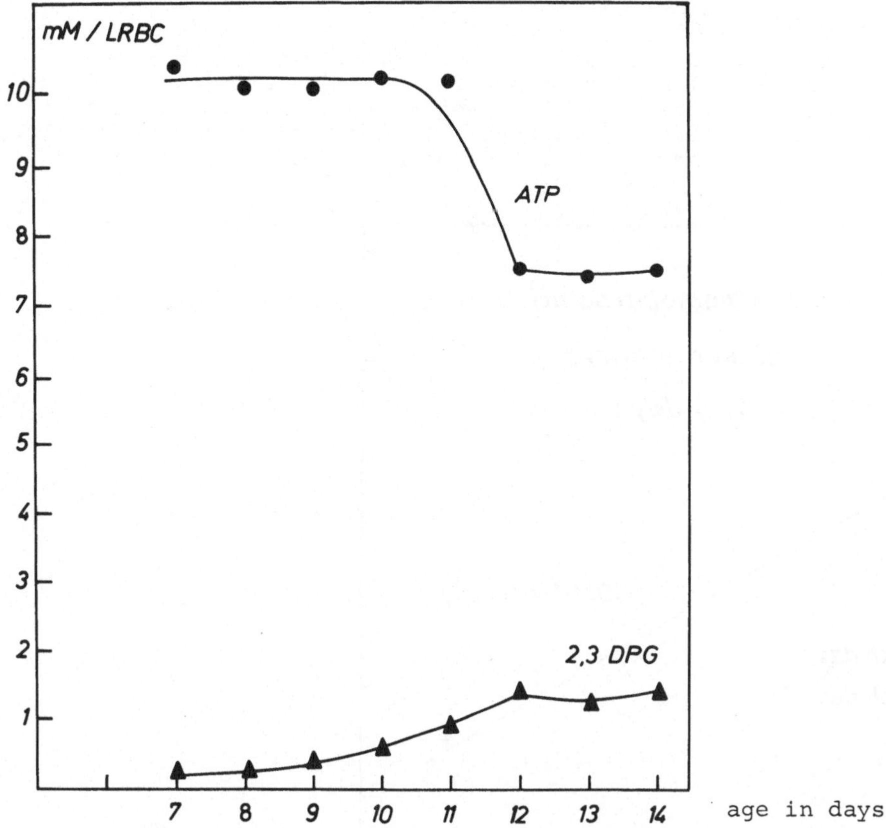

Fig. 2. Intraerythrocyte concentration of ATP and 2,3 DPG in embryonic chicken blood

Table 2. Oxygen affinity and Bohr effect of hemoglobin solutions from 5-, 6-, and 8-day-old chicken embryo compared with adult hemoglobin

		log P_{50} stripped	log P_{50} ATP	Δlog P_{50}/ ΔpH (stripped)	Δlog P_{50}/ ΔpH (ATP)	P_{50} ATP / P_{50} stripped
5 + 6 day	pH 6.7	0.65 (4.46)	1.38 (23.93)	- 0.36	- 0.58	5.38
	pH 7.2	0.47 (2.95)	1.09 (12.30)			4.16
8th day	pH 6.7	1.045 (11.1)	1.75 (56.23)	- 0.59	- 0.65	5.06
	pH 7.2	0.75 (5.62)	1.425 (26.60)			4.73
Adult	pH 6.7	1.225 (16.78)	1.95 (89.1)	- 0.45	- 0.63	5.30
	pH 7.2	1.00 (10.0)	1.635 (43.15)			4.31

Experiments with ATP were carried out at a tenfold excess of ATP over Hb_4. Values in brackets give P_{50} in mmHg. Stripped refers to hemoglobin solutions free of organic phosphates. T = 37°C, 0.05 M Bis Tris or Tris buffer. Total Cl^- concentration 0.16 M.

tion of 2,3 DPG in very low. The results are in good agreement with those of Bartlett and Borgese (1976).

d) Oxygen Affinity of Hemoglobin Solutions

Table 2 summarizes the values for P_{50}, Bohr coefficient, and effect of ATP on P_{50}, for hemoglobin solutions from 5-, 6-, and 8-day-old embryos and adult hemoglobin. In the absence of organic phosphates the O_2 affinity of solutions containing only embryonic hemoglobin (Fig. 3) is much higher than that of adult hemoglobin. The response toward

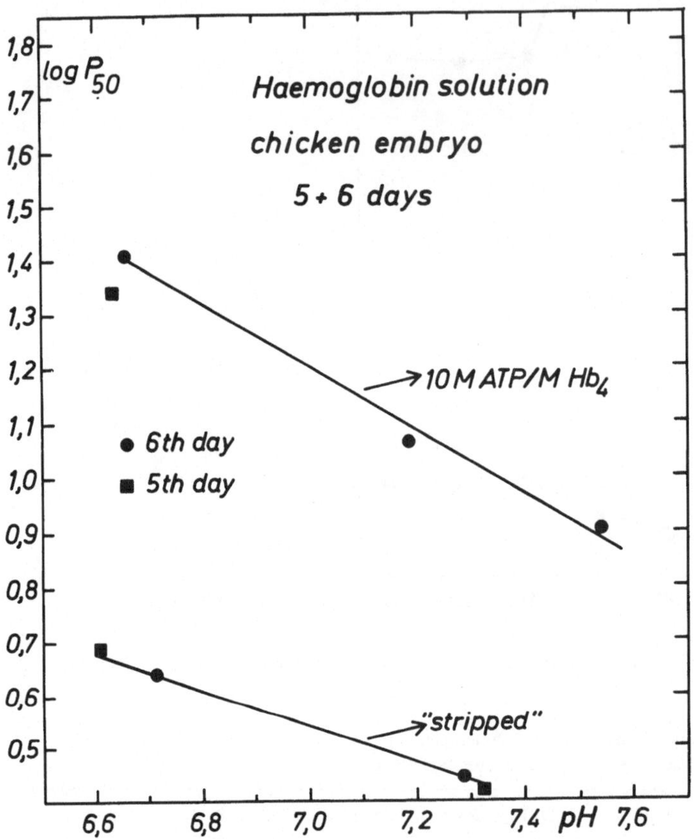

Fig. 3. Oxygen affinity of hemoglobin solutions from 5- and 6-day-old chicken embryos in dependence of pH and ATP

ATP is nearly the same for adult and embryonic hemoglobin solutions, resulting in a fourfold increase in P_{50} at pH 7.2. However, because of the initially low P_{50} of the embryonic hemoglobin solution, the absolute values for P_{50} of adult and embryonic hemoglobin are widely different in the presence of ATP.

4. Discussion

The present study corroborates the results of Lomholt (1975) and Tazawa et al. (1976) that the O_2 affinity of whole blood is lower at days 8-10 of incubation than at any other later stage of development.

The results obtained in hemoglobin solutions containing only embryonic hemoglobin are in contrast with those of Cirotto and Geraci. We find a higher O_2 affinity of embryonic hemoglobin compared to adult in the absence of organic phosphates, but a similar response towards ATP. The high concentration of ATP found between the 7th-10th day of incubation, when the amount of adult hemoglobin increases to about 60%, certainly helps to establish a low O_2 affinity. The decisive factor is, however, the low intracellular pH (pH_c 6.72 at day 8 versus 7.12 at day 16) at constant extracellular pH of 7.4. A quantitative analysis is however not possible, since in vivo adult and embryonic hemoglobins do not coexist within the same cell, but are synthesized by two different cell lines, the so-called primitive cell line synthesizing embryonic hemoglobin, and the difinitive cell line adult hemoglobin (Cirotto et al., 1975); Mahoney et al., 1977). In hemoglobin solutions as used by us, the different hemoglobins all coexist. This also implies that the values for ATP concentration and red cell pH obtained for day 8, only represent the mean average of both types of cells which are at this time present at about equal proportions (Bruns and Ingram, 1973), and do not necessarily represent the conditions encountered by adult and embryonic hemoglobin inside the erythrocyte. A further quantitative analysis of the factors regulating hemoglobin O_2 affinity during the period of transition from embryonic to adult hemoglobin requires the separation of definitive from primitive erythrocytes.

References

Bartlett, G.R., Borgese, T.A.: Phosphate compounds in red cells of the chicken and duck embryo and hatchling. Comp. Biochem. Physiol. 55 A, 207-210 (1976)

Bauer, Ch., Baumann, R., Engels, U., Pacyna, B.: The carbon dioxide affinity of various human hemoglobins. J. Biol. Chem. 250, 2173-2176 (1975)

Baumann, R., Bauer, Ch., Haller, E.A.: Oxygen-linked CO_2 transport in sheep blood. Am. J. Physiol. 229, 334-339 (1975)

Brown, J.L., Ingram V.M.: Structural studies on chick embryonic hemoglobins. J. Biol. Chem. 249, 3960-3972 (1974)

Bruns, G.A., Ingram, V.M.: The erythroid cells and haemoglobins of the chick embryo. Trans. Roy. Soc. Lond. ser. B Biol. Sci. 266, 225-305 (1973)

Cirotto, C., Geraci, G.: Embryonic chicken hemoglobins. Studies on the oxygen equilibrium of two pure components. Comp. Biochem. Physiol. 51A, 159-163 (1975)

Cirotto, C., Scotto di Tella, A., Geraci, G.: The hemoglobins of the developing chicken embryos. Fractionation and globin composition of the individual component of total erythrocytes and of a single erythrocyte. Cell Differentiation 4, 87-99 (1975)

Drysdale, J.W., Piergiorgio, R., Bunn, H.F.: The separation of human and animal hemoglobins by isoelectric focussing in polyacryamide gel. Biochim. Biophys. Acta 229, 42-50 (1971)

Ericson, A., De Verdier, C.H.: A modified method for the determination of 2,3-diphosphoglxcerate in erythrocytes. Scand. J. Cli. Lab. Invest. 29, 85-90 (1972)

Hall, F.G.: Hemoglobin function in the developing chick. J. Physiol. 83, 222-228 (1934)

Lomholt, J.P.: Oxygen affinity of bird embryo blood. J. Comp. Physiol. 99, 339-343 (1975)

Mahoney, K.A., Hyer, B.J., Chan, L.N.: Separation of primitive and definitive erythroid cells of the chick embryo. Sevel. Biol. 56, 412-416 (1977)

Sick, M., Gersonde, K.: Method for continuous registration of O_2-binding curves of hemoproteins by means of a diffusion chamber. Anal. Biochem. 32, 362-376 (1969)

Tazawa, H., Tsukase, O., Mochizuki, M.: Oxygen dissociation curve for chorioallentoic capillary blood of chicken embryo. J. Appl. Physiol. 40, 393-398 (1976)

The Effects of Restricted Gas Exchange on Embryonic Heart Rate

K. F. Laughlin

Summary

A technique for monitoring heart rate in a population of embryos during the last week of incubation has been used to examine the effects of restricting gas exchange across the eggshell. No correlation was found between embryonic heart rate and eggshell water vapour conductance (assumed to be a function of relative gas permeability). Application of PVC tape to the eggshell emphasized the natural trough in embryonic heart rate and restricting approximately two-thirds of the chorioallantoic shell surface also altered the timing of this event.

The temperature coefficient of heart rate was reduced in taped eggs.

1. Introduction

Chick embryo heart rate decreases gradually from 14 to 18 or 19 days of incubation (Cohn, 1925; Bogue, 1932; Cain et al., 1967) and then increases rapidly before hatching (Laughlin et al., 1976). One of several possible explanations for this pattern of heart rate is that it results from the decreasing efficiency of the chorioallantoic circulation prior to membrane penetration and the development of pulmonary respiration before hatching. The effects of restricting gas exchange over part of the eggshell on the respiratory properties of embryonic blood (Tazawa et al., 1971) and the pipping and hatching times (Visschedijk, 1968b) have previously been described.

This paper describes the effects on embryonic heart rate of natural variations and artificial reductions in the gas exchange across the eggshell of the chicken.

2. Materials and Methods

Three experiments were carried out. General procedures were similar for all three, detailed differences are noted below.

A total of 84 eggs taken from a single batch incubated for 14 days were divided at random into control or treatment groups and assigned to incubator positions. The gas exchange of the treated eggs was restricted by applying self-adhesive PVC tape (R.S. Components Ltd.).

Agricultural Research Council's Poultry Research Centre, King's Buildings, West Mains Road, Edinburgh EH9 3JS, Scotland.

This allowed equal application to a large number of eggs. The permeability of the tape to O_2, calculated from values for polymers (Lebovits, 1966), was 1.4×10^4 times less than the permeability of incubated eggshells (Tullett and Board, 1976). The average total surface area for the eggs was calculated and the chorioallantois was assumed to underlie 80% of this. Only portions of this area were covered with tape and care was taken not to apply tape close to the junction with the air space in order that it would not cover the air space, when the latter increased in size, or interfere with hatching.

The incubator was maintained at 37.5°C and 60% relative humidity and the heart rate of each embryo was measured every 2 h, until hatching, using the technique described previously (Laughlin et al., 1975, 1976).

a) Experiment 1 and 2

Eggs for these two experiments were taken, at flock ages of 35 and 38 weeks respectively, from heavy hybrid birds descended from Thornber 909 stock. In Experiment 1 eggs were weighed to the nearest mg at 3- or 4-day intervals during the first 14 days of incubation. The dewpoint of the incubator was monitored continuously with a lithium chloride hygrometer (Wallac Oy) and dry bulb measurements were taken with a mercury in glass thermometer. From these measurements the mean water vapour conductance was calculated for each egg.

A control and one treatment group were used in each experiment. For the treatment 30 cm^2 tape was applied to the shell; this covered approximately half the chorioallantoic membrane.

On days 16, 17, 18, and 19 of incubation, measurements of heart rate were also made at 36.5°C and 38.5°C. After changing the incubator temperature 2 h were allowed for equilibration before recording began.

During the hatching period eggs were exmained at roughly 6-h intervals and those pipped or hatched were noted. From these data mean pipping and hatching times were estimated for the control and treatment groups.

b) Experiment 3

Eggs from Ross 1 broiler breeders were incubated for 14 days and then allocated at random to a control or one of two treatment groups and to their incubator positions. Tape was applied to the shell over an estimated one-third and two-thirds of the chorioallantoic membrane, 18 cm^2 and 36 cm^2 respectively. Recordings of heart rate were made from 17 day to hatching. The condition of the eggs was noted on three occasions during the pipping and hatching period.

3. Results

a) Experiment 1 and 2

The pattern of changes in mean heart rate was similar in the two experiments. From day 16 the embryos in the taped eggs had consistently lower heart rates than those in the control eggs (Fig. 1, data for Experiment 2). The depression of heart rate on days 18 and 19 was greater in the taped eggs than in the controls; minimum values were 252 and 260 beats/min respectively. During the pre-hatch rise on day

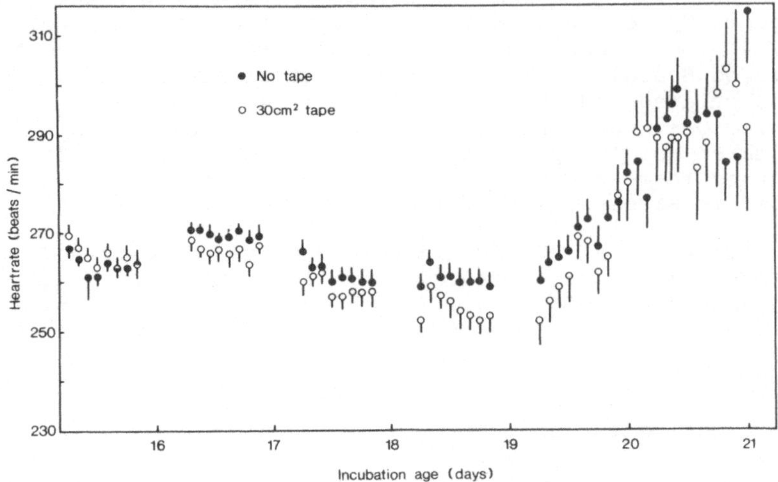

Fig. 1. Changes in the mean heart rate ± 1 SEM of control and taped embryos (n = 25 to 40) from 15 days to hatching

20 heart rates for embryos in the two groups of eggs rose to similar values.

In Experiment 1 the mean fresh egg weight was 58.19 g (Standard deviation, S.D., 3.52) and the mean water vapour conductance was 13.3 mg/mmHg/day (S.D. 1.81). The relationship between the water vapour conductance of the eggshells and the heart rate of the embryos they contained was examined in the control eggs for one record on each of days 16 to 18. No significant correlations were found.

The temperature coefficient of heart rate (beats/min/^{o}C) was consistently lower for embryos in taped eggs than the controls (Table 1).

Table 1. The temperature coefficient of heart rate (beats/min/^{o}C) in taped and untaped eggs on four days of incubation. Data for two experiments are presented. C, control; T, 30 cm^2 tape

Experiment number		1	2
Incubation age (days)			
16	C	24.14	25.20
	T	22.19	22.93
17	C	23.42	[a]
	T	21.93	
18	C	20.09	21.36
	T	19.86	19.54
19	C	20.65	17.20
	T	18.35	15.87

[a] No data available due to system failure.

Inspection of the raw data shows this to be due to the difference in heart rate between embryos in taped eggs and the controls being greater at 38.5°C than at 36.5°C. The depressive effect of egg taping on heart rate was emphasized at the higher temperatures.

For the small samples available measurements of mean pipping and hatching times are not accurate, but these events were estimated to be delayed in the taped eggs by 3 to 5 h and 7 to 8 h respectively.

b) Experiment 3

The pattern of changes in mean heart rate with embryo age in this experiment (Fig. 2) was similar to those in Experiments 1 and 2. The magnitude and timing of the changes were different.

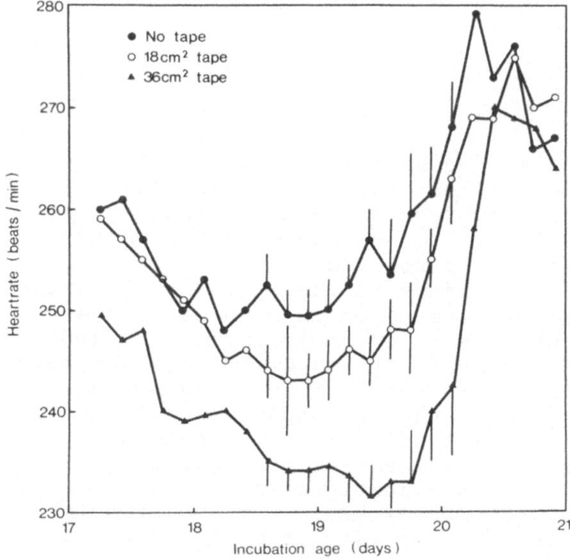

Fig. 2. Changes in the mean heart rate ± 1 SEM of control and two groups of taped embryos (n = 15 to 25) from 17 days to hatching

On day 17 the heart rate of embryos in the 36-cm^2 taped eggs (249.9, SEM 3.2) was already significantly lower (t-test, $P < 0.05$) than that in the control or 18-cm^2 groups (260.4, SEM 2.46 and 259.4, SEM 2.35). The overall minimum values for eggs in control, 18-cm^2 and 36-cm^2 taped eggs were 248, 243, and 232 beats/min respectively. This minimum occurred at a similar time, around 18 days 18 h, in the former two groups, but was delayed to 19 days 10 h in the 36-cm^2 taped eggs. The succeeding increase was, however, more rapid in the latter group so that all three had similar values, around 270 beat/min at about 20 days 12 h.

The record of pipping times (Table 2) shows little difference between the control and the 18-cm^2 taped eggs; the 36-cm^2 taped eggs were delayed. The number of hatched chicks at 21 days 0 h suggests some delay in hatching for 18-cm^2 taped eggs and considerable delay in the 36-cm^2 taped eggs. This delay in the latter is apparently much greater than the delay in pipping in the same group.

Table 2. The cumulative number of embryos pipped or hatched at three incubation ages in control and treated eggs

Incubation age (days, h)	20, 3	21, 0	21, 5
Number pipped			
control	10	20	22
18 cm^2 tape	7	20	20
36 cm^2 tape	5	17	19
Number hatched			
control	0	16	21
18 cm^2 tape	1	10	20
36 cm^2 tape	0	2	7[a]

[a] 12 embryos alive and pipped.

4. Discussion

The heart rate of individual embryos is consistent, relative to the population mean, throughout incubation, but Laughlin et al. (1976) found no simple relationship between embryonic heart rate and egg size or individual hatching times. Similarly in the experiments reported here it has not been possible to show any relationship between embryonic heart rate and water vapour conductance of the eggshell. If water vapour conductance values of eggshells are taken as a guide to their relative O_2 permeabilities then there is no evidence from this study that natural restrictions in shell permeability to O_2 affect embryonic heart rate. There are, however, wide variations between strains and flock ages of domestic fowl in egg water vapour conductance (Laughlin, in preparation). Eggs used in the present study were close to and above the average conductance for eggs of their weight. Further experiments are planned to examine embryonic heart rates in eggs from low water conductance strains and ages of chickens. Particular attention will be paid to the depression of heart rate prior to membrane penetration.

The present experiments showed that it was possible to reduce embryonic heart rate by artificially restricting gas exchange across the shell. The timing and size of these changes varied with the amount of restriction, but the major effect was to emphasize the natural decrease in embryonic heart rate around 18 days incubation. It is assumed (but not yet proved) that these minimum values occur just prior to the time of membrane penetration; after this any restriction of the chorioallentoic system becomes less important as pulmonary respiration takes over. Visschedijk (1968a) showed that for some time after membrane penetration the embryo was dependent on the chorioallantoic circulation which still accounted for 50% of the respiratory exchange at the time of pipping. This would explain the continued reduction in heart rate of taped eggs relative to controls even after the minimum values occurred in the present experiments.

Although the observations reported here were on relatively small numbers of eggs, in all cases restricting gas exchange by taping had a lesser effect on the timing of pipping than on hatching. In particular the response of the 36-cm^2 taped eggs in Experiment 3 could be explained by a delay in membrane penetration, due to general reduction in the developmental rate as a consequence of reduced gas exchange, followed by a relative advance in pipping time. The slower development would also delay hatching. It has previously been shown that severe reduction of the effective shell area delayed embryonic growth in the later stages of development (Tazawa et al., 1971). Visschedijk (1968b) produced a reduction in the time between lung ventilation and pipping by preventing gas exchange over the air space. In view of the continued importance of the chorioallantoic circulation during this period it might be expected that severe restriction over this surface would also shorten the parafoetal period.

The temperature coefficient of heart rate is reduced in taped eggs due to an increasing depression of heart rate relative to the controls as temperature increased. This suggests that the normal temperature response of the embryo is restricted by the reduced availability of O_2 in the taped embryos. A more detailed paper on the temperature responses of heart rate in normal embryos is currently being prepared (Laughlin and Lundy).

Embryonic heart rate can be reduced by artificially reducing the efficiency of the chorioallantoic gas exchange system, hence, it is possible that the natural reduction in heart rate during the last week of incubation is due to the natural reduction in the efficiency of the gas exchange system prior to the development of lung respiration. The fact that the embryos in taped eggs are less able to respond to the increased stress of high temperatures is an indication of the sensitivity of heart rate to O_2 supply.

References

Bogue, J.Y.: The heart rate of the developing chick. J. Exp. Biol. 9, 351-358 (1932)

Cain, J.R., Abbott, U.K., Rogallo, V.L.: Heart rate in the developing chick embryo. Proc. Soc. Exp. Biol. Med. 126, 502-510 (1967)

Cohn, A.E.: Physiological Ontogeny. A. Chicken Embryos. (V). On the rate of heart beat during the development of chicken embryos. J. Exp. Med. 42, 291-297 (1925)

Laughlin, K.F.: Water vapour losses from chicken eggs during incubation. (in prep.)

Laughlin, K.F., Lundy, H.: Chick embryo heart rate during the last week of incubation: the effect of temperature. (in prep.)

Laughlin, K.F., Lundy, H.:, Tait, J.A.: A method for monitoring embryonic heart rate during the last week of incubation. J. Physiol. (London) 224, 8P-9P (1975)

Laughlin, K.F., Lundy, H., Tait, J.A.: Chick embryo heart rate during the last week of incubation; Population studies. Brit. Poultry Sci. 17, 293-301 (1976)

Lebovits, A.: Permeability of polymers to gases, vapours and liquids. Modern Plastics pp. 139-213 (1966)

Tazawa, H., Mikami, T., Yoshimoto, C.: Effect of reducing the shell area on the respiratory properties of chicken embryonic blood. Respir. Physiol. 13, 352-360 (1971)

Tullett, S.G., Board, R.G.: Oxygen flux across the integument of the avian egg during incubation. Brit. Poultry Sci. 17, 441-450 (1976)

Visschedijk, A.H.J.: The air space and embryonic respiration. I. The pattern of gaseous exchange in the fertile egg during the closing stages of incubation. Brit. Poultry Sci. 9, 173-184 (1968a)

Visschedijk, A.H.J.: The air space and embryonic respiration. II. The times of pipping and hatching as influenced by an artificially changed permeability of the shell over the air space. Brit. Poultry Sci. 9, 185-196 (1968b)

List of Participants

ABDALLA, M.A.
Department of Veterinary Anatomy
University of Khartoum
Khartum / Sudan

AR, A.
Department of Zoology
Tel-Aviv University
Tel-Aviv / Israel

BALLINTIJN, C.M.
Zoologisch Laboratorium
Biologisch Centrum
Rijksuniversiteit Groningen
Haren (Groningen) / Netherlands

BAMFORD, O.S.
Department of Animal Physiology
University of Nairobi
Nairobi / Kenya

BANZETT, R.B.
Department of Physiology
School of Public Health
Harvard University
Boston, Massachusetts / USA

BARTELS, H.
Physiologisches Institut
Medizinische Hochschule Hannover
Hannover / FRG

BAUER, CH.
Physiologisches Institut
Universität Regensburg
Regensburg / FRG

BAUMANN, R.
Physiologisches Institut
Medizinische Hochschule Hannover
Hannover / FRG

BECH, K.
Department of Zoophysiology
University of Aarhus
Aarhus / Denmark

BERGER, M.
Landesmuseum fur Naturkunde
Münster (Westfalen) / FRG

BESCH, E.L.
Department of Metabolism
College of Veterinary Medicine
University of Florida
Gainesville, Florida / USA

BLACK, C.P.
Department of Physiology
Dartmouth Medical School
Hanover, New Hampshire / USA

BOUVEROT, P.
Laboratoire de Physiologie Respiratoire
Centre National de la Recherche
Scientifique
Strasbourg / France

BRACKENBURY, J.H.
Sub-Department of Veterinary Anatomy
University of Cambridge
Cambridge / England

BURGER, R.E.
Department of Avian Sciences
University of California
Davis, California / USA

BURNS, B.
John Hopkins University
Medical Institutions
Baltimore, Maryland / USA

BUTLER, P.J.
Department of Zoology and Comparative
Physiology
University of Birmingham
Birmingham / England

CALDER III, W.A.
Department of Ecology and
Evolutionary Biology
University of Arizona
Tucson, Arizona / USA

CAREY, C.
Department of EPO Biology
University of Colorado
Boulder, Colorado / USA

CRAWFORD Jr., E.C.
School of Biological Sciences
University of Kentucky
Lexington, Kentucky / USA

DAWES, C.M.
Department of Physiology
Royal Veterinary College
London / England

DEJOURS, P.
Laboratoire de Physiologie Respiratoire
Centre National de la Recherche
Scientifique
Strasbourg / France

DUNCKER, H.R.
Zentrum für Anatomie und Cytobiologie
Klinikum der Justus-Liebig Universität
Giessen / FRG

ESTAVILLO, J.
Department of Basic Medical Education
School of Medicine
Southern Illinois University
Carbondale, Illinois / USA

FARNER, D.S.
Department of Zoology
University of Washington
Seattle, Washington / USA

FEDDE, M.R.
Department of Anatomy and Physiology
College of Veterinary Medicine
Kansas State University
Manhattan, Kansas / USA

GANS, C.
Division of Biological Sciences
University of Michigan
Ann Arbor, Michigan / USA

GRAF, U.
Abteilung Tierphysiologie
Ruhr-Universität Bochum
Bochum / FRG

GRATZ, R.K.
Abteilung Physiologie
Max-Planck-Institut für experimentelle
Medizin
Göttingen / FRG

HEISLER, N.
Abteilung Physiologie
Max-Planck-Institut für experimentelle
Medizin
Göttingen / FRG

HOLLAND, R.A.B.
School of Physiology and Pharmacology
University of New South Wales
Kensington (Sydney) / Australia

HOYT, D.F.
Department of Physiology
State University of New York
at Buffalo
Buffalo, New York / USA

HUGHES, G.M.
Research Unit for Comparative
Animal Respiration
University of Bristol
Bristol / England

JOHANSEN, K.
Department of Zoophysiology
University of Aarhus
Aarhus / Denmark

JONES, D.R.
Department of Zoology
University of British Columbia
Vancouver, B.C. / Canada

KAWASHIRO, T.
School of Medicine
Keio University
Tokyo / Japan

KING, A.S.
Department of Veterinary Anatomy
University of Liverpool
Liverpool / England

KING, D.Z.
Department of Histology
Medical School
University of Liverpool
Liverpool / England

KUNZ, A.L.
Department of Physiology
Ohio State University
Columbus, Ohio / USA

LAUGHLIN, K.F.
Agricultural Research Council's
Poultry Research Centre
Edinburgh / Scotland

MALOIY, G.M.O.
Department of Animal Physiology
University of Nairobi
Nairobi / Kenya

MARIN, L.
Institut d'Embryologie
Centre National de la Recherche Scientifique
Nogent-sur-Marne / France

METCALFE, J.
Department of Medicine
Health Sciences Center
University of Oregon
Portland, Oregon / USA

MEYER, M.
Abteilung Physiologie
Max-Planck-Institut für experimentelle Medizin
Göttingen / FRG

MILLER, D.A.
Department of Physiology
Medical College of Georgia
Augusta, Georgia / USA

MILSOM, W.K.
Department of Zoology
University of British Columbia
Vancouver, B.C. / Canada

MOCHIZUKI, M.
Department of Physiology
School of Medicine
Yamagata University
Yamagata / Japan

MOLONY, V.
Department of Veterinary Physiology
Royal (Dick) School of Veterinary Studies
Edinburgh / Scotland

MONGIN, P.
Station de Recherches Avicoles
Institut National de la Recherche Agronomique
Tours-Nouzilly / France

MUNSHI, J.S. DATTA
Postgraduate Department of Zoology
Bhalgapur University
Bhalgapur (Bihar) / India

NILSSON, N.J.
Department of Clinical Physiology II
University of Göteborg
Göteborg / Sweden

OSBORNE, J.L.
Department of Physiology
University of California
Irvine, California / USA

PATTLE, R.E.
Chemical Defence Establishment
Porton Down (Salisbury) / England

PERRY, S.F.
Zentrum für Anatomie und Cytobiologie
Universität Giessen
Giessen / FRG

PETERSON, D.F.
Department of Pharmacology
Health Science Center
University of Texas
San Antonio, Texas / USA

PIIPER, J.
Abteilung Physiologie
Max-Planck-Institut für experimentelle Medizin
Göttingen / FRG

RAHN, H.
Department of Physiology
State University of New York at Buffalo
Buffalo, New York / USA

RAUTENBERG, W.
Abteilung Biologie
Ruhr-Universität Bochum
Bochum / FRG

RICHARDS, S.E.
Wye College
University of London
Ashford (Kent) / England

RÜFER, R.
Institut für Pharmakologie und Toxikologie
Fakultät für klinische Medizin
Mannheim, Universität Heidelberg
Mannheim / FRG

SCHEID, P.
Abteilung Physiologie
Max-Planck-Institut für experimentelle Medizin
Göttingen / FRG

SCHMIDT-NIELSEN, K.
Department of Zoology
Duke University
Durham, North Carolina / USA

SEYMOUR, R.S.
Department of Zoology
University of Adelaide
Adelainde, S. A. / Australia

SHELTON, G.
School of Biological Sciences
University of East Anglia
Norwich / England

STONE, D.
Department of Physiology
Ohio State University
Columbus, Ohio / USA

STURKIE, P.D.
Department of Environmental Physiology
Rutgers University
New Brunswick, New Jersey / USA

TALLMAN, R.D.
Department of Physiology
Ohio State University
Columbus, Ohio / USA

TAZAWA, H.
Department of Physiology
School of Medicine
Yamagata University
Yamagata / Japan

TORRE-BUENO, J.
Department of Zoology
Duke University
Durham, North Carolina / USA

TULLETT, S.G.
Agricultural Research Council's
Poultry Research Centre
Edinburgh / Scotland

ULTSCH, G.R.
Abteilung Physiologie
Max-Planck-Institut für
experimentelle Medizin
Göttingen / FRG

VISSCHEDIJK, A.H.J.
Department of Veterinary Physiology
School of Veterinary Medicine
University of Utrecht
Utrecht / Netherlands

WEST, N.H.
Department of Zoology
University of British Columbia
Vancouver, B.C. / Canada

WHITE, F.N.
Physiological Research Laboratory
Scripps Institution of Oceanography
University of California
San Diego
La Jolla, California / USA

WRIGHT, P.G.
Department of General Physiology
School of Dentistry
University Witwatersrand
Johannesburg / South Africa

Subject Index

Zoomorphologie

An International Journal of Comparative and Functional Morphology

Avian Physiology

Editor: P. D. Sturkie

3rd edition. 1976. 106 figures. XIII, 400 pages
(Springer Advanced Texts in Life Sciences)
ISBN 3-540-07305-1

The first edition of this important book, published in 1954, was acknowledged to be the most comprehensive and authoritative onevolume account of the avian physiology available. The third edition has been enlarged, and brought up to date by an outstanding international group of experts, including many new authors.

Avian Physiology is still the only single-volume work in any language that is devoted to the physiology of aves. It is a well-balanced account of the physiology of most of the organ systems of aves and may be used as a textbook or a reference.

Contents: Nervous System. – Sense Organs. – Blood. – Heart and Circulation. – Heart. – Respiration. – Regulation of Body Temparature. – Energy Metabolism. – Alimentary Canal. – Secretion of Gastric and Pancreatic Juice. – Carbohydrate Metabolism. – Protein Metabolism. – Lipid Metabolism. – Kidney, Urine, and Extra-Renal Salt Secretion. – Hypophysis. – Reproduction in the Female and Egg Formation. – Reproduction in the Male, Fertilization, and Early Embryonic Development. – Thyroids. – Parathyroids. – Adrenals. – Pancreas.

Springer-Verlag
Berlin
Heidelberg
New York